The Dark Matter Discoverer's Guidebook

Exploring WIMPs, Axion-Like Particles, and the Dark Sector

Jodi Cooley

Stephen Jacob Sekula

YBK Publishers

New York

The Dark Matter Discoverer's Guidebook:
Exploring WIMPs, Axion-Like Particles, and the Dark Sector

YBK Publishers, Inc.
39 Crosby Street
New York, NY 10013
www.ybkpublishers.com

ISBN:978-1-936411-95-5

Manufactured in the United States of America for distribution in
North and South America or in the United Kingdom or Australia
when distributed elsewhere.

For more information, visit www.ybkpublishers.com.

Dedication

For Ann, Rich, Annetta, and Stephen...our greatest teachers, mentors, and friends.

Acknowledgements

We are extremely grateful to the organizers of the Les Houches Summer School on Dark Matter for reaching out to Jodi in 2019, requesting that she teach on the subject of the direct detection of dark matter at the 2020 school. It was that event, more than anything else, that led to this book.

Due to the global SARS-CoV-2 pandemic raging in 2020 and 2021, the school was postponed to 2021. The delay, rather than being a frustration, afforded a key opportunity...one provided by our then-home physics department at SMU in Dallas, TX.

The spring of 2021 facilitated the chance to create a one-of-a-kind special topics course entitled "The Higgs and Dark Matter." The first half was taught by our colleague, Dr. Ryszard Stroynowski, on the subject of the Higgs particle, quantum field theory, and particle physics experiments. After seven weeks of instruction, Jodi took over the course, dedicating the remaining half of the term to the subject of dark matter. The notes for that second portion of the course are the basis for this manuscript. That subsequently aided in the final development of her lecture notes for the long-delayed Les Houches Summer School on Dark Matter in August, 2021. Those lecture notes became the basis for chapter four of this book. The continuity of the book's entirety was developed from the set of slides Jodi created for the SMU special topics course.

Jodi would like to acknowledge the students enrolled in that course. To Lucas, Brandon, Noah, and Jared, she expresses her deep appreciation for their strong and active participation. She is especially grateful for their questions, comments, and conversation during the term, inspiring her to work harder, think more, and dedicate herself to a deeper understanding of the difficult and broad subject of dark matter more than at any earlier time in her career. Thanks also go to the students who sat in on the course—Ishwita and Jasmine—for providing her a broader audience of young future dark matter hunters. Their participation benefited this text in broadening this subject for a future audience.

Stephen's part in the story, as co-author, arises in parallel with the development of the course. Inspired by similar examples of past special courses deployed for the purpose of both educating students and advancing the field, he saw an opportunity in the department's teaching schedule for spring of 2021 and used that to encourage the creation of the course. Models for this, where the lectures were actually a vessel for fundamental progress, were Robert Serber's "Los Alamos Primer" and Sir Andrew John Wiles's special course at Princeton, "Calculations on Elliptic Curves," which disguised a presentation of his proof of Fermat's Last Theorem intended for his colleague, Nick Katz.

Throughout the development of the course, during the teaching term, and in the summer of 2021, Stephen encouraged both instructors to prepare their lectures as manuscripts. Jodi consented to have her lecture notes transcribed to book form by Stephen, which he did over the summer months and into the fall of 2021. During the process of converting the lecture materials, Stephen added a narrative structure to the whole of the work and fleshed out a large number of points, steps, and ideas that had been necessarily compressed for the special topics course and especially the lecture materials.

This provided a means for him to develop and increase his familiarity with astrophysics. Paraphrasing the late Steven Weinberg, if one wishes to become expert in something, start by teaching a course (ideally more than once) with the aim of collecting the lecture materials, insights, and experiences into a book on the subject.

Jodi would like to express her thanks to her co-author. She self-describes as "not a writer" and is grateful for Stephen's relentless (here, she would like to pause and over-emphasize the word *relentless*) effort in strong-arming her into writing down all of this material. Stephen would like to thank his co-author for her patience and brilliance.

Stephen would also like to express his gratitude to his former student, Dr. Nicole Hartman, for her work on low-energy neutrino scattering which appears briefly in this book as a private communication, including some unpublished conclusions.

We are grateful to our families, especially our parents Ann, Rich, Annetta, and Stephen. We are heartbroken that Rich did not live to see so many things, including the publication of this, Jodi's first book. We dedicate this text to his memory. Rich helped raise Jodi right, made sure she spent a lot of time in school, and taught her the importance of frequent breaks from routine. We are also very grateful to our siblings Jackie, Jerry, Jolene, and Kate, who teach us not to take ourselves too seriously while still encouraging us to be serious.

We are grateful to our scientific mentors, who shaped us into practicing physicists. For Jodi, her deepest thanks go to Professors John Friedman (University of Wisconsin-Milwaukee), Sharon Morsink (University of Alberta), Albrecht Karle (University of Wisconsin-Madison), Francis Halzen (University of Wisconsin-Madison), Kate Scholberg (Duke University), and Blas Cabrera (Stanford University). For Stephen, his deepest thanks go to Dr. Charles Dumais (Executive Director at Cooperative Educational Services), Professors Michael Schmidt (Yale University), Sau Lan Wu (University of Wisconsin-Madison), Yibin Pan (University of Wisconsin-Madison), Gabriella Sciolla (Brandeis University), Dick Yamamoto (MIT), Klaus Honscheid (The Ohio State University), Richard Kass (The Ohio State University), Harris Kagan (The Ohio State University), and K. K. Gan (The Ohio State University).

Finally, but certainly not least, we would like to express our deepest thanks to those who provided assistance and technical support, including manuscript review during the writing process. This resulted in a much stronger book.

We thank Professor Yoni Kahn (University of Toronto) for his feedback on material related to dark matter scattering theory and on future directions for this field. We are especially grateful for his encouragement to help our audience understand the shape of the recoiling dark matter spectrum in direct detection experiments. We note below his additional contributions to future directions in dark matter detection.

We thank Professor Joel Meyers (Southern Methodist University) for his feedback on material related to astrophysics and cosmology. We are especially grateful for his encouragement of the inclusion of big bang nucleosynthesis. It is an important, independent test of the baryon density parameter of the universe. We benefited immensely from his expertise especially on the cosmic microwave background.

We thank Professor Steven Robertson (Institute for Particle Physics Scientist and Professor at the University of Alberta) for his feedback on material related to both indirect detection of dark matter (through dark matter annihilation or decay) and especially on the use of particle colliders to search for dark matter. Stephen would like to personally express his gratitude to Prof. Robertson for being one of his earliest mentors when each of them worked together on the BaBar Experiment.

We thank Professor Tarek Saab (University of Florida) for his feedback on material related to direct detection of dark matter and future directions for this field. We are especially grateful for his attention to detail and his encouragement, along with similar encouragement from Yoni Kahn and Katelin Schutz (see below), to further subdivide the future directions discussion into technologies oriented at low energy thresholds, directional detection, and weak chemical bonds. We are exceptionally grateful for his contribution of two updated sets of graphs. The first was for comparison experiment and theory regarding the energy deposition mechanisms in germanium. The second was his contribution of updated dark matter scattering constraint graphs for the conclusion of Chapter 4, as he is a lead developer of the community tool for generating such graphs [Tar24].

We thank Professor Katelin Schutz (McGill University) for her feedback on material related to axion physics and cosmology and for future directions for this field. We are especially grateful for the extra time she took to have a deeper face-to-face conversation about the subject material with excellent suggestions for improving and sharpening the discussion.

We are also extremely grateful to our colleague, Dr. Aaron Vincent (Queen's University), for his assistance in a detailed understanding of the "freeze-in" mechanism for creating a relic abundance of dark matter particles (see Section 3.6). His lectures during the inaugural SNOLAB Underground Science Institute (SuSi) lecture program in the summer of 2024 also had a strong impact on the final discussion of early-universe dark matter population dynamics. We extend gratitude also to Dr. Edward Laird (Lancaster University) for his input and comments on our explanation of the axion concept, which shaped how we communicate this framework in Chapter 6.

Finally, we are extremely grateful to the research and academic institutions who have been host to us over the last 15 years. First, we thank Southern Methodist University, especially our colleagues in the Department of Physics, for their insights, encouragement, and collegiality. We thank SNOLAB, where we presently work, for their support of excellence in underground science and especially for the effort to better nourish the human spirit in addition to the mind. Science is nothing without the people who conduct it. Finally, we thank Queen's University, our new academic home, for their dynamic engagement in and support of physics and people. Though we are far from Queen's while working at SNOLAB, we feel the light of our colleagues every day.

The image on the cover was generated using artificial-intelligence-powered software. As physicists who have utilized neural networks to generate non-parametric functions that describe complex relationships since the 1990s, we enjoy the new creative directions opened by these tools. We used the *Bing Image Creator* tool and the prompt, "Silhouetted young male and female detective examining a puddle reflecting a galaxy with dark matter using a magnifying glass under the light of a street lamp, photorealistic."

Contents

Preface

This book is intended as a guide for any rising physicist having the ambition, to make great discoveries. Such a book, however, cannot be unlimited in its scope. It is the authors' intent to first introduce our audience to a bit of the history of the subject of dark matter, with an emphasis on those parts of physics and astronomy that played a role in its past development (Chapter 2). We then aim to broadly survey the major lines of evidence for dark matter's existence, both now and much earlier in the universe (Chapters 2 and 3). Finally, we will survey the wide range of discovery technologies, pointing out the kinds of dark matter candidates to which they could be sensitive (Chapters 4-6).

Twenty years ago, when this field began to gain attention, it might have been possible to explore both the dark matter candidates and the experiments that seek to find them in a single textbook. Those days are behind us. The concepts that might explain dark matter have exploded in number along with the ways in which unknown particle types can be detected. This is a golden era of both ambition and uncertainty.

It is probably a good thing that we cannot fit all of the ideas and technologies into this one book. That might leave the intrigued reader with the impression that all that can be known on the subject is already known.

That, certainly not being the case, we will show our audience the ideas themselves and the physics concepts behind them; reinforce those concepts where possible, and to finally leave the reader with those key insights extant in the field as it stands today.

As no one knows the answer to the riddle of dark matter, this book should be used to stimulate thinking, to practice basic ideas, to discover new ones, and to serve as intellectual fuel for advancing the reader's own career.

A reader completing this text should be able to do two things: (1) engage in an investigative area of their choosing at the start of a long and, hopefully, prosperous career and (2) begin to generate new ideas and new technologies beyond the scope of this book to advance the study of dark matter. We provide additional fodder for this in the concluding Chapter 7.

The reader should be an individual with basic formal training in physics, having completed its core subjects (these are usually classical and quantum mechanics, electromagnetism, and statistical mechanics/thermodynamics) as well as having completed core mathematics (calculus, up to and including multivariable approaches; linear algebra;

differential equations). As part of classical mechanics, we expect a reader to have familiarity with Lagrangian and Hamiltonian dynamics.

The philosophy of this book is quite different from a standard physics textbook. We will not abandon our audience, leaving them to guess what is meant or intended. We anticipate that many of our readers are only part way along the path to acquiring the tools that are necessary in solving technical problems using mathematics. We will therefore avoid skipping any steps during mathematical discussions.

As we demonstrate a key idea, we will itemize the individual steps; step by step. Many textbooks ignore those steps and simply present a final equation, leaving the reader to guess, or to spend much wasted time working out how to go about arriving at the same conclusion. We will show our work, very much in the manner in which many of our readers may be familiar—just as was expected when we all were students.

This book should be a resource for the reader. We relied on a large number of references to write it. These are cited in an organized bibliography at the back of the book. The bibliography is sorted using an alphabetic tag whose form is [<AUTHOR ABBREVIATION><YEAR>]. A single-author paper will have an author abbreviation of one or three letters, using the beginning of the last name (or of a collaboration). A multi-author paper with two or three authors will use the first initials of the last names of the authors. For more than three authors, the beginning of the first author's last name is used, followed by a plus sign to signify a multi-author work. The year code is the last two digits of the publication year. This approach should make it easier to find resources later if one is familiar with authors' names.

Oh! A surprise for you! We provide homework. Also, each problem is worked out and shown in detail in the appendix. We invite you to struggle... but not to get lost in the dark.

Jodi A. Cooley   Stephen Jacob Sekula

Sudbury, Canada, October, 2024

CHAPTER 1

Introduction

The material in this brief chapter is meant to help establish conventions for discussing the subject of dark matter. A student with some experience in astronomy and astrophysics can likely skip much of this chapter. A student with less or no such experience will find this material useful for understanding basic units of measure and other fundamental concepts needed to follow the rest of the book.

1.1 Astrophysical Units

A basic unit of distance measure, appropriate for scales at the level of a solar system or larger, is the *Astronomical Unit* (AU):

$$1 \text{ AU} = 1.496 \times 10^{13} \text{ cm.} \tag{1.1}$$

This is the distance between Earth and the Sun. For example, the planet Venus—next-closest to the sun—is, on average, 0.72 AU from the sun during its elliptical orbit. Mars, the next-farthest planet from the sun, is, on average, 1.52 AU from the sun. Jupiter, first of the outer planets, and also the largest in the solar system, is, on average, 5.2 AU from the sun, while the farthest planet, Neptune, is about 30 AU from the sun. The farthest planet-scale objects are located in the Kuiper Belt. This is about 50 AU from the sun. The Oort Cloud, a population of objects believed to be the source of comets, extends outward from the sun to distances as far as 100,000 AU.

We can see that the AU is a convenient unit of measure for solar system-scale measurements. However, the AU becomes inconveniently small when considering interstellar or galactic distances. Instead, a more useful measure is the *light-year*:

$$1 \text{ light year (ly)} = 6.324 \times 10^{5} \text{ AU} = 9.461 \times 10^{17} \text{ cm.} \tag{1.2}$$

A light-year corresponds to the distance that light, within empty space, traveling at 2.998×10^{8} m/s, will travel in one year.[1] The closest star to the sun, Proxima Centauri,

---

[1] It is convenient to remember that, in seconds, one year of time can be approximated to better than 0.5% accuracy using the number $\pi \times 10^{7}$ s.

is just over 4 ly from our central star. Stars described to be in the "neighborhood" of our solar system are typically stated to be within about 100 ly of our sun.

Another astrophysical unit about the same size as the light-year is the *parsec*:

$$1\,pc = 3.26\,ly = 2.06 \times 10^5\,AU = 3.09 \times 10^{18}\,cm. \tag{1.3}$$

The parsec is related to the AU geometrically and is short for "parallax arcsecond."

Consider two objects in the sky separated by 1 AU perpendicular to the line of sight. If the two objects are 1 parsec away and separated by 1 AU, they will subtend 1 arcsecond on the sky. For example, an observer 1 parsec (3.26 ly) from the earth and sun, which are separated by 1 AU, will observe the two bodies to subtend 1 arcsecond of angle. This unit is a consequence of considering parallax motion of distant observed objects as the earth orbits the sun (Fig. 1.1).

The parsec is a standard unit for describing the cosmic distance between interstellar, galactic, or intergalactic objects. For example:

- 1 pc = interstellar distances

- 1 kpc = 10^3 pc = distances in a galaxy

- 1 Mpc = 10^6 pc = distances between galaxies

- 1 Gpc = 10^9 pc = scale of the observable universe

Within just a few multiples of 1000 we are able to range from galactic distances to ones that span the visible universe. It is useful to remind ourselves of the various prefixes for various powers of ten. A range of powers of 10 will be found in use throughout this text. A list of the most common ones are as follows, from largest to smallest.

Exa (E): 10^{18}, typically used to describe the highest energies observed in the current universe (e.g., exa-electron-Volts).

Peta (P): 10^{15}, again, typically associated with very high-energy cosmic processes.

Tera (T): 10^{12}, typically used to describe the highest energies available in terrestrial, human-made particle accelerators (e.g., tera-electron-Volts).

Giga (G): 10^9, typically marks the beginning of the scale used with mass-energy of the heaviest subatomic particles, beginning with the proton and neutron. In terms of distances and times, this prefix appears in typical descriptions of the size of the visible universe and in the age of the universe (e.g., gigaparsecs and gigayears).

Mega (G): 10^6, typically used for energy scales at the level of nuclear processes (e.g., fission and fusion and binding energy in nuclei). In distances, this prefix is used to define the distances between galaxies and galaxy clusters (e.g., megaparsecs).

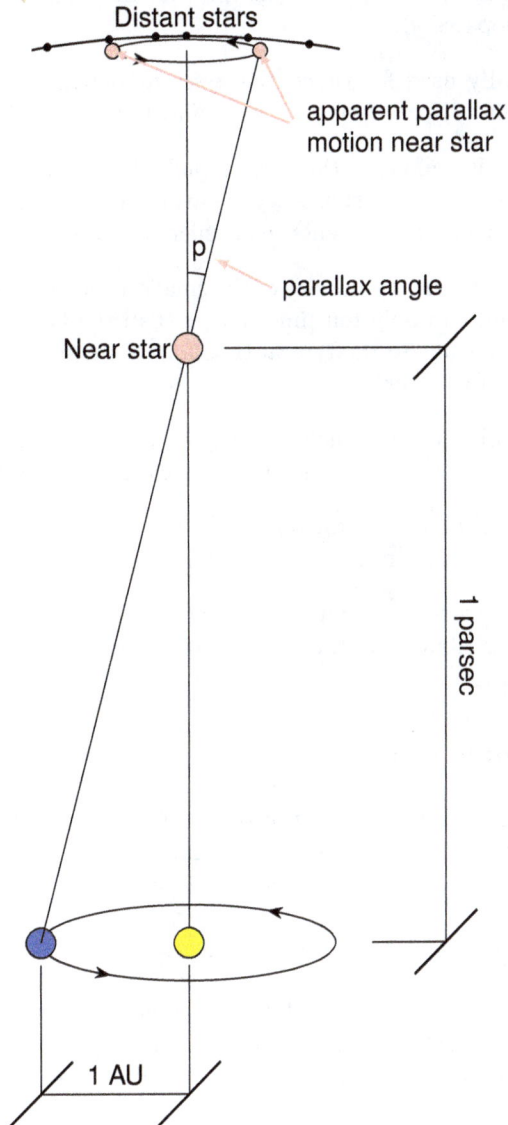

Figure 1.1: A geometric representation of the parsec as a unit of distance, a consequence of parallax of nearby objects (such as stars) against much-more-distant background objects. Based on an example from Ref. [use19].

kilo (k): 10^3, typically used for energy scales at the level of the strongest bonded atomic electrons (e.g., those closest to the nucleus in heavy atoms like lead, which are usually kilo-electron-volts). In distances, this prefix is used to define distances within galaxies (e.g., kiloparsecs).

milli (m): 10^{-3}, typically used for timescales associated with software-based decision-making algorithms typical of modern experiments (e.g., milliseconds).

micro (μ): 10^{-6}, typically marking the size of biological cells. In time, this prefix is associated with the typically shortest time available to make algorithmic decisions in hardware in modern experiments (e.g., microseconds).

nano (n): 10^{-9}, typically marking the size of the smallest biological cell components (e.g., the cell membrane) and only ten times larger than the typical size scales of atoms. In time, this corresponds to the typical time for a light signal to travel about 1 foot (about one-third of a meter).

pico (p): 10^{-12}, typically marking the size between the atomic and nuclear size scales. In time, this typically marks the shortest timescales available to modern electronics.

femto (f): 10^{-15}, typically marking the size scale of the nucleus (e.g., femtometers). In time, this represents the timescale of a chemical reaction (e.g., femtoseconds).

atto (a): 10^{-18}, typically associated with the current limit of knowledge of subatomic size scales (e.g., the distribution of the electron's electric charge is known to be only to about one attometer).

1.2   Particle Physics Units

In particle physics, it is standard to let the speed of light, c, and the reduced Planck's Constant, $\hbar$, define the maximum speed and smallest unit of action, respectively. Thus it is standard to redefine all units by setting $c = \hbar = 1$. This system is typically called *natural units*.

Units of energy are conventionally stated in electron-Volts (eV) and related powers of ten of this unit. An energy of 1 eV is the scale of energy associated with atomic physics while 1 keV is associated with x-rays, and 1 MeV with nuclear physics. Subnuclear physics scales are at the level of GeV and TeV. The highest energy particles observed in the universe to date are cosmic rays, having energies as high as PeV (10^{15} eV) and even EeV (10^{18} eV).

In natural units it is also conventional to set Boltzmann's constant, k_B, to $k_B = 1$, so that energy and temperature are the same (by way of $E = k_B T$). Energy and time are related through one of the statements of the Heisenberg uncertainty principle, $\Delta E \Delta t \geq \hbar/2$[2]. In natural units, since $\hbar = 1$, it will be true that $E \propto 1/t$. In natural units,

---

[2]Alternatively, the same conclusion can be drawn from one of the de Broglie relationships, $E = \hbar\omega$, where ω is an angular frequency with units s^{-1}.

time is measured in eV^{-1}. Energy is thus also directly related to distances in space by the special theory of relativity, since invariant intervals in a geometrically flat space-time are given by $s^2 = c^2 \Delta t^2 - \Delta \vec{r} \cdot \Delta \vec{r}$. Since energy and time are related, and time and space are related, space also has natural units of eV^{-1}.

The combination of particle physics units with energy units is to harmonize the description of momentum, energy, and mass. Consider Einstein's special relativistic energy-momentum-mass relationship in meters-kilograms-seconds (MKS) units:

$$E^2 = p^2 c^2 + m^2 c^4. \tag{1.4}$$

Here, energy will have units of eV; momentum units of eV/c; and mass units of eV/c^2. However, applying the convention $c = 1$ allows all of these quantities to be measured in units of just eV. A helpful conversion factor is that $1 \, kg \approx 5.7 \times 10^{26} \, GeV$ in natural units. The proton mass in natural units is $m_p = 0.93827208816 \, GeV \approx 1 \, GeV$ and provides a useful baseline for comparing dark matter candidate masses to more familiar particles.

The standard unit of charge is the elementary charge, $e = 1.602 \times 10^{-19} \, C$. In natural units, it is also conventional to set $e = 1$ and measure all charges relative to the fundamental unit.

The standard unit of cross-sectional area in particle physics is the *barn*, where $1 \, b = 10^{-24} \, cm^2$. In quantum mechanics, cross-sectional area is related to the probability of an interaction (e.g., via scattering theory). Therefore, cross-section in physical processes is synonymous with interaction probability. The smaller the cross section of a process, the smaller the probability for that process to occur. The cross-sectional area of a uranium nucleus is approximately 1 b. The interaction probability associated with the strong nuclear interaction is also about this same scale. The weak nuclear interaction has cross-sections that are more at the level of nanobarns or picobarns, where $1 \, nb = 10^{-36} \, cm^2$ and $1 \, pb = 10^{-39} \, cm^2$. The comparison yields the expectation, consistent with observation, that strong nuclear processes dominate over weak nuclear processes owing to the much higher probability for the former to occur.

Exercise 1

In the United States, the standard distance unit is the mile. There are exactly 5280 feet to a mile, and exactly 12 inches to a foot. If 1.00 inch = 2.54 cm, how many kilometers are there in a mile?

Exercise 2

Earth is a nearly round planet that can be modeled as a sphere having a radius of 3958.8 miles. How many kilometers is that?

Exercise 3

In the system of natural units, the speed of light (c) and Planck's reduced constant, $\hbar \equiv h/(2\pi)$, are chosen to represent the fundamental units of speed and angular momentum, respectively. Therefore, they are set to $\hbar = c = 1$. This then scales all angular momenta and speeds so that they are defined *relative* to $\hbar$ and c, respectively. In natural units, the product $\hbar c$ is thus also equal to 1. What is this equal to in electron-volts and meters (which are familiar as part of the more traditional meter-kilograms-seconds, or MKS, unit system)?

Exercise 4

There are approximately 10^{80} atoms in the cosmos. If 76% of these atoms are hydrogen (with mass $m_H = 1.67 \times 10^{-27}$ kg) and 24% are helium (with mass $m_{He} = 6.65 \times 10^{-27}$ kg), what is the total mass of atomic matter in the universe?

Exercise 5

A neutron star is the result of the death of a very heavy star, one with a core at least 1.5 times heavier than our sun's. If a typical neutron star is approximately the shape of a sphere having a diameter of 25 km, but a mass of 3.0×10^{30} kg, what is the density of a typical neutron star? (BONUS: compare this to the density of lead.)

Exercise 6

The galactic center of the Milky Way is marked by a supermassive black hole called Sagittarius $A^\star$. This black hole is so far away that it takes light 26,673 years to reach Earth. In kilometers, this distance is immense. Astronomers prefer to quote such vast distances in units of *kiloparsecs*, where 1 parsec is the distance traveled by light in 3.26 years. How many kiloparsecs is it from Earth to Sagittarius $A^\star$?

Exercise 7

A potentially habitable planet orbits the star nearest to the sun and the earth, Proxima Centauri. The planet, Proxima Centauri b, is approximately 4.017×10^{13} km from Earth.

- A beam of light (e.g. a radio signal) travels at the speed of light in the mostly empty space between earth and Proxima b. At that speed ($c = 2.998 \times 10^8$ m/s), and assuming straight-line travel, how long (in years) does it take for a light signal to go from Earth to Proxima b?

- The longest-traveling and fast-moving instruments we have ever put into space are the Voyager probes, launched in 1977. The fastest of these moves, relative to our sun, at a speed of 3.803×10^4 mph. How long, in years, would it take Voyager to reach Proxima b?

1.3 The Standard Model

We do not provide a detailed review of the Standard Model of Particle Physics (which shall henceforth be abbreviated as SM). We do, however, summarize the salient features of this model here.

The SM is a quantum field theoretical framework that describes the known constituents of matter and the forces that bind them together or alter one form of matter into another (e.g., the *decay* process). The framework incorporates mathematical symmetry. A very basic example of a transformation that could have an associated symmetry would be to shift a potential energy function by some offset, e.g. $V(x)' = V(x) + V_0$. A theory that possesses a symmetry would be invariant under such a transformation, implying that the Lagrangian (or Lagrangian density, in the case of a quantum field) describing the system remains invariant. Symmetries like this can be described by the properties of mathematical groups, and are referred to as "symmetry groups." A "gauge symmetry" is present if one can apply a transformation to a field at every point in spacetime and the theory remains invariant under such a change.

The matter particles in the standard model are described by fields that obey Fermi-Dirac statistics and are known as *fermions*. As a class, their distinguishing feature is that they possess half-integer quantum spin angular momentum, henceforth referred to as *spin*. The force-carrying particles (e.g., the photon) are consequences of the gauge symmetries, obey Bose-Einstein statistics, and are called *bosons*. Their distinguishing feature is that they possess integer spin. Matter particles are mathematically described using wave functions, denoted $\psi(\vec{r}, t; \mathcal{Q})$, where $\vec{r}$ and t refer to space and time coordinates of the function and $\mathcal{Q}$ refers to the set of quantum numbers (such as charge and spin) associated with the particle.

There are also discrete mathematical transformations that can be applied to wave functions. These transformations may have physical consequences. The three discrete transformations most often discussed in the context of the SM are the parity (P), time reversal (T), and charge conjugation (C) operations. Consider a wave function with spacetime coordinates and a charge, q. The three operations described above would be written mathematically as

$$P\psi(\vec{r}, t, q) = \psi(-\vec{r}, t, q); \tag{1.5}$$
$$T\psi(\vec{r}, t, q) = \psi(\vec{r}, -t, q); \tag{1.6}$$
$$C\psi(\vec{r}, t, q) = \psi(\vec{r}, t, -q). \tag{1.7}$$

A wave function that remains unchanged by any one of these discrete transformations, or any combination of them, is said to be *invariant* under that operation. A phenomenon is referred to as *chiral* if its parity-inverted form is not the same as its non-inverted form.

For example, one can define the projection of spin onto linear momentum for a particle, $\lambda = \vec{s} \cdot \vec{p}/(|s||p|)$, called "helicity." A left-handed particle is one in which $\lambda = -1$, while a right-handed particle is one in which $\lambda = +1$. Spin is an angular momentum and thus can be described by a vector composed from two other vectors; classically, mechanical angular momentum is defined as $\vec{L} = \vec{r} \times \vec{p}$. Both $\vec{r}$ and $\vec{p}$ reverse under P, but their vector product remains invariant, so angular momentum is unaffected by a parity inversion. A vector that is not invariant under P is simply referred to as a "vector" (like $\vec{r}$ and $\vec{p}$) while one that is invariant under P is a "pseudovector" or "axial vector" (like $\vec{L}$ or $\vec{s}$). We see that helicity is not invariant under P, since it is the product of a regular vector and an axial vector.

Particles can have both *chirality* and *helicity*, and these need not be the same. The former refers to how the wave function itself transforms under the Lorentz symmetry group, while the latter refers to the spin-momentum product described above. In the limit of massless particles, these two quantities become the same, but otherwise are distinct.

Matter is organized in three generations. The first generations contains two *quarks* (the up and down) and two *leptons* (the electron e^- and the electron-type neutrino, ν_e). Up and down quarks are the main constituents of protons and neutrons (e.g., a proton is uud while a neutron is udd). The second generation of matter is organized similarly, with the charm and strange quarks and the muon and muon-type neutrino. As these are generally heavier than the first generation counterparts, matter and interactions involving them are more rare in nature as they exist only in higher-energy processes. The third generation, consisting of the top and bottom quarks and the tau lepton and tau-type neutrino, are the heaviest particles and thus the rarest in the universe (they are only known to have been directly produced in high-energy particle collisions due either to human-made instrumentation or cosmic rays).

There are three forces that are described in the SM: the electromagnetic force, the weak force, and the strong force. We use *force* and *interaction* interchangeably to describe these phenomena. The electromagnetic (EM) force is familiar as the one responsible for macroscopic phenomena like static electricity and electric current, as well as the reason atoms exist (they are bound states of protons with electrons). This force is infinite in range (the force is never zero between two electrically charged objects unless they are an infinite distance apart from each other). This is due to the fact that the photon is massless. Electric charge is the origin of the EM interaction and comes in two kinds, denoted *positive* (+) and *negative* (−). The quantum theory of the electromagnetic interaction is known

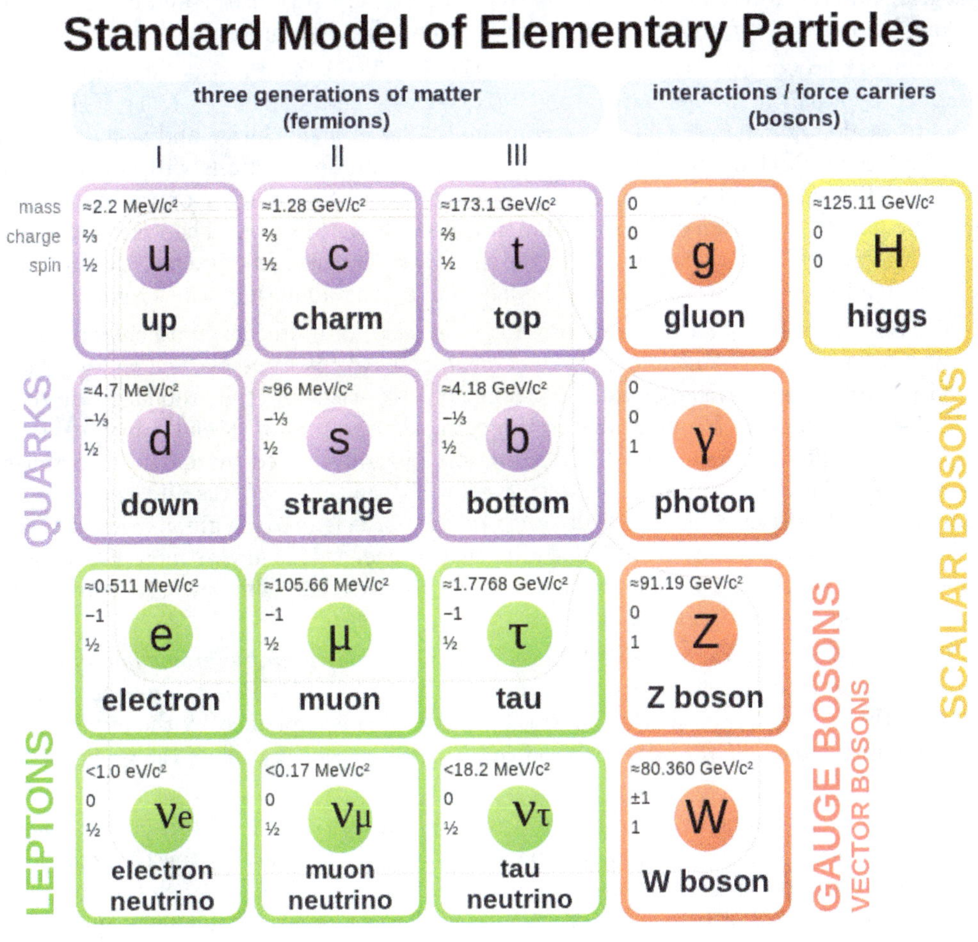

Standard Model of Elementary Particles

Figure 1.2: A table organizing the particles that describe matter (left three columns) and interactions between matter (right two columns). Figure from Ref. [Wik24e].

as *quantum electrodynamics*, or QED. The known properties of the particles described by the SM are illustrated in Fig. 1.2.

The weak interaction is less familiar, but responsible for important macroscopic phenomena such as radioactive decay and stellar fusion. At the microscopic scale, the weak force acts to change one kind of quark (or lepton) into another (e.g., $d \rightarrow u$) and underlies the process of radioactive *beta decay*, which results in the emission of high-energy electrons from atomic nuclei. The weak force is short-ranged, limited to the approximate size scale of a proton or neutron (about 1 fm). This is due to the fact that the particles that transmit this force, the *weak bosons* $W^{\pm}$ and Z^0, are very heavy and cannot exist for long before they decay into other particles or are absorbed in an interaction.

Symmetry groups are a key idea built into the SM. The EM and weak interactions are described by a pair of groups: $U(1)_Y$ and $SU(2)_L$, written as $SU(2)_L \times U(1)_Y$. Here, Y refers to weak hypercharge, which is a combination of electric charge and weak isospin, and L refers to the fact that the weak interaction favors interactions with left-handed particles and right-handed anti-particles over interactions with their opposites, which naturally leads to violation of parity symmetry of the wave function. The weak force is a very chiral phenomenon. For example, in the SM since neutrinos only interact through the weak force all neutrinos are expected to be left-handed and all antineutrinos right-handed.

The strong nuclear force binds quarks together (e.g., into protons and neutrons) and protons and neutrons to one another. Since protons have the same electric charge, they would naturally repel each other. The strong force overwhelms that repulsion at very short distances, distances comparable to the diameter of a proton (about 1 fm), causing them to bind into atomic nuclei. The strong force is transmitted by massless particles called *gluons* that act on objects possessing color charge. Only quarks and gluons have color charge. The reason for the short range of this force is because gluons can interact both with quarks and themselves. This limits their range, as it is always more likely for a gluon to have another strong interaction than to continue traveling away from another quark or gluon as a free particle.

The theory of the strong sector of the SM is known as *quantum chromodynamics*, or QCD, in a nod to its predecessor theory, QED. The heart of QCD is the mathematical fact that the interaction of quarks under the strong force is explained by the application of a symmetry group, $SU(3)$ (often labeled $SU(3)_{color}$), the generators of the group being identified as gluons.

At the current temperature of the universe, quarks are found only in bound states with one another. The primary rule for quark bound states is that they must be colorless. This means their color charges, denoted by "red," "green,", and "blue" labels, can only be found in groups like $rgb = colorless$ or $r\bar{r} = b\bar{b} = g\bar{g} = colorless$). The most common of these are triplets of quarks called *baryons*. The proton is a color-neutral baryon made from an rgb combinations of quarks, uud. The neutron is a similar color-neutral combination, udd. The lightest bound states made from quark-antiquark pairs (e.g., $u\bar{u}$) are the *pions*, but these are unstable and decay to other particles in about 1 billionth of a second.

The Higgs particle (H^0) is the consequence of spontaneously broken gauge symmetry in the SM. Specifically, breaking the electroweak gauge symmetry yields masses for some of the force-carrying particles (the bosons that carry the weak force) but leaves the photon and the gluons (the particles that transmit the strong force) massless. Fermions also appear to acquire their masses in this process, except in the case of the neutrinos. Neutrinos are known to have small, non-zero masses but the origin of this attribute is not yet satisfactorily explained. Neutrino mass is generally considered to be explained by something outside the SM.

Gravity is not described by the SM but is instead only known to be accurately modeled by the General Theory of Relativity, which is not a quantum field theory. Attempts to make a quantum field theory of gravity have not succeeded, but that work is not considered complete.

Feynman Diagrams

A common tool in particle physics is the use of a special diagram to summarize the mathematical rules for combining particles, for example from an initial set of particles that can interact and produce, through intermediaries, a final set of particles. These images are called *Feynman diagrams*.

Each direction in the diagram has meaning, as does each line and each intersection of more than two lines. A well-written Feynman diagram can be converted into a mathematical calculation of a matrix element representing the transition from an initial to a final state. The example shown in Fig. 1.3, described below, is one contribution to the transition matrix $\langle gg \, | \mathcal{H} | \, \gamma\gamma \rangle$, where $\mathcal{H}$ is the Hamiltonian of the system that can connect the initial di-gluon state to the final di-photon state.

Implicit in each Feynman diagram is a space axis and a time axis. Generally, space will be the vertical axis and time the horizontal axis. The condensed notation for the elements of a diagram in use in this book is as follows:

- A solid line: this represents a fermion, generally a quark or lepton. If a specific fermion is involved, the line will be labeled. If the diagram admits any possible fermion, the line will remain unlabeled.

- A dashed line: this represents a spin-0 particle;

- A wavy line: this represents a photon, W boson, or Z boson (labels will indicate which);

- A curly line: this represents a gluon.

A line represents the propagation of a particle between two spacetime points in the diagram. A *vertex* is the intersection of more than two lines at a common spacetime point. A vertex represents an interaction. A vertex is allowed if conserved quantum numbers

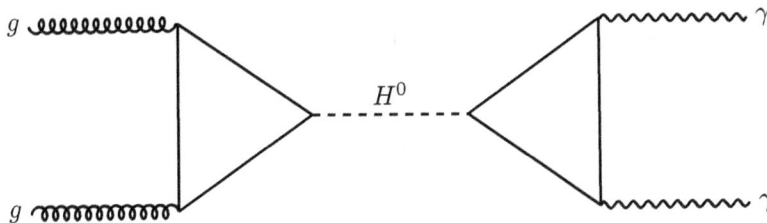

Figure 1.3: An example of Feynman diagram. This one represents the interaction of two gluons, via an intermediate "loop" containing quarks, which then produce a Higgs particle. The Higgs decays through another loop to photons.

are respected by writing the intersection (for example, conservation of electric or color charge, or both). When a gauge boson is part of a vertex, the "strength" of the coupling to the boson (proportional to the observed degree of this interaction in nature) enters through a coupling constant, α. A factor of $\alpha \propto g^2$, where g is the intrinsic strength of an interaction ($g = e$, the elementary charge, for the electromagnetic interaction of electrons via a photon), will appear at each such vertex.

The electromagnetic coupling constant, α_{EM}, is also known as the *fine-structure constant* owing to its role in the details of atomic structure, and is famously $\alpha_{EM} \approx 1/137 \approx 10^{-2}$. This dimensionless number is related to the strength of the electric charge, $\alpha_{EM} = e^2/(4\pi\varepsilon_0\hbar c)$, where ε_0 is the permittivity of free space in electromagnetic theory. The strong coupling constant is $\alpha_s \approx 1$ under similar conditions (e.g. when considering atomic or nuclear phenomena), while the weak coupling constant is $\alpha_W \approx 10^{-6}$.

We show an example of a Feynman diagram in Fig. 1.3. The process represented there is the interaction of a pair of gluons, via a fermion loop, into a Higgs particle. The Higgs particle then decays, also through a fermion loop, into a pair of photons. The gluon loop can only involve quarks, since color and electric charge must be conserved at the gluon-fermion vertexes. Gluons carry net non-zero color charge, so the fermion pair that stems from the gluon-fermion vertex must not be colorless. A lepton pair would carry net-zero color charge and is forbidden in the left-hand loop.

The right-hand loop, after the Higgs decay, can contain either leptons or quarks. The Higgs particle carries a net-zero color and net-zero electric charge, so any lepton or quark can be present in the subsequent loop so long as the combination sums to zero. This requires the loop to contain quark-antiquark or lepton-antilepton combinations only.

The fermions in the right-hand loop must contain electric charge or they cannot couple to the photons. Therefore, this narrows the contributions in the right-hand loop to any quark or any charged lepton; neutrinos are forbidden in the right-hand loop.

A single such diagram can convey a wealth of information. Conversely, assumptions about the nature of a process can constrain the ways in which one can draw such diagrams.

PART I

OBSERVATIONS AND EVIDENCE

CHAPTER 2

The Evidence for Dark Matter

2.1 Overview

This chapter will review some of the evidence for dark matter. This is by no means intended as a complete historical record, nor a compendium of all such evidence. For that, we recommend reviews like Ref. [BH18]. Instead, we will focus on the major elements of the case positing the existence of nonluminous gravitating matter.

The history of unseeable objects in astronomy is long, but key insights are often highlighted in discussions of the history of astronomy. John Michell's conceptualization, based on Isaac Newton's formulation of mechanics, of a star so massive that its gravity prevents light from escaping its surface (a "dark star") would later find even deeper mathematical and physical purchase in the general theory of relativity. Ultimately, these concept seeds (scattered and planted by many scientists over almost two centuries) would lead to the prediction of black holes. These were first observed in the latter half of the twentieth century. Black holes—stellar corpses so massive that there is a region around them from which light cannot escape—are now tools for studying the larger cosmos, in addition to they being, themselves, objects of study [Abb+23b].

If dark stars were permissible even by Newton's laws of mechanics, then it would be possible for there to be a large inventory of unseeable objects littering the universe. Over centuries, astronomers sought to catalog the unseeable, using gravitational influence (and, later, wavelengths undetectable by the human eye). In this chapter, we highlight two key players in these efforts. These two individuals are recent participants in the scientific story of cataloging unseen mass and are important to the modern era of dark matter hunting. This chapter will highlight the astronomical and physical aspects of this work, using it to motivate the case for an unseen astronomical form of matter present in the universe today.

2.2 Galaxy Clusters

The most famous early evidence for the existence of dark matter was put forth by the astrophysicist, Fritz Zwicky. Zwicky was born in Bulgaria, later conducting his university studies at the Swiss Federal Polytechnic (ETH Zürich) in Switzerland's capital, Zurich.

In 1925, he immigrated to the United States to work with the physicist, Robert Millikan, at the California Institute of Technology (Caltech).

Zwicky (Fig. 2.1) made several major contributions to both astronomy and physics, including the concept of the "supernova" as an exploding star; the idea of "neutron stars," collapsed dead stars supported from further collapse only by quantum mechanics; and the observation that there may be substantial nonluminous, but gravitationally affecting matter in the universe.

His observation of what came to be known as "dark matter" began with investigations of a large cluster of galaxies known as the Coma Cluster (Fig. 2.2) [Zwi33; Zwi37]. Drawing on astrophysical scales, recall that Megaparsecs (Mpc) correspond to sizes among galaxies, while kiloparsecs correspond

Figure 2.1: Fritz Zwicky [Wik22a]

to distances within galaxies. In the Coma Cluster, we are dealing with size scales at the level of Mpc. Formally, the Coma Cluster is designated as object Abell 1656, referring to

Figure 2.2: A false-color image of the central region of the Coma Cluster compiled in 2007. The fuzzy objects, of which there are thousands, are galaxies. The smaller fuzzy objects are dwarf galaxies, while the two larger fuzzy objects are a pair of dominant elliptical galaxies. Image from NASA / JPL-Caltech / L. Jenkins (GSFC).

the identification of this astronomical object in the catalogue compiled in 1958. [ACO89]. The cluster is located an average of 99 Mpc from Earth, or about 321 million light-years.

Zwicky was able to establish a number of properties of the galaxies in the Coma Cluster using basic physics ideas and detailed astronomical observations. For example, consider the Doppler-shifting of light emitted by galaxies in the Coma Cluster due to their motion along the line of sight—the radial Doppler shift. If light is emitted from hydrogen-rich stars in a galaxy of the Coma Cluster at $\lambda_{H-\alpha} = 656.3$ nm (corresponding to frequency $f_{H-\alpha} = 4.568 \times 10^{14}$ Hz, also known as the *Balmer line* in the hydrogen spectrum), but the cluster is moving along the line of sight at a velocity, u, then the shift of its frequency, according to the relativistic Doppler effect, will be given by[1]:

$$f_{obs} = \frac{f_{H-\alpha}}{\gamma(1 + u/c)} \xrightarrow{c = \lambda f} \lambda_{obs} = \lambda_{H-\alpha} \times \gamma(1 + u/c) \qquad (2.1)$$

Here, $\gamma = (1 - u^2/c^2)^{-1}$ is the relativistic *gamma factor*. It is convention to define all velocities relative to the speed of light[2] and to denote such relative velocities as $\beta \equiv u/c$. We will use this convention.

If $u > 0$, corresponding to a galaxy receding from our view (moving away from us along the line of sight), we see that $f_{obs} < f_{H-\alpha}$. The Balmer line appears more red-shifted (lower in frequency, longer in wavelength) than it would if the hydrogen-emitting source were at rest along the line of sight. Conversely, if $u < 0$, and the galaxy is approaching the observer along the line of sight, then the Balmer light will appear more blue-shifted (higher in frequency, shorter in wavelength). With the proper instrumentation, it is possible to observe the spectrum of light emitted from distant stars and galaxies and, using measurements of the wavelength of key emission or absorption lines in the spectrum, determine the Doppler shift away from the nominal rest-value of the light emission. This, in turn, allows astronomers to infer the radial velocity of distant objects.

Using this approach, Zwicky established that the galaxies of the Coma Cluster have radial speeds of about 1000 km/s, or about 0.33% the speed of light. The average radial velocities of the galaxies, $\bar{u}$, can be related to the gravitational mass, M_{motion}, that must be holding those galaxies together in the cluster by application of the virial theorem[3]. This leads to the approximate relationship that

$$M_{motion} \approx \frac{R\bar{u}^2}{G} \qquad (2.2)$$

where G is Newton's gravitational constant ($G = 6.674 \times 10^{-11}$ N $\cdot$ m^2/kg^2) and R is the average radius of the cluster.

---

[1]You will find this formula written in one of at least a few ways, depending on either the textbook or the paper you are reading. A common way of writing this is $\lambda_{obs} = \lambda_{rest} \frac{(1+\beta)}{(1-\beta)}$, where $\beta = u/c$. The number of identities and relationships between γ (see Eqn. A.8.2) and β in special relativity is a deep, fun rabbit hole. We invite you to explore this rabbit hole, which we consider to be very good practice with algebra.

[2]In the system of units introduced in Section 1.2, the speed of light is redefined as $c = 1$. We see that, in this case, all velocities become fractions of 1, and so, $\beta = u/c \xrightarrow{c=1} \beta = u$. In other words, β *is* the same as velocity in this system of units.

[3]The word "virial" is derived from the Latin word *vis* meaning "force" or "energy."

Zwicky's clever approach was to then measure the mass in the cluster using an alternative and independent means and compare the two numbers. A reliable establishment of the physical properties of a system is best accomplished by at least two independent methods, if they are available. If there is a complete understanding of the system under study, one expects the independent measurements to yield the same value (within uncertainty on the observations) of the property. If one's knowledge is incomplete, the comparison may reveal significant disagreements in the numbers. This is an invitation to improve one's knowledge of the system under study.

Zwicky used an independent fact about light and mass to infer the mass in the Coma Cluster. Galaxies are made from stars and hot gas. This is the primary means by which light is emitted from galaxies; the gas, itself, is expected to be primarily heated by the action of the stars. Therefore, if there is a relationship between stellar mass and the amount of light emitted by the stars, one can use that relationship to infer mass from light. Specifically, one can measure the brightness ("luminosity") of galaxies and, using the mass–light relation, infer the amount of mass tied up in stars and gas. One can refer to this as the "luminous mass" of the cluster.

The mass-luminosity relationship for a single star results from a complex set of factors and is given approximately by:

$$L \approx \frac{1}{15} \frac{2\pi^3}{9 \cdot 5^5} \frac{G^4 \, m_e^2 \, m_n^5}{\alpha^2 \hbar^5} M^3 \qquad (2.3)$$

where m_e is the mass of the electron, m_n is the average mass of the nuclei in the star, $\alpha \approx 1/137$ is the fine-structure constant, and $\hbar$ is the reduced Planck's constant. We can make a hand-waving argument that motivates this relationship without wading into the specific derivation of this particular formula.

Radiation in a star results from nuclear fusion. However, for nuclear fusion to occur, the nuclei must be forced into very close proximity, or made to collide at high speeds, such that the Coulomb barrier, owing to the repulsion of protons between colliding nuclei, can be overcome. Initiating fusion requires immense pressure. This is provided by the gravitational collapse of the hydrogen and helium present in a protostar, which relentlessly draws the matter of the star in toward its center of mass. This compression is therefore going to be dependent on the mass of the star. A heavier star will generate more pressure, and, thus, more rapid fusion. A lighter star will generate less pressure, and thus a slower rate of fusion. We would then expect the luminosity of a single star to depend proportionally on some power of the mass and Newton's gravitational constant, as well as the mass of the subatomic particles (e.g., nuclei) being driven into fusion and thus generating the radiant energy. An equation of the form shown in Eqn. 2.3, where luminosity increases with mass (more mass yields more gravitational pressure, increasing the rate of fusion) and with the mass of the nuclei involved (since the energy of fusion tends to grow as the mass of the nuclei increase), provides the appropriate description.

We can conclude that there would naturally be a relationship between the observed luminosity of an overall galaxy and the mass present in that galaxy as stellar material. In addition, gas heated by stellar activity will emit light, so the radiance of the stars also drives radiation from the gas in galaxies not itself bound up in stars. Zwicky used the observed relationship between mass and luminosity to make an independent estimate of the luminous mass in the Coma Cluster.

The comparison of his two results—the gravitational mass from the galaxy dynamics and the luminous mass from the mass-to-light ratio—led to a stunning inconsistency. From the two numbers, Zwicky was led to the conclusion (assuming no fatal flaw in either or both approaches) that the amount of gravitating mass was ten times greater than the mass inferred from light.

Exercise 8

Using the observation that a galaxy in the Coma Cluster is receding along the line-of-sight at a velocity of $u = 1000$ km/s, estimate the percent shift in the wavelength of the Balmer series hydrogen α spectral line due to this motion.

The Virial Theorem

Zwicky utilized the virial theorem to infer mass from gravitation. It is worth spending some time on this theorem. It is essentially a statement of the relationship of gravitational potential energy and kinetic energy in a closed system (free from any meaningful external forces) in equilibrium.

A student with sufficient preparation in mathematics, and with a basic understanding of conservation of energy, will find the derivation of the virial theorem useful. We are interested in relating the motion of galaxies to the gravitational potential energy of the overall cluster. If we can make such a relationship (which is what the virial theorem provides), a measurement of motion allows us to infer the mass in a gravitating system.

Consider a system of N galaxies. Let the ith galaxy have a position, measured relative to the center of mass of the cluster, given by $\vec{r}_i$. Let its mass be denoted m_i. Let us also assume the center of mass (COM) of the cluster is in a state of uniform motion. These assumptions are consistent with the premise: a system in equilibrium free from significant external forces, $\vec{F}$, e.g., $\sum_k \vec{F}_k = d\vec{p}_{COM}/dt = 0$, implies that the COM moves at constant speed.

We can then write down the moment of inertia (rotational inertia) of the system:

$$I = \sum_i m_i \vec{r}_i \cdot \vec{r}_i = \sum_i m_i \vec{r}_i^2. \tag{2.4}$$

This equation assumes that the rotation of the system of galaxies occurs about a spatial axis and that we are measuring distances to the galaxies from that axis. In anticipation of needing these later, let us compute some useful time derivatives of the rotational inertia:

$$
\begin{aligned}
\frac{dI}{dt} &= \frac{d}{dt}\left(\sum_i m_i \vec{r}_i \cdot \vec{r}_i\right) \\
&= \sum_i m_i\left[\left(\frac{d}{dt}\vec{r}_i\right)\cdot\vec{r}_i + \vec{r}_i\cdot\left(\frac{d}{dt}\vec{r}_i\right)\right] \\
&= \sum_i m_i\left[\dot{\vec{r}}_i\cdot\vec{r}_i + \vec{r}_i\cdot\dot{\vec{r}}_i\right] \\
&= 2\sum_i m_i\dot{\vec{r}}_i\cdot\vec{r}_i
\end{aligned}
\tag{2.5}
$$

$$
\begin{aligned}
\frac{d^2 I}{dt^2} &= \frac{d}{dt}\left(2\sum_i m_i\dot{\vec{r}}_i\cdot\vec{r}_i\right) \\
&= 2\sum_i m_i\left(\ddot{\vec{r}}_i\cdot\vec{r}_i + \dot{\vec{r}}_i\cdot\dot{\vec{r}}_i\right) \\
&= 2\sum_i m_i\ddot{\vec{r}}_i\cdot\vec{r}_i + 2\sum_i m_i\dot{\vec{r}}_i\cdot\dot{\vec{r}}_i
\end{aligned}
\tag{2.6}
$$

The rotational kinetic energy of any single galaxy in the cluster will be given by

$$
T_i = \frac{1}{2}m_i\dot{r}_i^2
\tag{2.7}
$$

such that $T = \sum_i T_i$ is the total kinetic energy of the galaxies. We can rewrite the second derivative of the rotational inertia as

$$
\frac{d^2 I}{dt^2} = 2\sum_i m_i\ddot{\vec{r}}_i\cdot\vec{r}_i + 4T.
\tag{2.8}
$$

The time-average of any quantity $y(t)$ from $t = 0$ to a time $t = \tau$ is defined as

$$
\langle y\rangle = \frac{1}{\tau}\int_0^\tau y(t)\,dt.
\tag{2.9}
$$

We can use this to calculate the time-average of the second derivative of the rotational inertia:

$$
\begin{aligned}
\left\langle\frac{d^2 I}{dt^2}\right\rangle &= \frac{1}{\tau}\int_0^\tau\left(2\sum_i m_i\ddot{\vec{r}}_i\cdot\vec{r}_i + 4T\right)dt \\
&= \frac{2}{\tau}\int_0^\tau\sum_i m_i\ddot{\vec{r}}_i\cdot\vec{r}_i\,dt + \frac{4}{\tau}\int_0^\tau T\,dt \\
&= 2\sum_i m_i\langle\ddot{\vec{r}}_i\cdot\vec{r}_i\rangle + 4\langle T\rangle.
\end{aligned}
\tag{2.10}
$$

Let us assume not only that this system (the Coma Cluster) is in equilibrium (free of net external linear forces and torques), but is also closed and isolated in all other ways. This assumption allows us to impose the conservation of linear and angular momentum. This is a reasonable assumption for a cluster like this one, which, while bound to itself by gravity, would only be comparatively weakly influenced by the gravitational forces of exterior entities (e.g., other clusters).

In that case, one can demonstrate that the second-derivative of the moment of inertia of the entire system necessarily must vanish.[4] We require that $\ddot{I} = 0$, which then (based on Eqn. 2.8) requires that:

$$\left\langle \frac{d^2 I}{dt^2} \right\rangle = 0 = 2 \sum_i m_i \langle \ddot{\vec{r}}_i \cdot \vec{r}_i \rangle + 4\langle T \rangle \tag{2.11}$$

The left term in the above equation contains what is known as "the virial" of the system, $m_i \langle \ddot{\vec{r}}_i \cdot \vec{r}_i \rangle$. We can recognize this product as the dot product of a force ($m\ddot{\vec{r}}$) by Newton's second law, with a position vector ... very much like the definition of work done by an external force, $W = \vec{F} \cdot \vec{r}$. In fact, owing to the relationship between work done by a conservative force on matter, and the change in potential energy of matter in the field associated with that conservative force, we can expect that this equation will result in a statement relating the average energy of motion (kinetic energy) to the average energy of configuration (potential).

We can unpack this by beginning with Newton's second law for the ith galaxy in the cluster:

$$m_i \ddot{\vec{r}}_i = \sum_{j,\, j \neq i} \vec{F}_{ij} \tag{2.12}$$

where $\vec{F}_{ij}$ is the force on the ith galaxy exerted by the jth galaxy. From Newton's law of gravity:

$$m_i \ddot{\vec{r}}_i = \sum_{j,\, j \neq i} -\frac{Gm_i m_j}{|\vec{r}_i - \vec{r}_j|^3}(\vec{r}_i - \vec{r}_j) \tag{2.13}$$

Now construct the virial:

$$m_i \ddot{\vec{r}}_i \cdot \vec{r}_i = \sum_{j,\, j \neq i} -\frac{Gm_i m_j}{|\vec{r}_i - \vec{r}_j|^3}(\vec{r}_i - \vec{r}_j) \cdot \vec{r}_i \tag{2.14}$$

and sum over all the galaxies in the Coma Cluster:

$$\sum_i m_i \ddot{\vec{r}}_i \cdot \vec{r}_i = \sum_i \sum_{j,\, j \neq i} -\frac{Gm_i m_j}{|\vec{r}_i - \vec{r}_j|^3}(\vec{r}_i - \vec{r}_j) \cdot \vec{r}_i \tag{2.15}$$

---

[4]Consider a counter-example, wherein the system in fact satisfies that $\ddot{I} = C_0 \neq 0$. If that is true, then $\dot{I} = C_0 t + C_1$, where neither C_0 nor C_1 are guaranteed to be zero (and where C_0 has been explicitly assumed to be non-zero). The conservation of angular momentum in a system dictates that $\dot{L} = 0$. Using $L = I\omega$, where ω is the angular velocity of the rotating system and is non-zero, this would require that $\dot{L} = \dot{I}\omega + I\dot{\omega} = 0$ for conservation to hold. The second term is identifiable as $I\alpha = \tau$, the torque acting on the system. For a system in equilibrium, this must vanish. That would leave $\dot{L} = \dot{I}\omega = C_0 t\omega + C_1\omega$. Even if we impose that $C_1 = 0$, this equation is nevertheless explicitly not zero at all times. Therefore, for conservation of angular momentum to hold in such a system, it must be true that $C_0 = 0$, imposing the requirement that $\ddot{I} = 0$.

Since the indices are arbitrary, swap $i \leftrightarrow j$:

$$\sum_j m_j \ddot{\vec{r}}_j \cdot \vec{r}_j = \sum_j \sum_{i,\, i \neq j} -\frac{Gm_j m_i}{|\vec{r}_j - \vec{r}_i|^3}(\vec{r}_j - \vec{r}_i) \cdot \vec{r}_j \tag{2.16}$$

Now sum Eqns. 2.15 and 2.16:

$$\sum_i m_i \ddot{\vec{r}}_i \cdot \vec{r}_i + \sum_j m_j \ddot{\vec{r}}_j \cdot \vec{r}_j = \sum_i \sum_{j,\, j \neq i} -\frac{Gm_i m_j}{|\vec{r}_i - \vec{r}_j|^3}(\vec{r}_i - \vec{r}_j) \cdot \vec{r}_i + \tag{2.17}$$

$$\sum_j \sum_{i,\, i \neq j} -\frac{Gm_j m_i}{|\vec{r}_j - \vec{r}_i|^3}(\vec{r}_j - \vec{r}_i) \cdot \vec{r}_j \tag{2.18}$$

and simplify this to

$$2\sum_i m_i \ddot{\vec{r}}_i \cdot \vec{r}_i = \sum_i \sum_{j,\, j \neq i} -\frac{Gm_i m_j}{|\vec{r}_i - \vec{r}_j|^3}\left[(\vec{r}_i - \vec{r}_j) \cdot \vec{r}_i + (\vec{r}_j - \vec{r}_i) \cdot \vec{r}_j\right]. \tag{2.19}$$

This looks complicated, but we are very close to a final relationship between gravitational potential energy and the virial. Concentrate on those last terms in the square brackets:

$$(\vec{r}_i - \vec{r}_j) \cdot \vec{r}_i + (\vec{r}_j - \vec{r}_i) \cdot \vec{r}_j = (\vec{r}_i - \vec{r}_j) \cdot \vec{r}_i - (\vec{r}_i - \vec{r}_j) \cdot \vec{r}_j \tag{2.20}$$

$$= (\vec{r}_i - \vec{r}_j) \cdot (\vec{r}_i - \vec{r}_j) \tag{2.21}$$

$$= |(\vec{r}_i - \vec{r}_j)|^2 \tag{2.22}$$

Thus:

$$2\sum_i m_i \ddot{\vec{r}}_i \cdot \vec{r}_i = -\sum_{ij,\, j \neq i} \frac{Gm_i m_j}{|\vec{r}_i - \vec{r}_j|^3}|(\vec{r}_i - \vec{r}_j)|^2 \tag{2.23}$$

$$= -\sum_{ij,\, j \neq i} \frac{Gm_i m_j}{|\vec{r}_i - \vec{r}_j|} \tag{2.24}$$

$$\sum_i m_i \ddot{\vec{r}}_i \cdot \vec{r}_i = -\frac{1}{2}\sum_{ij,\, j \neq i} \frac{Gm_i m_j}{|\vec{r}_i - \vec{r}_j|} \tag{2.25}$$

The gravitational potential energy of a single galaxy, i, due to the gravitational field exerted by another galaxy, j, is given by:

$$U_{ij} = -\frac{Gm_j}{|\vec{r}_i - \vec{r}_j|} \tag{2.26}$$

The gravitational potential energy of all unique pairs of galaxies is:

$$U = \sum_i U_i = -\frac{1}{2}\sum_{ij,\, i \neq j} \frac{Gm_i m_j}{|\vec{r}_i - \vec{r}_j|} \tag{2.27}$$

Substituting Eqn. 2.27 back into Eqn. 2.25,

$$\sum_i m_i \ddot{\vec{r}}_i \cdot \vec{r}_i = U \tag{2.28}$$

If we then time-average the potential energy:

$$\langle U \rangle = \frac{1}{\tau} \int_0^\tau \left(\sum_i m_i \ddot{\vec{r}}_i \cdot \vec{r}_i \right) dt \tag{2.29}$$

$$= \sum_i m_i \langle \ddot{\vec{r}}_i \cdot \vec{r}_i \rangle \tag{2.30}$$

We finally put all the pieces together and arrive at the virial theorem:

$$0 = 2 \sum_i m_i \langle \ddot{\vec{r}}_i \cdot \vec{r}_i \rangle + 4\langle T \rangle$$

$$0 = \sum_i m_i \langle \ddot{\vec{r}}_i \cdot \vec{r}_i \rangle + 2\langle T \rangle$$

$$0 = \langle U \rangle + 2\langle T \rangle \tag{2.31}$$

where in the second line of the above sequence of equations we have merely divided both sides by 2, which changes nothing about the equation. Equation 2.31 is referred to as *the virial theorem*. Any system that achieves this condition is said to be *virialized*. This term means that the system has achieved stability because it is in equilibrium.

Applying the virial theorem to the Coma Cluster

We can now use the virial theorem to estimate the total mass of the Coma Cluster. For example, if we assume a spherical galaxy cluster with N galaxies, each of mass m with a typical root-mean-square velocity, $\nu \equiv \sqrt{\langle v^2 \rangle}$, then, given the typical kinetic energy of a single galaxy, we can approximate the kinetic energy of all N galaxies:

$$T_i = \frac{1}{2} m v_i^2 \longrightarrow \langle T \rangle = \frac{1}{2} N m \nu^2 \tag{2.32}$$

We can imagine this galaxy is part of a spherically symmetric group of galaxies. For a single galaxy a distance R_i from the center of the distribution, where the sphere of radius R_i encloses a total mass M, the potential energy is:

$$U_i = -\frac{3}{5} \frac{GM}{R_i^2} \longrightarrow \langle U \rangle = -\frac{3}{5} \frac{GM}{R^2}. \tag{2.33}$$

In both cases, we have assumed that there is some typical galaxy in the cluster with velocity v_i, and at radius R_i, whose speed and distance represent the typical (e.g., mean) values of galaxies in a similar region of the distribution. We insert these into the virial theorem:

$$0 = \langle U \rangle + 2\langle T \rangle = -\frac{3}{5} \frac{GM}{R^2} + N m \nu^2 \tag{2.34}$$

The total mass of the cluster is given by $Nm = M$, so:

$$0 = -\frac{3}{5}\frac{GM}{R^2} + M\nu^2 \longrightarrow M = \frac{5R\nu^2}{3G} \tag{2.35}$$

We need to briefly deal with the issue of galaxy velocities and how they would be distributed in three dimensions in the cluster. The dispersion of velocities in the cluster is the distribution of the velocities about the mean value. The velocity-squared in one dimension (for example, the radial direction) will be equivalent to the radial velocity dispersion, σ_r: $\nu_{1D}^2 = \sigma_r^2$.

If the motion of the galaxies in the cluster is isotropic, then there is nothing special about the velocity in the radial direction compared to any other direction. For example, there is nothing special about the velocity component pointing toward the earth compared to any other direction of motion. Thus, the three-dimensional velocities of galaxies in the cluster will have an average squared velocity of $\nu_{3D}^2 = 3\nu_{1D}^2 = 3\sigma_r^2$.

Let us apply this thinking—a spherically symmetric cluster of N galaxies with isotropic motion—to the virial theorem.

$$0 = \langle U \rangle + 2\langle T \rangle = -\frac{3}{5}\frac{GM}{R^2} + Nm\nu^2 = -\frac{3}{5}\frac{GM}{R^2} + 3M\sigma_r^2 \tag{2.36}$$

We would then predict that

$$M = \frac{5R\sigma_r^2}{3G}. \tag{2.37}$$

We can employ information from Zwicky's measurements of the Coma Cluster to reproduce his calculations. He estimated the radius of the Coma Cluster to be 613 kpc. He observed the average line-of-sight galaxy velocity to be 236 km/s. From this, we can estimate the mass of the Coma Cluster from kinetic information, $M = 7.8 \times 10^{43}$ kg $= 3.9 \times 10^{13}$ M$_\odot$. Here, M$_\odot$ is the mass of our own sun (1.99×10^{30} kg), so, from this, we can see that the Coma Cluster galaxies consist of about forty trillion solar masses. We know from modern astronomical measurements that galaxies like the Milky Way or Andromeda consist of about a trillion stars. The Coma Cluster consists of about 1000 galaxies of various sizes.

The average mass of a galaxy in the Coma Cluster would then be about 3.9×10^{10} M$_\odot$. The luminosity of a typical galaxy in the Coma Cluster is about $8.5 \times 10^7 L_\odot$, where $L_\odot$ is the luminosity of our own sun. One would expect, taking the mass-to-luminosity ratio, that, if the Coma Cluster's mass is due to stars, then $M/\text{M}_\odot \approx L/L_\odot$.

We can see that, in fact, the mass-to-luminosity ratio is about 460 for the Coma Cluster. This implies that there is about 460 times more gravitational than luminous mass.

Exercise 9

The galaxy cluster A2142 is estimated to contain 994 individual galaxies, with a cluster radius of 1371 kpc and a line-of-sight velocity dispersion of 1062 ± 70 km/s. The baryonic mass is estimated, using x-ray imaging of the cluster's gas (assumed to be the dominant mass component) and a separate estimate of the mass of the stars, and is found to be $\log(M_{bar}) = 14.142 \pm 0.003$ where the underlying mass units are solar masses, $M_\odot$. Estimate the ratio of gravitational to visible mass using the virial theorem.

Considerations in the Application of the Virial Theorem

An astrophysicist, looking at the evaluation so far, would be right to raise a number of concerns with such a basic analysis. First, one needs to be sure which galaxies are part of the cluster, and which are not. The individual velocities of the galaxies can have an influence on the overall average velocity of galaxies in the cluster; for example, faster moving galaxies will skew $\langle v \rangle$.

There are more subtle issues. For example, if two neighboring galaxies are found in the cluster, since they are strongly gravitationally influenced by one another, should one treat their motions separately or group them together? A less subtle issue is that of the velocity relations between one and three dimensions, which relied on assumptions of a spherical cluster with uniform motion. Deviations from spherical, which are fairly likely in real clusters, would impact such assumptions. We have also implicitly made the assumption that the cluster has no overall expansion or contraction, limiting the velocities to being primarily about a center of mass and not toward or away from it.

In fact, such considerations have entered the study of galaxy clusters in the decades since Zwicky's original work. While improvements in the analysis of the motion of clusters like the Coma Cluster has resulted in closure of the gap between the gravitational and luminous masses, even in a modern analysis, these still differ by a factor of $\mathcal{O}(10)$.

2.3 Galaxy Rotation Curves

Zwicky's work on the Coma Cluster is among the earliest evidence for nonluminous gravitating matter. As a device to explain the discrepancy between the two masses in the Coma Cluster, Zwicky invoked [Zwi33; Zwi37] the term *Dunkle Materie*—dark matter—to suggest that there is simply a large amount of matter that isn't absorbing, reflecting, or emitting optical light.

As later decades would inform, there are potentially many standard explanations for how this could take place without invoking a novel new form of matter beyond the SM. The unseen matter might have simply turned out to be unseen gas and dust both in the galaxies and between them, or perhaps due to large populations of stellar-scale objects (brown dwarf star or black holes). A particle physics explanation could have been that there are many more neutrinos in the universe than expected from just ongoing processes;

perhaps the big bang itself produced excessively large numbers of primordial neutrinos. Those ideas ultimately fell out of favor owing to a variety of problems in using them to explain nonluminous gravitating matter.

It may also be that there are simply too many flaws in applying so many physical assumptions to so few astronomical observables to be able to draw large conclusions from them. Perhaps the Coma Cluster is just too far away, and too hard to understand, even by reasonable assumption or approximation. We can look to other places in the universe for independent measurements of motion and luminosity. For the next case, we will look at galaxy rotation curves and how they independently suggest that, even for galaxy-sized objects, there is a disconnect between luminous and gravitational mass.

Vera Rubin (Fig. 2.3) was an American astronomer who pioneered work on galaxy rotation rates, resulting in measurements of those rates referred to as "rotation curves." She received her undergraduate degree in astronomy at Vassar College and did graduate studies at Cornell University and Georgetown University. Among her many distinctions, she was elected to the National Academy of Sciences. After she passed away in 2018, a project known as the Large Synoptic Survey Telescope

Figure 2.3: Vera Rubin [Wik23i]

was renamed in her honor, a nod to her contributions to the study of dark matter and to her outspoken advocacy for the equal treatment and representation of women in science.

One of her key contributions was to have made meticulous measurements of the light spectrum of stars within single galaxies, and to do so as a function of the location of the light in the galactic structure. A galaxy, one like the Milky Way, consists of hundreds of billions or trillions of stars, bound together by gravity as they circle around the center of the galaxy. Our solar system is about 8 kpc (26,000 ly) from the center of the Milky Way. It is estimated that our sun makes one revolution around the galaxy every 230 million years. Our sun is about five billion years old, so we've made almost twenty-two trips around the galaxy since its birth.

All stars in the Milky Way orbit the galactic center. The sun does so at about 1/1300 of the speed of light, or about 230 km/s. Stars closer to the galactic center orbit at different speeds than does the sun. Because we can think of galaxies as large collections of non-interacting stellar point masses[5], bound together only by gravity, it is possible

---

[5]It can be jarring to think of a star as a "point mass" when compared to a galaxy. After all, stars are so much bigger than planets or people. To build comfort with this casual assumption, let us run some numbers. Our own sun, a mid-sized star, has a diameter of 1,392,684 km, or about 1.4×10^9 m. The Milky Way has a diameter, including most of its stars and gas, that is something like 100,000 ly or about 9.4×10^{20} m. We already see that a star like the sun pales in comparison to the size of its host galaxy.

to predict what a galactic rotation curve should look like by describing the relationship between the speed of the star and the radius of its orbit around the galactic center.

Models of Rotation

In introductory physics we learn how to relate angular velocity, linear velocity, and distance from an axis of rotation for points on a rigid rotating body. For example, consider a solid wheel rotating around an axis passing through the center of the wheel perpendicular to the plane of the wheel itself. In this case, for a point of displacement $\vec{r}$ from the rotation axis, with linear velocity $\vec{v}$, the angular velocity will be

$$\vec{\omega} = \frac{\vec{r} \times \vec{v}}{r^2} = \frac{\hat{r} \times \vec{v}}{r}. \tag{2.38}$$

All points on the rotating wheel have the same angular velocity owing to the rigidity of the body. The cross-product of $\vec{r}$ and $\vec{v}$ will be $rv \sin\theta$, where θ is the angle from the first vector to the second. Velocities will always be perpendicular to the displacement vector from the axis of rotation ($\theta = \pi/2$ or $-\pi/2$) for this case, so the magnitude of the angular velocity will be

$$\omega = \frac{rv \sin\theta}{r^2} \rightarrow \frac{v}{r} \tag{2.39}$$

allowing us to write the linear speed of the point at any distance r from the axis of rotation:

$$v = r\omega \tag{2.40}$$

Here, $\vec{v}$ is always transverse to $\vec{r}$.

The problem with this model, when applied to galaxies and stars, is that galaxies are not rigid bodies. Each star is bound by the gravitational influence of all stars inside its orbital distance, $\vec{r}$, from the center of the galaxy (assuming that a flat, circular disk of stars can be used as an approximation of a galaxy). We might be tempted to use a planetary model to describe galactic rotation. Let's explore this model briefly so we can understand why this is not a good representation of a galaxy.

A planetary model imagines a central star and orbiting planets. As a general rule of thumb, the mass, M, of the central star will exceed the mass of any single planet, m_i, where i labels a planet [6]. We can treat the star, and its planets, as a closed and isolated system. To a first approximation, owing to $M \gg m_i$, the planets will have much less influence on each other compared to the influence of the central star. Therefore, we can

---

Let's get some useful numbers from this. The diameter of our sun is a size scale that is just 1.5×10^{-12} that of the Milky Way. As a way of comparing size scales, a single carbon atom has a size that is about $10^{-10} - 10^{-11}$ that of the human body. Certainly, when comparing atoms to macroscopic objects like a person, we are comfortable treating atoms as points. We can see that an atom in relation to the body is actually a bit closer to the size scale of the body than is a star in comparison to its host galaxy, but the two comparisons are roughly comparable. Atoms are to people what stars are to galaxies, and this justifies the point-like approximation.

[6] For example, in our own solar system the mass of the sun accounts for more than 99% of the mass of the solar system.

also approximate for each planet that the only force influencing each planet is the central star.

We can then use Newton's second law to relate gravity to the acceleration of the planet:

$$\vec{F}_i = m_i \vec{a}_i \tag{2.41}$$

$$G\frac{Mm_i}{r_i^2}(-\hat{r}) = m_i \vec{a}_i \tag{2.42}$$

If we treat the orbital shape as approximately circular, then we can identify the acceleration as being centripetal in nature:

$$G\frac{Mm_i}{r_i^2}(-\hat{r}) = m_i \vec{a}_i \tag{2.43}$$

$$G\frac{Mm_i}{r_i^2}(-\hat{r}) = m_i \frac{v_i^2}{r_i}(-\hat{r}) \tag{2.44}$$

This allows us to relate the speed of the planet (transverse to the displacement from the star to the planet) to the radius of the orbit:

$$v_i = \sqrt{\frac{GM}{r_i}} \tag{2.45}$$

We learn from this that, in a planetary model with a central force binding the planet to the star, the transverse speed of the planet relates as $1/\sqrt{r}$ of the orbital radius.

In either the rigid rotating disk model, or the planetary model, we see that speed varies with distance from the axis (or center) of rotation. In a rigid model, the dependence goes as r, while in the planetary model it relates as $1/\sqrt{r}$. In the former, it increases with distance, and, in the latter, it decreases with distance.

Neither of these is a great model for a galaxy. While galaxies are now known to have central supermassive black holes, with masses that range between $100,000$ M$_\odot$ and a few billion M$_\odot$, they also contain hundreds of billions, to trillions, of stars. So, while there is a massive central object (like a star in the planetary model), there are also millions, billions, or even (for those with large galactic orbital radii) trillions of stars exerting a gravitational influence on a single star.

We can begin from the planetary model and adjust it slightly to take into account that it is the mass *enclosed* by a star's orbit that influences its speed[7]. We merely need to change $M \rightarrow M(r)$. In that case:

$$v_i = \sqrt{\frac{GM(r)}{r_i}} \tag{2.46}$$

---

[7]The fact that it is only the mass *enclosed* about the center of rotation that exerts a force on a test mass, m_i, is entirely due to Gauss's law. For mass and gravitational fields, this is expressed (in differential form) as $\vec{\nabla} \cdot \vec{g} = -4\pi G\rho$, where G is Newton's gravitational constant, ρ is the mass density, $\vec{g}$ is the gravitational field (force per unit mass), and $\vec{\nabla} \cdot$ is the divergence operator.

To obtain a mathematical function for $M(r)$ we need to adopt some model of how mass is distributed in a galaxy. A simple model would be a homogeneous sphere of uniform density, ρ. In that case:

$$\rho = \frac{M}{V} \longrightarrow M(r) = \rho V = \rho \frac{4}{3}\pi r^3 \tag{2.47}$$

This only applies for $r \leq R$, where R is the outermost radius of the "galactic sphere" in this model. For $r > R$, the enclosed mass will always be the total mass of the entire galaxy, $M(r) = M$. Finally, we arrive at a key relationship from this model:

$$v_i(r \leq R) = \sqrt{\frac{4\pi\rho G r_i^3}{3 r_i}} = \sqrt{\frac{4\pi\rho G}{3}} \, r_i. \tag{2.48}$$

Again, v_i is the magnitude of the velocity transverse to the displacement vector from the center of the galaxy to the star. Once we get to R, we run out of stars in the galaxy. For some point-like "test star" of mass m_i placed in orbit at $r_i > R$, the transverse speed of that star would be given by the planetary model result:

$$v_i(r > R) = \sqrt{\frac{4\pi\rho G R^3}{3 r_i}} = \sqrt{\frac{4\pi\rho G}{3}} \, r_i^{-\frac{1}{2}} \tag{2.49}$$

Test stars imagined to be placed in a stable orbit at these distances would have a decreasing orbital speed the farther we position them from the edge of the galaxy. This is known as the *Keplerian fall off*, honoring the discovery that Johannes Kepler made about the laws of planetary motion.

How do we apply this thinking to galaxies and the orbital speed of stars about the galactic center? If galaxies are made only from stars and gas, with the mass dominated by the stars, then we expect to see a rise in stellar orbital speed as we look farther and farther from the center of the galaxy. That will continue until we get to the edge of the galaxy, which we can define as the radius at which most of the stars in the galaxy are already found within that orbital radius. Stars beyond that boundary will then have lower and lower speeds as the vast majority of the mass of the galaxy should already be enclosed within their orbital radii. We should see, therefore, a downturn in stellar orbital speed as we get to, and then beyond, the edge of the galaxy's primary population of stars. That is the prediction for a galaxy dominated by luminous matter.

Rubin's Observations of Galaxy Rotation Curves

In practice, rotational velocities of stars in galaxies are determinable by observing galaxies using telescopes fitted with spectrometers. This allows the light from stars to be split into its constituent colors. As Zwicky did with galaxies in the Coma Cluster, rotational velocities along the line of sight can be determined by looking for red- or blue-shifts in key parts of the stellar light spectrum. In fact, this is precisely how Vera Rubin and her collaborator, Ken Ford, designed their astronomical experiments to measure galactic

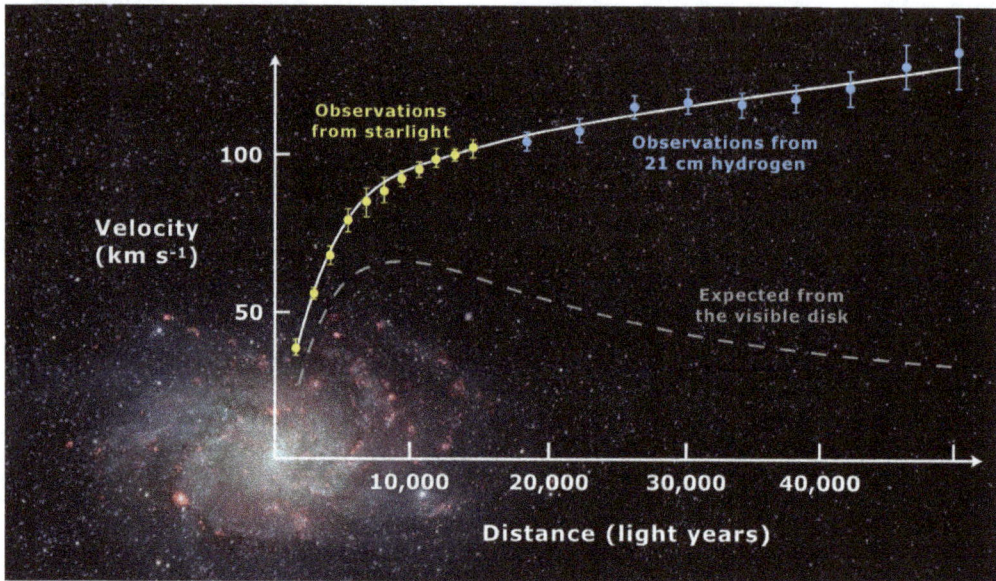

Figure 2.4: Rotation curve of spiral galaxy Messier 33 (data points with error bars), and a predicted rotation curve taken from the distribution of only the visible matter (smooth dashed line). The data and the model predictions are from Ref. [CS00]. Image is from Ref. [Wik21].

rotation speeds. By focusing first on stars on one side, then closer to the center, then on the other side of the galaxy, they were able to build precise maps of orbital speeds as a function of orbital radius (or, at least, the distance from the center of the observed galaxy). A modern example of this, using data from before 2000, is shown in Fig. 2.4.

What they observed defied the expectation from Newtonian gravity applied to a galaxy model in which luminous matter (in the shape of a "visible disk") determines rotational speed. Stars (and even gas) at large radii from the galactic center were observed to be moving as fast as stars nearly halfway between the galactic center and the visible edge of the galaxy. In other words, the rotation curve of speed vs. radius appeared to flatten out. Such "flat rotation curves" imply that the mass enclosed by these outer stellar orbits was in excess of what one would expect from the fall-off of the luminous matter (or, conversely, to the flattening of the cumulative distribution of luminous matter). The rotation curves suggested that, as the orbital radius increased, such far-positioned stars were enclosing an $M(r)$ that *continued to grow with r*, even as enclosed luminous mass visibly appears to flatten out to a constant. In Fig. 2.4, the location where most of the visible mass is enclosed occurs around 10,000 ly from the galactic center of M33.

Numerically, this implied that there was about ten times more mass than was luminous in individual galaxies. Measurements across many galaxies served only to cement this observation: in galaxies, gravitational mass exceeds luminous mass. We can put this in more firm mathematical terms by considering a spherical halo of matter with a density profile that increases linearly with radius. For example:

$$\rho(r) = \frac{1}{4\pi r^2} \frac{dM(r)}{dr}$$

Recall that, for the case in which we increase orbital radius and also continue to allow the enclosed mass to increase,

$$v_i = \sqrt{\frac{GM(r)}{r_i}}$$

where i labels a star. This implies that

$$M(r) = \frac{v_i^2 r}{G} \tag{2.50}$$

We can thus determine that

$$\rho(r) = \frac{1}{4\pi r^2} \frac{dM(r)}{dr} = \frac{1}{4\pi r^2} \frac{d}{dr}\left(\frac{v_i^2 r}{G}\right) \tag{2.51}$$

$$= \frac{v_i^2}{4\pi r G} \tag{2.52}$$

This would be valid for large orbital radii. If there is a halo of nonluminous matter, and if we are considering the motions of stars far from the galactic center, this would then tell us about the shape of the nonluminous matter halo that includes the galaxy, but extends beyond the luminous boundary of that galaxy.

Exercise 10

Our sun is a star located approximately 24,000–28,000 ly from the galactic center of the Milky Way. Our solar system orbits the galactic center with a speed of 240 km/s.

1. Estimate the gravitational mass enclosed by the orbit of the sun about the galactic center.

2. Make the assumption that luminous mass (stars, gas) explain all the mass of the Milky Way, and, noting that there is far less luminous material between our solar system and intergalactic space (as compared to between our orbit and the galactic center), predict the speed of a star at twice the distance from the galactic center.

3. Compare your previous answer to the observed speeds of stars three times our distance from the galactic center (note that stars have been observed beyond such distances; e.g., ULAS J001535.72+015549.6).

2.4   Inferring the Structure of Halo Models

A commonly used parameterization for pseudo-isothermal halos that can also account for the velocity-radius behavior observed closer to the galactic center, as well as far away, is of the form

$$\rho(r) = \frac{\rho_0}{1 + \left(\frac{r}{a}\right)\gamma'} \tag{2.53}$$

where ρ_0 is known as the "finite central density," describing the density profile of matter near the galactic core; a is the characteristic distance scale of the linear system (the critical radius of that core); and the coefficient γ' is a free-scaling parameter that must be determined by constraining the model using observed galaxy data. A value of $\gamma' \approx 2$ is determined to provide a good description of the observed characteristics of most galactic halos.

The state-of-the-art of galactic dark matter halo models has pushed far beyond these more simplistic ad hoc models. For example, the Navarro–Frenk–White (NFW) profile [NFW96] is determined from numerical simulations of cosmic structure formation. These simulations evolve the universe from very early times with quantum fluctuations from the Big Bang causing over-densities and under-densities of matter, especially dark matter. These, much later, become the seeds of galaxy formation as luminous matter (baryonic matter) in-falls on these dark matter wells. The NFW profile is determined empirically from these models and is routinely used when interpreting experiments trying to directly or indirectly observe dark matter particles:

$$\rho(r) = \frac{\rho_{crit}\,\delta_c}{\left(\frac{r}{r_s}\right)\left(1 + \frac{r}{r_s}\right)^2} \tag{2.54}$$

where r_s is, again, a scale radius indicating the approximate size of the halo core; δ_c is a dimensionless density parameter that is known as the "characteristic," and $\rho_{crit} = 3H^2/8\pi G$ is the critical density for the closure of the universe based on cosmological considerations (H is the Hubble constant)[8]. This model is fairly "universal" in that it appears to work well to describe a wide range of halo masses from those of individual galaxies to those of galaxy clusters.

While the NFW model is often used as a baseline for comparing interpretations of experimental searches for dark matter, it is not considered a state-of-the-art model and is no longer favored as the most likely description of dark matter halos. For example, in diffuse galaxies whose observed brightness (seen from Earth) is lower than the average brightness of the night sky, this model appears to poorly describe the dynamics of the stars. This leads to what is known as the *core-cusp problem*, wherein NFW would predict dark matter halos having sharp cusps at $r = 0$ (where the density blows up to infinity) while some classes of galaxies seem to prefer a profile with a finite-sized (and density) core. This represents an ongoing area of investigation. We will provide a few alternative

---

[8]A more thorough introduction to the Hubble constant, the critical density, and other essential concepts is provided in Chapter 3.

models, including NFW, in a more detailed discussion of dark matter density and velocity distributions in Section 2.7.

2.5 Non-Visible Baryonic Components of Galaxies and Clusters

The discussion so far has been tightly focused on a narrow portion of what we know about baryonic matter (matter whose mass is dominated by bound-states of quarks, such as neutrons and protons). For example, both of the previous reviews of galaxy cluster dynamics and galactic rotation curves have looked extensively at stellar mass as the key indicator of mass in galaxies or clusters of galaxies.

However, this neglects at least two things:

- We know that baryonic matter is capable of emitting or reflecting more than just visible wavelengths of light. Would we draw the same conclusion from investigations of galaxies and clusters using other wavelengths, such as radio, x-rays, or gamma rays?

- We know that there is more than just stars that comprise the baryonic matter of galaxies. For example, our own galaxy is rich in dust and gas, as evidenced by the fact that observations of its galactic center depends on long-wavelength astronomical methods (e.g., radio astronomy) to penetrate the intervening gas and dust that otherwise block shorter-wavelength visible light.

X-ray astronomy has been developing over many decades, focusing recently on satellite-based instrumentation to avoid observations through the x-ray-absorbing atmosphere of the earth. Intergalactic hot gas can extend multiple galactic radii from the central galaxy. That gas can have temperatures of millions of degrees Kelvin, more than sufficient to generate light, but not in the visible spectrum. In fact, you might argue, doesn't this help explain away the kind of "halo of nonluminous matter" we saw evidenced in our rotation curve work on galaxies?

This is precisely the point of such multi-wavelength investigations of astronomical objects. X-ray astronomy, for example, provides a crucial tool to disentangle the potential effects of hot gas from other kinds of matter, perhaps non-baryonic in nature. An example of the scale and scope of hot gas in galaxies is shown using x-rays in Fig. 2.5. These kinds of gas halos are very much a potential confounding effect in the effort to understand the mass composition of galaxies and galaxy clusters. In addition, assuming that the masses of galaxies and clusters still turn out to be more than just the sum of gas and stars, the gas itself may provide a natural target for dark matter–baryonic matter interactions detectable on Earth. We will return to this point later.

X-ray emissions should be relatable to the mass of gas producing those emissions. X-ray radiation will be an effect of both the temperature of the gas (hotter means more emission, and at higher average energies) and the mass of the gas (more gas permits more emission). To explore the relationship, we can employ Euler's fluid dynamics equation,

$$\rho \frac{d\vec{v}}{dt} = -\nabla p - \rho \nabla \Phi \qquad (2.55)$$

Figure 2.5: Chandra telescope x-ray observations of the giant elliptical galaxy NGC 6482 (center). It is surrounded by a cloud of hot gas (the large-diameter, fainter, fuzzier glow). The gaseous region extends 700,000 ly across and has an average temperature of about 10 million Kelvin. Image from NASA/CXC/SAO.

where p is the fluid pressure, ρ is the fluid density, Φ is the gravitational potential, and $\vec{v}$ is the velocity of the fluid flow. Let us approximate a gas halo like the one seen in Fig. 2.5 as a spherical system in hydrostatic equilibrium. In this model the acceleration of the fluid will necessarily be zero, $d\vec{v}/dt = 0$. In that case, we have:

$$\nabla p = -\rho \nabla \Phi. \tag{2.56}$$

For a spherically symmetric distribution, it must be the case that $dp/d\phi = dp/d\theta = 0$. In the gradient operator, in spherical coordinates, this leaves only a radial derivative. From Newton's law of gravitation, we can then write the gradient of the gravitational potential to obtain:

$$\frac{dp}{dr} = -\frac{GM(r)}{r^2}\rho. \tag{2.57}$$

Here, $M(r)$ is the mass *enclosed* in a radius, r, centered on the halo. We can combine the above fluid equation with the ideal gas law to learn about the relationship of temperature and mass, trading pressure for temperature. In doing so, we are necessarily assuming no, or minimal, interactions among the atoms in the gas. However, under this assumption we can still learn something. Begin by employing $pV = nk_BN_AT$ (where n is the number of moles of gas, k_B is Boltzmann's constant, and N_A is Avogadro's number). The volume of the gas can be related to its total mass, and its density, using $\rho = M/V$. The number of moles of gas is given by the number of gas particles divided by Avogadro's number, $n = N/N_A$. The number of gas particles is simply the total mass of the gas divided by the mass of a single gas constituent, $N = M/m$. Thus

$$pV = \frac{M}{m}\frac{1}{N_A}k_BN_AT = \frac{M}{m}k_BT \tag{2.58}$$

We can then use our substitution for volume to find

$$pV = p\frac{M}{\rho} = \frac{M}{m}k_BT \longrightarrow p = \frac{\rho k_BT}{m} \tag{2.59}$$

We are now ready to combine both pieces of information,

$$\frac{dp}{dr} = -\frac{GM(r)}{r^2}\rho \tag{2.60}$$

$$\frac{d}{dr}\left(\frac{\rho k_BT}{m}\right) = -\frac{GM(r)}{r^2}\rho \tag{2.61}$$

Let us now use this to derive the mass of gas enclosed in a radius, r, in terms of density and temperature:

$$\frac{d}{dr}\left(\frac{\rho k_BT}{m}\right) = -\frac{GM(r)}{r^2}\rho \tag{2.62}$$

$$\frac{k_B}{m}\frac{1}{\rho}\frac{d}{dr}(\rho T) = -\frac{GM(r)}{r^2} \tag{2.63}$$

$$\frac{k_B}{m}\left(\frac{T}{\rho}\frac{d\rho}{dr} + \frac{dT}{dr}\right) = -\frac{GM(r)}{r^2} \tag{2.64}$$

$$\frac{k_B}{m}\left(T\frac{d\ln\rho}{dr} + T\frac{d\ln T}{dr}\right) = -\frac{GM(r)}{r^2} \tag{2.65}$$

$$-\frac{k_BT}{Gm}r^2\left(\frac{d\ln\rho}{dr} + \frac{d\ln T}{dr}\right) = M(r) \tag{2.66}$$

By application of the chain rule and the relationships:

$$\frac{du}{dx} = \frac{1}{dx/du} \tag{2.67}$$

$$\frac{d\ln u}{dx} = \frac{1}{u}\frac{du}{dx} \tag{2.68}$$

we obtain

$$M(r) = -\frac{k_B T}{Gm}r\left(\frac{d\ln\rho}{d\ln r} + \frac{d\ln T}{d\ln r}\right) \tag{2.69}$$

This tells us that, if we can measure both the temperature and the density profile, $\rho(r)$, of the gas as a function of r, we can estimate the mass of the gas.

We can model the gas as a body at thermal equilibrium that absorbs all incoming energy (e.g., radiation from stars). This is what is known as a *black body*. Such a body will re-emit energy having a particular spectrum (intensity of emission at various wavelengths and at a given temperature). This *black body radiation* is described by Planck's equation,

$$B_f(E) = \frac{2f^2}{c^2}\frac{hf}{e^{hf/k_B T} - 1}. \tag{2.70}$$

Here, E is the energy of a particle, $B_f(E)$ is the distribution of energy among the particles (known as the *spectral radiance*) per units solid angle, area, and time, and T is the temperature of the gas. The remaining numbers are the familiar Boltzmann's constant, k_B ($k_B = 8.31446261815324$ J$\cdot$K$^{-1}\cdot$mol^{-1}), and the speed of light, c. From this, it is possible to answer the question, *what is the most prominent frequency of light emitted by the gas?* The peak frequency (or wavelength) of the x-rays emitted from the gas can be obtained from Wien's displacement law, from which we can then infer the temperature of the gas at a given radius from the center of the distribution:

$$\lambda_{peak} = \frac{b}{T} \tag{2.71}$$

where $b = 2897.771955$ μm$\cdot$K. An analysis of the x-ray intensity distribution will permit the temperature to be determined at any r within the resolution of the instrument. The density profile can be inferred from the intensity of x-rays vs. r, assuming intensity follows density of gas. If we assume the gas is dominated by hydrogen, such that $m = m_p$, the proton mass, then the gas halo mass can be determined.

What, then, are the typical gas halo masses that have been determined from x-ray observations? Existing data tells us that the masses of such gas coronas, even extending far beyond the visible stars and gas in the galactic disk, can account for only a fraction of a percent (at most) of the mass of stars in the same galaxy (c.f. Ref. [Ras+09]). This is insufficient to explain the enclosed mass that must be present to determine the observed motion of stars and gas at large radii from the galactic center.

We will return to this technique later, when analyzing galaxy cluster collisions, to allow for gas, stars, and non-baryonic matter to be separated from one another by the collision dynamics.

Exercise 11

Derive Wein's displacement law (Eqn. 2.71) from the black body spectrum equation (Eqn. 2.70).

2.6 Gravitational Lensing of Galaxies and Galaxy Clusters

Gravitational lensing is a technique that takes advantage of the fact that light will follow, traveling on geodesics[9], the curvature of spacetime around regions in spacetime that are significantly bent by large energy densities. Modern astronomy routinely observes this effect.

The popular depiction of the phenomenon is often represented by a consequence of what is called *strong lensing*. This occurs when the lensing mass, between the observer and the object, is so large that multiple images (either clear or slightly distorted) of the object appear to the observer. If the lensing mass is precisely centered on the true line of sight between the object and the observer, and if the lensing mass is spherically symmetric, what results is an infinite number of images of the object appearing in a ring around the lensing mass—an Einstein ring. A true example of an Einstein ring (albeit, one not quite meeting the exacting conditions of perfection as stated above) is shown in Fig. 2.6. Strong lensing observations are rare, but when discovered, they provide an excellent local constraint on the gravitational potential of the lens object itself.

---

[9]Courses in general relativity are not common at the undergraduate level, and even at the graduate level are not always available. The concept of a *geodesic* derives from the study of the mapping of the surface of the earth, or *geodesy*. As a mathematical concept, a geodesic is the generalization of the concept of a *line* in flat Euclidean geometric space. While this is not a wholly accurate summary, a geodesic can be thought of in certain cases as the shortest path between two points in a space.

Figure 2.6: An Einstein ring (technically, a horseshoe Einstein ring given its incompleteness) of a background galaxy lensed by an intermediate foreground galaxy (central bright fuzzy object). This image was taken by the Hubble Space Telescope.

More common, and more difficult to detect and measure, are *weak lensing* distortions. These occur when the lensing mass is much smaller. This causes subtle distortions in the image of the background object. For example, if a cluster of galaxies lies between the observer and an object, the object's image will be distorted as its light passes through the cluster and is warped by alterations in a straight-line path by the local curvature of spacetime.

It is possible to use lensing within a single galaxy (e.g., the Milky Way) to detect nonluminous local collections of mass. This technique, known as *micro lensing*, is possible with powerful telescopes and algorithmic approaches to detect and interpret the local distortion of objects by lensing masses in our own galactic neighborhood. For example, large collections of black holes or other *massive astrophysical compact halo objects* (MACHOs), which might be made from normal matter, be nonluminous, and neatly explain dark matter, can be detected by this technique. The difference among these three lensing scales—strong, weak, and micro—depends upon the positions of the source, lens, and observer, as well as the mass and shape of the lens.

Gravitational lensing is a powerful tool for conducting mass surveys. Since any energy density will curve spacetime, all mass—luminous and nonluminous—can be accounted for using this approach. By combining this technique with other approaches, such as dynamical methods (motion of luminous matter) or mass-luminosity relationships, it is possible to build detailed surveys of the amount and kind of matter in a target.

The deflection of light by gravity is not a new idea. It predates Einstein and falls well within the purview of classical Newtonian mechanics. What distinguishes the general relativistic approach from the classical approach is merely accuracy: Einstein's general theory of relativity matches the observed degree of deflection of light while Newton's mechanics gets it wrong. This, of course, is what distinguishes a more complete theory of nature from a less complete or incorrect theory of nature. The key distinction in these two approaches is the axiomatic description of spacetime itself: in Newton's mechanics, time is assumed to be the same for all observers, and space and time are assumed to be rigid, flat, and immutable; the opposite is true in general relativity, where space and time are key players in the dynamics of the universe and neither is experienced the same for observers in different inertial or non-inertial reference frames (owing to their propensity for curvature).

The first written account of the bending of light by gravity was made by John Soldner in 1804 in his treatise, *On the Deflection of a Light Ray from its Straight Motion due to the Attraction of a World Body which it Passes Closely*. Newtonian mechanics recognizes that all objects in a uniform gravitational field fall at the same rate, independent of their mass. Introductory physics students are usually introduced to this observation by using drop experiments—crumpling paper into a ball (to reduce air drag) and dropping it from the same height as a tennis ball, or an iron brick. Students observe that the objects strike the ground at nearly the same time, and a discussion then ensues about the small effects of air resistance. This is usually followed by the dramatic footage of an APOLLO astronaut on the moon dropping a hammer and a feather at the same time from the same height. They strike the lunar regolith at the same moment, and, since the moon has no atmosphere, we have removed the effect of the drag force to reveal what Galileo Galilei proved, with careful experiments, to be true almost 500 years ago.

Students are then treated to an explanation from Newton's laws of motion. Newton's second law is familiar to all:

$$\sum_i \vec{F}_i = m\vec{a} \tag{2.72}$$

The sum of the forces acting on a mass, m, results in a net acceleration, a, of the mass. An object in free fall in the uniform gravitational field of the earth near its surface experiences a force we call *weight*, which is given merely by the product of its mass and the acceleration due to gravity. If this is the only force, then

$$m\vec{g} = m\vec{a} \longrightarrow \vec{g} = \vec{a} \tag{2.73}$$

Mass plays no role in this statement, but this approach also doesn't give us a completely satisfying picture. It is when we employ Newton's law of gravity that we get to the more satisfying version of all of this:

$$G\frac{M_{earth}m}{R_{earth}^2}(-\hat{r}) = m\vec{a} \longrightarrow \vec{g} \equiv G\frac{M_{earth}}{R_{earth}^2}(-\hat{r}) = \vec{a} \tag{2.74}$$

We learn that the acceleration due to gravity is not only a constant and points toward the center of the earth, but it is entirely determined from Newton's gravitational constant, the mass of the earth, and the radius of the earth at which we drop the object (assuming the height of the drop, h, satisfies $h \ll R_{earth}$).

This is usually where the discussion stops in introductory physics, but going just one small step forward would open up students' eyes to a whole world of possibility. We learn in second semester introductory physics that light is a wave-like phenomenon (Maxwell's equations) whose mass is entirely zero. Nevertheless, since mass plays no role in the above statement about acceleration due to a local gravitational field, one is forced to hypothesize that light, too, will fall with an acceleration $\vec{g}$ in the gravitational field of Earth!

The reason this never comes up is the impracticability of observing such an effect, even if it is real. The speed of light, c, is so great that over the length of, say, an introductory physics laboratory room (maybe $L = 10$–20 m), the deflection of a laser beam from a purely horizontal original orientation would be (by a classical calculation)

$$\Delta y_{laser} \approx -\frac{1}{2}g\left(\frac{L}{c}\right)^2 \approx -10^{-14} \text{ m.} \tag{2.75}$$

The width of the laser beam alone would confound any attempt to measure such a small vertical displacement and the equipment required to observe such an effect is non-standard for most laboratory classes. However, there are far more massive objects in the universe than Earth. What if we apply this classical thinking to a light beam passing close to the surface of the sun?

In that case, and over the same distance ($L \approx 10$m), the classical deflection transverse to the radial direction would be

$$\Delta y_{laser} \approx -\frac{1}{2}G\frac{M_\odot}{R_\odot^2}\left(\frac{L}{c}\right)^2 \approx -10^{-13}\text{m} \tag{2.76}$$

Bigger, but not that much bigger. Of course, when considering the deflection of a beam of light by the sun, one is not confined to deflections over just several meters. The light beam will enter the influence of the sun's gravitational field, experience a rather non-uniform gravitational field as it traverses that field and accelerates, and then eventually exit the influence of the field having been deflected by some angle from its original trajectory. Soldner worked this out and found that:

$$\hat{\theta} = \frac{2GM_\odot}{c^2 R_\odot} \qquad (2.77)$$

which yields a net deflection of 0.83 arcsecond, where 1 second of arc is 1/3600 of a degree. If anyone had been able to test this prediction before about 1920, they would have found that Newtonian mechanics gave the wrong answer by a factor of about two. The inability to measure such a subtle deflection bought time for Albert Einstein to work out the details of his general theory of relativity and arrive at what turned out to be the correct prediction:

$$\hat{\theta} = \frac{4GM_\odot}{c^2 R_\odot} = 1.74 \text{ arcsecond.} \qquad (2.78)$$

Let's work out the geometry of gravitational lensing and then apply it to simple archetypes.

Lensing and Geometry

Figure 2.7 provides the geometric notation used in the derivation in the following paragraphs. This notation is standard for the field. The observer is taken as the

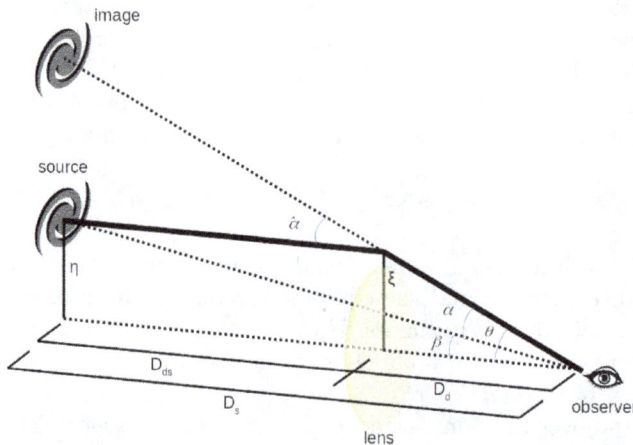

Figure 2.7: A diagram illustrating the standard notation for angles involved in gravitational lensing calculations. Created by the authors based on Ref. [Wik20b].

origin of the geometric system. The distances involved in the construction of a lensing calculation are:

- D_s: the distance from the observer to the plane containing the object (the "source") on a line that connects the observer and the center of the lensing system (the lens-observer line). All planes are transverse to this line.

- D_d: the distance from the observer to the plane of the lensing system.

- D_{ds}: the distance from the source plane to the lens plane.

- η: the distance from the lens-observer line to the source in the source plane. (The true source position.)

- ξ: the distance from the lens-observer line to the light ray from the source in the plane of the lens. (The position of the source image in the lens plane.)

Angles involved in the calculations will be:

- θ: the angle from the lens-source line to the apparent image of the source.

- α: the angle between the source and the image of the source.

- β: the angle between the lens-observer line and the source.

Some relationships become immediately apparent:

$$\theta = \alpha + \beta \qquad (2.79)$$

$$D_d = D_{ds} + D_d \qquad (2.80)$$

If we apply a small-angle approximation, then a few other relationships can be established:

$$\beta \approx \frac{\eta}{D_s} \qquad (2.81)$$

$$\theta \approx \frac{\xi}{D_d} \qquad (2.82)$$

Finally, we can define the *deflection angle* $\hat{\alpha}$ as the angle between the source's apparent and true positions.

Now that we have defined the geometry of the problem and the notation, we can use a bit of algebra to derive what is known as *the lens equation*. We begin with the small angle approximation once again, which implies that

$$\cos\theta \approx 1, \qquad (2.83)$$

$$\sin\theta \approx \theta, \text{ and} \qquad (2.84)$$

$$\tan\theta = \frac{\sin\theta}{\cos\theta} \approx \theta. \qquad (2.85)$$

Figure 2.8: Quasar QSO 0957+561 A and QSO 0957+561 B, the "Twin Quasar," appear as a pair of bright blue objects near the center of the image. The lensing galaxy is the large, diffuse, object in the center. Figure modified from Ref. [Wik23e]. Credit: ESA/Hubble & NASA.

Trigonometry then tells us that

$$D_s \tan \theta \quad \approx \quad D_s \tan \beta + D_{ds} \tan \hat{\alpha} \tag{2.86}$$

$$D_s \theta \quad = \quad D_s \beta + D_{ds} \hat{\alpha} \text{ (small angle approx.)}. \tag{2.87}$$

This then allows us to relate

$$\theta - \beta \quad = \quad \frac{D_{ds}}{D_s} \hat{\alpha} \tag{2.88}$$

$$\alpha(\theta) \quad = \quad \frac{D_{ds}}{D_s} \hat{\alpha}(\theta). \tag{2.89}$$

This yields the *lens equation*:

$$\beta = \theta - \alpha(\theta) \tag{2.90}$$

A lensed image is observed when the lens equation is satisfied.

An example of a lensed object other than the Einstein ring in Fig. 2.6 is shown in Fig. 2.8, the "Twin Quasar"—a quasi-stellar object (active galaxy) with a twin image caused by lensing. The lens is a red galaxy intermediate between us and the quasar. This was first detected by Dennis Walsh, Robert Carswell, and Ray Weyman in 1979 using the Kitt Peak National Observatory [WCW80]. It is the first multiply imaged source due to gravitational lensing. The first gravitational lens ever discovered was our own sun, when it was observed to bend starlight during a famous 1919 expedition to observe a total solar eclipse and test the general theory of relativity [GP19][10].

The two quasars are measured using decomposition of their atomic spectra and found to be identical. Both appear to be about 4400 Mpc from us, while the lensing galaxy appears to be 1400 Mpc from us. There is evidence of time delays in the two objects, when an optical event that occurred in one (B) then occurs later in the second (A), but about 400–500 days later [Kun+97], owing to the different lengths in optical paths between the source and the image.

The Twin Quasar has been observed in all wavelength ranges from radio to x-ray. This emphasizes a useful point: gravitational lensing has the same effect on light regardless of the wavelength. This is unlike an optical (e.g., glass) lens, which bends light of different wavelengths at different angles, causing them to spread out (as in a rainbow). A concrete realization of this is Eqn. 2.78: the equation depends only on the speed of light in vacuum but not individually on frequency or wavelength. All light travels at the same speed in vacuum regardless of wavelength. In a sense, a gravitational lens is a perfect lens and provides no chromatic distortion (aberration), unlike optical lenses based on the passage of light through matter.

The Einstein Radius

The lens equation allows us to establish the condition for the Einstein ring and its observed radius about the center of the lens. We can begin with the general theory of relativity's conclusion concerning the angle of deflection of light about a lensing system of mass distribution, $M(r) \equiv M(\xi)$ (Eqn. 2.78). Combining this with Eqn. 2.89 we obtain

$$\alpha(\theta) = \frac{D_{ds}}{D_s} \frac{4GM(\xi)}{c^2 \xi}. \tag{2.91}$$

We then assert the condition in the lens equation:

$$\begin{aligned} \beta &= \theta - \alpha(\theta) & (2.92) \\ &= \theta - \frac{D_{ds}}{D_s} \frac{4GM(\xi)}{c^2 \xi} & (2.93) \\ &= \theta - \frac{D_{ds}}{D_s} \frac{4GM(\xi)}{c^2 (D_d \theta)} & (2.94) \\ &= \theta - \frac{D_{ds}}{D_s D_d} \frac{4GM(\xi)}{c^2 \theta} & (2.95) \end{aligned}$$

where we have employed from the small-angle approximation that $\tan \theta \approx \theta = \xi/D_d$.

---

[10]We are grateful to Professor Joel Meyers (Southern Methodist University) for the "aha!" moment we earned when he noted that, technically, the first gravitational lens ever observed was our own sun!

The Einstein ring will form under the very special condition that the source lies directly along the line of sight between the observer and the center of the lens, or when $\beta = 0$. Employing this special condition we find:

$$\theta_{Einstein} = \sqrt{\frac{D_{ds}}{D_s D_d} \frac{4GM(\xi)}{c^2}}. \tag{2.96}$$

This provides the Einstein radius, in radians, of the resulting ring image of the source.

Exercise 12

The distances to the source (and, approximately, the image) of the "Twin Quasar," as well as to the gravitational lens, are given in the text. The source quasar and its image are observed to be separated by 5.7 arcseconds. Assuming that the singular quasar that becomes the "twin quasar" lies directly behind the lensing object along the line of sight, estimate the mass of the lens.

Point-Like Lens Geometry, $M(\xi) = \delta(\xi - 0)$

We can use Eqn. 2.96 in combination with Eqn. 2.95 to obtain the relationship

$$\beta = \theta - \frac{\theta_E^2}{\theta} \longrightarrow \theta^2 - \beta\theta - \theta_e^2 = 0. \tag{2.97}$$

This implies two solutions for θ that satisfy the equation:

$$\theta_\pm = \frac{1}{2}\left(\beta \pm \sqrt{\beta^2 + 4\theta_E^2}\right). \tag{2.98}$$

The *magnification* (μ) of an object by a lens, in classical optics, is defined as the ratio of the heights of the image and the source; in a thin lens system, this can also be related to the image and source distances from the lens, which in our notation will be

$$\mu = -\frac{D_{ds}}{D_d} = \frac{h_{image}}{h_{source}}. \tag{2.99}$$

This will be given equivalently by the ratio of the solid angles occupied by the image and the source,

$$\mu = \frac{\theta \, d\theta}{\beta \, d\beta}. \tag{2.100}$$

For the case where the mass of the lens system is entirely concentrated at $\xi = 0$, we find that we always obtain two images of the source; one that has a radius inside the Einstein radius, and one with a radius outside the Einstein radius. As the angle of the source becomes more and more non-zero, $\beta > 0$, one image will approach the lens and become faint while the other will approach the true source position and tend toward a magnification of 1. This makes physical sense. As we move the source from behind the lens and then farther from the possibility of the gravitational spacetime distortions of the lens system influencing light rays from the source, we observers should "recover" the simple act of seeing the undeflected source at its true position. The same effect occurs as $M(\xi) \to 0$ (the lens mass vanishes, meaning the lens ceases to exist).

An Extended Gravitational Lens

In reality, lensing systems are not generally small and compact, nor are they necessarily spherically symmetric, even when extended. A distributed system is less likely to cause the kind of strong-lensing optical effects we have been inferring from special cases, and are more likely to impose more subtle and weaker distortions on the image of the source.

To handle distributed sources causing a weaker lensing effect on the source, let us assume that the total deflection angle can be obtained by a linear sum of the deflection angles induced by each of the component sources in the extended and distributed lensing object:

$$\hat{\alpha} = \sum_i \hat{\alpha}_i. \tag{2.101}$$

We also assume that the object is far behind the lensing system such that the physical size of the lensing system is small compared to the distance between the source and the lens. This is the gravitational equivalent of the thin-lens approximation. We assume that all lensing elements are precisely located a distance D_d from the observer. If we then define ξ_i as the projected position of mass element m_i in the lens plane, and ξ again as the position of the light ray from the source in the lensing plane, then

$$\hat{a}(\xi) = \sum_i \frac{4Gm_i}{c^2} \frac{\xi - \xi_i}{|\xi - \xi_i|^2}. \tag{2.102}$$

Let us imagine now that the distribution of lensing masses in the lens plane can be described by a single surface density of lensing masses, σ, which has units of mass per unit area. In general, the total mass of the lensing system will be given by $M = \int \sigma(\xi')d^2\xi'$, where we assume the surface density is a function of the position of the masses within the lens, ξ', and that we need to integrate over the area of the lens plane to obtain the total mass.

Using this idea, we can recast our sum as an integral:

$$\alpha = \frac{D_{ds}}{D_s}\hat{a} \tag{2.103}$$

$$= \frac{D_{ds}}{D_s} \sum_i \frac{4Gm_i}{c^2} \frac{\xi - \xi_i}{|\xi - \xi_i|^2} \tag{2.104}$$

$$= \frac{D_{ds}}{D_s} \int \frac{4G}{c^2} \sigma(\xi') \frac{\xi - \xi'}{|\xi - \xi'|^2} d^2\xi'. \tag{2.105}$$

We can perform a transform of coordinates in the integration by again employing the small-angle approximation, such that $\theta = \xi/D_d$. This implies that $d\theta/d\xi = 1/D_d$, and thus $d\xi = D_d\, d\theta$. Therefore:

$$\alpha = \frac{D_{ds}}{D_s}\frac{4G}{c^2}\int \sigma(D_d\theta')\frac{D_d\,(\theta - \theta')}{D_d^2|\theta - \theta'|^2}D_d^2 d^2\theta' \tag{2.106}$$

$$= \frac{D_{ds}D_d}{D_s}\frac{4G}{c^2}\int \sigma(D_d\theta')\frac{(\theta - \theta')}{|\theta - \theta'|^2}d^2\theta' \tag{2.107}$$

We can then define a *critical surface mass density* such that any lens whose $\sigma > \sigma_{critical}$ will generate multiple images of the source:

$$\sigma_{critical} = \frac{c^2 D_s}{4\pi G D_d D_{ds}}. \tag{2.108}$$

This allows us to write:

$$\alpha = \frac{1}{\pi}\int \frac{\sigma}{\sigma_{critical}}\frac{(\theta - \theta')}{|\theta - \theta'|^2}d^2\theta' \tag{2.109}$$

$$= \frac{1}{\pi}\int \kappa\frac{(\theta - \theta')}{|\theta - \theta'|^2}d^2\theta'. \tag{2.110}$$

The ratio κ is known as the *gravitational convergence*. It tells us the strength of the lens. As a rule of thumb,

- $\kappa > 1$: This is the *strong-lensing* domain, in which multiple images of the object will be possible, as well as a strong magnification and strong distortions of the object in forming the image. The lens strength will also, of course, depend on the relative distance of the observer and source to the lens plane.

- $\kappa \ll 1$: This is the *weak-lensing* domain, in which the net effect of the lens on the object will be a series of weak distortions with little magnification.

It is on weak lensing that we wish to focus attention, especially as it pertains to assessing the mass content of the lens system itself. In many cases, the lens is not strong enough or concentrated enough to form multiple images or arcs. Nevertheless, while the effect of the lens on the source will be small, it will yield an image similar to the source, but with a pattern of distortions. The kinds of distortions we can expect are stretching (shearing) of the shape of the background object and magnification (convergence) of the shape of the background object. Using shear and convergence to deduce properties of the lens would work well if all sources were already well-characterized in un-lensed size and shape. Normally, only information about average properties can be or is known (e.g., the average shapes, characterized by how elliptical they appear, of galaxies in the cosmos).

An example of how the lensing system will distort the objects behind it is shown in Fig. 2.9. One set of images shows the unlensed population and the lensed population when the original population has no "shape noise" (no natural ellipticity), while the second set shows a more realistic case where the unlensed population already has shape noise. There are subtle but important differences between the two lensing outcomes. If we are to measure mass distributions as accurately as possible, these must be considered.

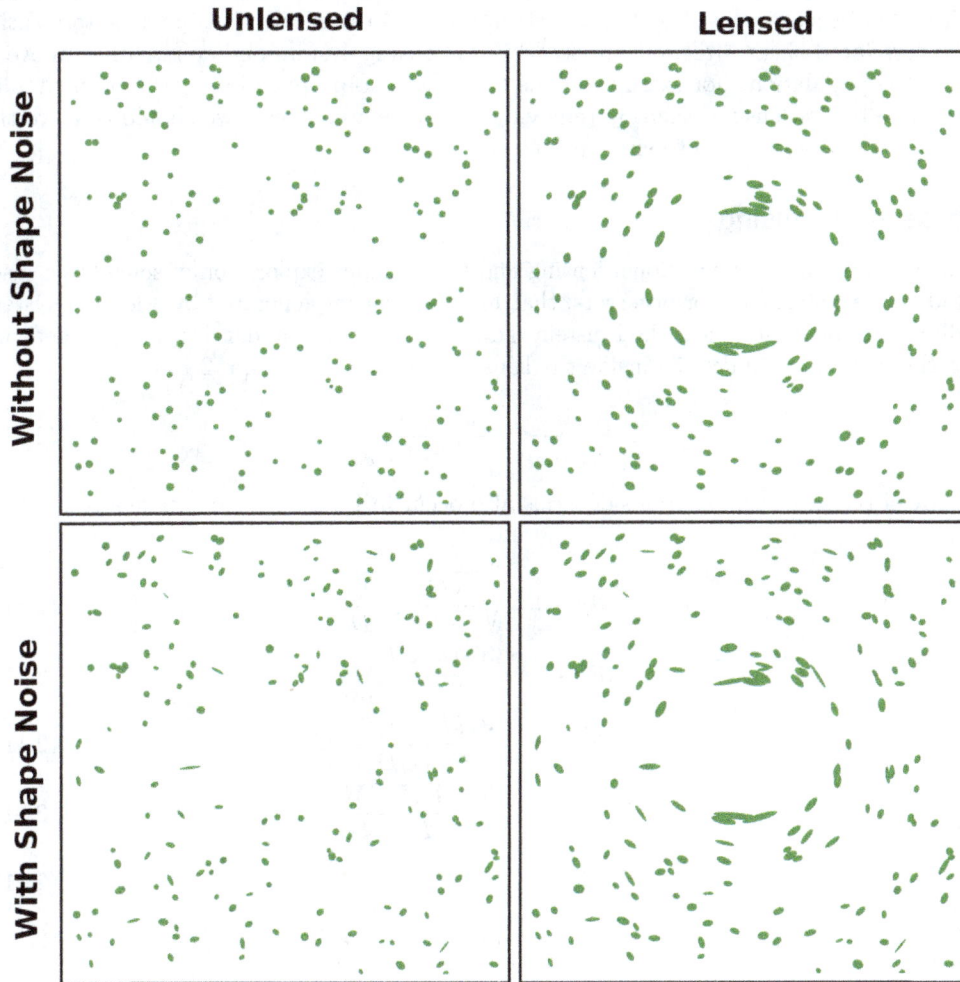

Figure 2.9: Weak gravitational lensing of galaxies in astronomy, with and without "shape noise"—that galaxies are not circular, and their own natural ellipticity will have to be factored into any shape distortions also caused by gravitational lensing. Figure from Ref. [Wik23g]

One example of how to do this is to look for correlated shearing effects on the population of lensed objects. The objects themselves may have random ellipticities, oriented in uncorrelated and unpredictable directions. However, the *lensing* of those objects will impose, on top of that natural variation, noise, as well as an overall preferred direction of shear imposed by the lens structure. Algorithmically, it is possible to identify the correlated shear direction imposed by the lensing system and coherently remove it from the population. For example, if the background objects had all really been circular in shape before lensing, then by removing the correlated shear, we should observe the objects to all be restored to circular once again.

Mass from Lensing

The implications of gravitational lensing and the previous exploration of some basic cases lead to the conclusion, for observers, that multiple images generated by a lensing system will tend to be located near the Einstein radius, θ_E. The critical density, $\sigma_{critical}$, depends on the *mean surface density* inside the Einstein radius. From Eqn. 2.108,

$$\frac{1}{\sigma_{critical}} = \frac{4\pi G}{c^2} \frac{D_s}{D_d D_{ds}}. \tag{2.111}$$

Many of the same combinations of terms also occur in Eqn. 2.96. Let's take advantage of that:

$$\theta_E = \sqrt{\frac{4GM}{c^2} \frac{D_s}{D_d D_{ds}}} \tag{2.112}$$

$$\theta_E^2 = \frac{4GM}{c^2} \frac{D_s}{D_d D_{ds}} \tag{2.113}$$

$$(\pi D_d)\theta_E^2 = \frac{4GM}{c^2} \frac{D_s}{D_d D_{ds}}(\pi D_d) \tag{2.114}$$

$$\pi D_d \theta_E^2 = \frac{4G\pi}{c^2} \frac{D_s D_d}{D_{ds}} \frac{M}{D_d} \tag{2.115}$$

$$\pi D_d^2 \theta_E^2 = \frac{M}{\sigma_{critical}} \tag{2.116}$$

$$M = \pi(D_d\theta_E)^2 \sigma_{critical} \tag{2.117}$$

One can estimate the Einstein radius from observations of a lensing mass in a lensing system. The critical density can be obtained from red shifts of the spectra of the objects in the lens system and the lensed objects, since it depends only on those distances. With current methods, mass estimates obtained from this approach can be accurate to the level of a few percent.

Applications of Kinematics, Luminosity, and Lensing: The Bullet Cluster

The Bullet Cluster, object 1E0657-56, is one of the most famous examples of the composition of mass and light information about a dynamic, complex astrophysical

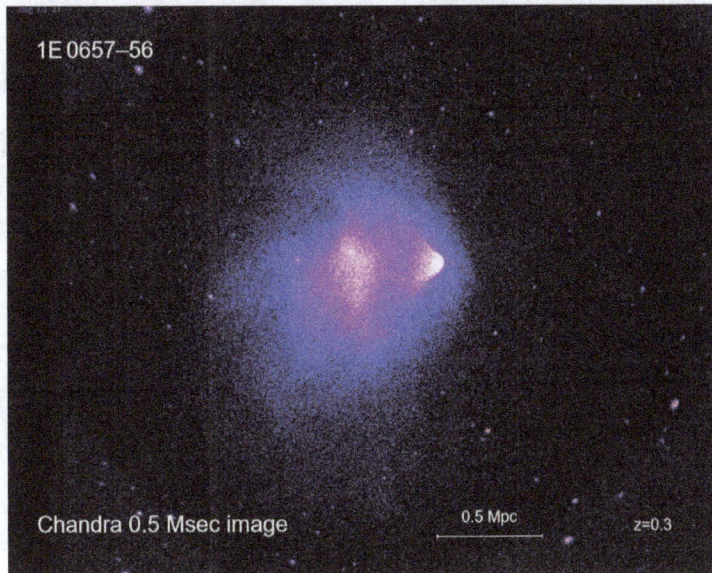

Figure 2.10: An x-ray image of the Bullet Cluster (1E0657-56) by the Chandra X-ray Observatory. This image was built from a 140-hour exposure. The scale is shown in Mpc. The colliding clusters are about 4 billion ly from Earth. Figure from Ref. [Wik23a] and originally provided by NASA.

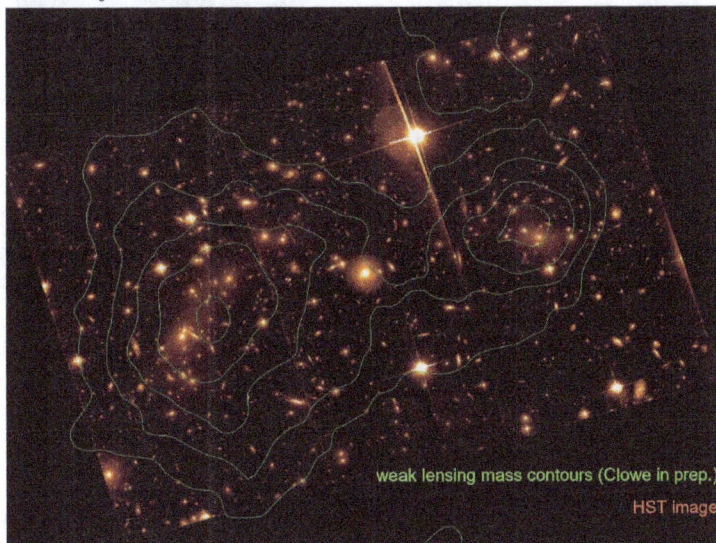

Figure 2.11: A visible-light image by the Hubble Space Telescope of the Bullet Cluster with the density contours derived from weak gravitational lensing analysis overlaid on the image. Figure from Ref. [Wik20a] and originally from NASA.

Figure 2.12: A composite image of the Bullet Cluster. The x-ray image (two bright central fuzzy shapes, shaped red in color image) is superimposed, along with the lensing image (the fainter fuzzy lobes to the left and right of the central bright shapes, shaded blue in color image), onto a visible-light image of the cluster. Credit: NASA/CXC/M. Weiss [Wik24a]

structure. It is actually the result of a pair of objects—galaxy clusters—that, long ago, passed through one another and have now separated again.

The "bullet" in the Bullet Cluster is visible in x-ray imaging (Fig. 2.10). This reveals the envelopes of hot gas surrounding the galaxies in each of two colliding clusters. This envelope has the appearance of a shock wave resulting from one object impacting another. As we saw in Section 2.5, such envelopes of hot gas can extend many galactic radii beyond their central galaxies. The two galaxy clusters in the Bullet Cluster are not the same; one (the "bullet") is smaller than the other. The smaller cluster has punched through the larger cluster. This collision occurred around 100 million years ago.

The x-ray image (Figs. 2.10 and 2.12) reveals that a shock front has formed in the gas clouds, with the smaller one appearing as the "bullet" that has created a front in the envelope of the larger cloud. This dynamical appearance tells us that the gas has experienced a strong collision.

The galaxies, which can be seen in visible light (Fig. 2.11), are much more compact than the surrounding halos of hot gas. These appear to have largely escaped collisions, and observations of their kinematics using astrometry indicate that the two groups involved in the collision were largely unaffected by each other. Just as stars are so much smaller than their host galaxies, and can be treated as points, galaxies are much smaller than galaxy clusters. The density of galaxies in such a cluster is so low that, in the case that two clusters pass through one another, the chance of a direct galaxy collision or gravitational interaction is quite small.

However, as we have seen, the gas coronas surrounding galaxies extend far beyond their hosts while nevertheless containing little overall mass (less than 1% of the mass of the stars in that galaxy). Nevertheless, their much larger extent means that, even if the galaxies miss each other, the gas may not. Indeed, from the bullet cluster we see that the luminosity of the two groups suggests that, while the galaxies (defined by their stars) have missed each other, their gas envelopes have lagged behind. This separates these two populations—gas and stars—from one another. The hot gas collisions generate a lot of light in the form of x-rays, while the stars produce most of the visible light.

Systems like the Bullet Cluster offer a unique laboratory for studying the question of whether gas halos explain nonluminous matter in galaxies and galaxy clusters. If the excess mass in galaxies and clusters is due to the hot gas halos, we expect the mass density to follow the gas. If, on the other hand, the unseen mass component of galaxies and clusters is not tied to the gas halos, we might expect the mass to follow the galaxies. Lensing, agnostic as to the kind of mass (luminous or nonluminous), will tell us the answer.

Lensing analysis (Fig. 2.12), combined with x-ray and visible-light imaging of the Bullet Cluster, shows us that the mass density (more leftward and rightward, fainter, blue-shaded regions) follows the galaxies (shown in their visible-light imaging) and not the gas (the more central, intense, red-shaded regions). Color intensity is proportional to density. The highest concentrations of gas are not associated with the highest concentrations of mass. Those appear to follow the density of galaxies in the post-collision clusters.

This single-cluster collision is not the only one observed. To date, at least four such colliding galaxy clusters have been analyzed using visible light, x-rays, and gravitational lensing [Wik23j]. A systematic analysis across these astronomical objects consistently reveals that the main mass densities do not follow the gas halos. The gas halos strongly interact and become entangled. The galaxies are compact compared to the size of the cluster and tend to avoid collisions, passing through unchanged. The mass density follows the galaxies. This confirms again what visible-light astronomy has revealed: the luminous mass of galaxies is insufficient to explain their total mass. There is an unseen mass envelope that extends far beyond the visible stellar and gas populations. This allows galaxies to grow larger than would otherwise be expected.

Whatever holds the galaxies together and influences the overall dynamics of galaxy clusters is not the electromagnetically responsive atomic matter that fills the voids between the stars and the galaxies. The hot gas, while massive, does not explain the total mass distribution of the galaxies and the clusters. Hot gas, pulled away from host galaxies by the collision process, does not also pull the mass distribution with it; rather, the mass distribution follows the galaxies. It is as if the enveloping halo of unseen matter is also relatively collisionless.

2.7   Dark Matter Density and Velocity Distributions

Bullet Cluster-like events offer us a baseline model for dark matter: a (nearly) collisionless gas. From this model, we can begin to infer the properties of dark matter halos (those surrounding luminous galaxies).

Let us consider a collection of N particles contained in some volume, each with some position, velocity, and mass. We can describe a complex N-body system like this by thinking not about each constituent, but about the *phase space* of the system: the mathematical representation of all possible *states* of the system. The phase-space distribution contains the complete information for statistically describing a system of particles.

The probability that, at time t, a specific dark matter particle is in the phase-space volume $d^3x\, d^3v$ around the location $(\vec{x}, \vec{v})$ is given by a function we shall write as $f(\vec{x}, \vec{v}, t)$. The latter function is known as a *probability density function* and tells us the probability of finding the particle-per-unit volume in the space of all possible positions and velocities.

The probability of finding the particle somewhere in the entire phase space must be unity (100%). Therefore, we obtain the normalization condition for this probability density function,

$$1 = \int f(\vec{x}, \vec{v}, t) d^3\vec{x} d^3\vec{v}. \tag{2.118}$$

The Boltzmann equation provides us with a method of describing a dynamical system like this that is not in equilibrium (e.g., is in a process of flowing due to affects on the system). This equation, in the framework of Hamiltonian mechanics, states that

$$\hat{L}[f] = \hat{C}[f] \tag{2.119}$$

where $\hat{L}$ is the *Liouville operator* which describes how, in time, the phase-space density will change; and $\hat{C}$ is the *collision operator*, which describes the number of particles per phase-space volume lost or gained per unit time. The Boltzmann equation relates the activity of these two operators acting on the probability density function, essentially stating that changes in the phase-space density in time are directly related to the change in the number of particles per phase-space volume in that time.

In the non-relativistic limit (whose applicability will be clear later), the Liouville operator can be written

$$\hat{L}[f] = \frac{\partial f}{\partial t} + \dot{\vec{x}}\frac{\partial f}{\partial \vec{x}} + \dot{\vec{v}}\frac{\partial f}{\partial \vec{v}}. \tag{2.120}$$

In principle, the collisions described by the Boltzmann equation can grow quite complex, depending on the nature and strength of the interactions involved in the gas. However, we will consider a (near-) collisionless gas of nonluminous massive particles as an explanatory mechanism for dark matter. On the galactic or cluster scale, this seems a practical model given the evidence. This allows us to write:

$$\hat{L}[f] = \frac{\partial f}{\partial t} + \dot{\vec{x}}\frac{\partial f}{\partial \vec{x}} + \dot{\vec{v}}\frac{\partial f}{\partial \vec{v}} = \hat{C}[f] = 0. \tag{2.121}$$

Jeans's theorem allows us to determine steady-state solutions (those properties of the system that have no time dependence) to the Boltzmann equation. Without proving this, Jeans's theorem states that any steady-state solution of the collisionless Boltzmann equation depends on the phase-space coordinate through integrals of motion $I(\vec{x}, \vec{v})$ that satisfy

$$\frac{d}{dt}\left(I(\vec{x}, \vec{v})\right) = 0. \tag{2.122}$$

Integrals of motion are defined as any functions of only the phase-space coordinates that are constant along an orbit. Some common examples would be angular momentum, $\vec{L} = \vec{r} \times \vec{p}$ or the time-independent Hamiltonian, $H(\vec{x}, \vec{v}) = \frac{1}{2}v^2 + \Phi(\vec{x})$. Here, Φ is a generic potential term. In the case of a steady-state collection of collisionless particles forming a halo, the phase space is solely a function of energy,

$$f(\vec{x}, \vec{v}) = f(\mathcal{E}) \tag{2.123}$$

where $\mathcal{E} = \Psi - \frac{1}{2}v^2$ and Ψ is specifically the gravitational potential, $\Psi = -GM(r)/r$, due to a spherically symmetric mass distribution. Consider the density distribution of a phase-space distribution for an isotropic steady-state halo of $f(\mathcal{E}) = \exp(\mathcal{E}/\sigma^2)$ where σ is the RMS velocity dispersion. The associated density distribution is then

$$\rho(r) \propto \int_0^\infty dv\, v^2 f(v) = \int_0^\infty dv\, v^2 \exp\left(\frac{\Psi - v^2/2}{\sigma^2}\right) \propto e^{\Psi/\sigma^2} \tag{2.124}$$

In other words, in this model, since we are dealing with a collisionless gas in which the velocity dispersion about any value will be distributed in a purely uncorrelated way (Gaussian-distributed), we can write the probability density function for the phase space in terms of a Gaussian function, where the deviations from the mean are written in terms of the expectation that gravitational potential determines orbital velocity (and thus kinetic energy). The density as a function of r then simply becomes akin to the expectation value of v^2.

We can find a relationship between radius and density distribution starting from the equations of hydrostatic equilibrium and assuming that the gravitational potential plays the role of the potential in the scenario,

$$\nabla p = -\rho \frac{d\Psi}{dr}. \tag{2.125}$$

For an ideal gas, $p = k_B T/V$. Recalling the definition of density, $\rho = M/V$, and employing spherical coordinates, we can take the derivative of the pressure with respect to the radial coordinate (for a spherically symmetric gas under the influence of a gravitational potential, this is the only non-zero component of the gradient):

$$\frac{dp}{dr} = \frac{k_B T}{m}\frac{d\rho}{dr} \tag{2.126}$$

We then recognize that the derivative on the right-hand side relates to the radial derivative of the gravitational potential,

$$\frac{k_B T}{m}\frac{d\rho}{dr} = -\rho\frac{d\Psi}{dr} \tag{2.127}$$

$$\frac{k_B T}{m}\frac{d\rho}{dr} = -\rho\frac{GM(r)}{r^2} \tag{2.128}$$

We can rearrange the terms and continue:

$$\frac{k_B T}{m}\frac{d\rho}{dr} = -\rho\frac{GM(r)}{r^2} \tag{2.129}$$

$$r^2\frac{1}{\rho}\frac{d\rho}{dr} = -\frac{mGM(r)}{k_B T} \tag{2.130}$$

$$r^2\frac{d\ln(\rho)}{dr} = -\frac{mGM(r)}{k_B T} \tag{2.131}$$

Take the derivative with respect to r of both sides:

$$\frac{d}{dr}\left(r^2\frac{d\ln(\rho)}{dr}\right) = -\frac{mG}{k_B T}\frac{dM(r)}{dr}. \tag{2.132}$$

The mass enclosed in an orbit of radius r will be given by:

$$M(r) = \int_0^{2\pi}\int_0^{\pi}\int_0^r \rho(r)r^2 dr\sin\theta d\theta d\phi = 4\pi\int_0^r \rho(r)r^2 dr \tag{2.133}$$

which implies

$$dM(r) = 4\pi r^2\rho(r)dr \longrightarrow \frac{dM(r)}{dr} = 4\pi r^2\rho(r). \tag{2.134}$$

Therefore,

$$\frac{d}{dr}\left(r^2\frac{d\ln(\rho)}{dr}\right) = -\frac{4\pi Gm}{k_B T}r^2\rho(r). \tag{2.135}$$

For our isotropic steady-state halo model, $\rho(r) = \rho_1 e^{\Psi/\sigma^2}$ (Eqn. 2.124). Inserting this into the above equation results in $\ln(\rho(r)) = \ln(\rho_1) + \Psi/\sigma^2$. The first term is a constant, and its derivative with respect to any variable is zero. This allows us to use Poisson's equation[11] to relate the velocity dispersion, σ^2, to other information in the system,

$$\nabla^2\Psi = 4\pi G\rho(r) \tag{2.136}$$

$$\frac{1}{r^2}\frac{d}{dr}\left(r^2\frac{d\Psi}{dr}\right) = 4\pi G\rho(r) \tag{2.137}$$

---

[11]Poisson's equation involves the Laplace operator, ∇^2, which in Cartesian coordinates is $(\partial^2/\partial^2 x + \partial^2/\partial^2 y + \partial^2/\partial^2 z)$, but when transformed into spherical coordinates is more complicated. It is instructive to review this and to note that, when acting on a spherically symmetric function, the only term that survives from Laplace's operator is $(1/r^2)\partial/\partial r(r^2\partial/\partial r)$. This is crucial to the mathematics in this section.

$$\frac{d}{dr}\left(r^2\frac{d\Psi}{dr}\right) = 4\pi Gr^2\rho(r) \tag{2.138}$$

$$\frac{d}{dr}\left(r^2\frac{1}{\sigma^2}\frac{d\ln(\rho(r))}{dr}\right) = 4\pi Gr^2\rho(r) \tag{2.139}$$

$$\frac{d}{dr}\left(r^2\frac{d\ln(\rho(r))}{dr}\right) = 4\pi G\sigma^2r^2\rho(r) \tag{2.140}$$

This allows us to identify, by combining this with Eqn. 2.135, that, in this case, $\sigma^2 = k_BT/m$.

Let us now find a descriptive solution to our equation assuming that it has a more general form $\rho(r) = Cr^{-b}$. In this case,

$$\frac{d}{dr}\left(r^2\frac{d\ln(\rho)}{dr}\right) = -\frac{4\pi G}{\sigma^2}r^2\rho(r) \tag{2.141}$$

$$\frac{d}{dr}\left(r^2\frac{1}{\rho}\frac{d\rho}{dr}\right) = -\frac{4\pi G}{\sigma^2}r^2\rho(r) \tag{2.142}$$

$$\frac{d}{dr}\left[r^2\frac{1}{Cr^{-b}}\left(-bCr^{-b-1}\right)\right] = -\frac{4\pi G}{\sigma^2}r^2Cr^{-b} \tag{2.143}$$

$$\frac{d}{dr}\left(-br\right) = -\frac{4\pi G}{\sigma^2}r^2Cr^{2-b}. \tag{2.144}$$

By inspection, we can determine that

$$-b = \frac{4\pi G}{\sigma^2}Cr^{(2-b)} \tag{2.145}$$

is a condition that forces $b = 2$ in order for b to have no r dependence. That, in turn, forces $C = \sigma^2/(2\pi G)$. We can then write our solution:

$$\rho(r) = \frac{\sigma^2}{2\pi Gr^2}. \tag{2.146}$$

We have employed a simple model for a spherical isotropic halo in steady state that fits the evidence, assuming an inverse-square dark matter density and phase-space distribution described by a Maxwell–Boltzmann distribution. In this case, $\rho(r) \propto 1/r^2$ and $f(v) \propto v^2\exp(-v^2/(2\sigma^2))$. This is exactly what is expected for a self-gravitating isothermal gas sphere. Rotation curves tell us, experimentally, that we should have expected a dependence like $\rho(r) \propto 1/r^2$—marking a remarkable concordance between a simple collisionless model for dark matter and observational evidence for the shape of its density profile necessary to explain galactic dynamics.

However, it is important to also consider the shortcomings of this model. There is a divergence in the total mass of the system described by the density profile above. The rotation curves of galaxies are not flat all the way to $r \to \infty$, which would be true in our model—the dark matter density profile must cut off someplace far beyond the core of the galaxy. The presumption of a steady-state system is not a realistic assumption.

Galaxies merge and as such are not virialized. This affects both the spatial density and the velocity distribution. Structure and substructure are expected to result from the mergers of galaxies, which are quite common.

The Standard Galactic Halo Model

Considering the compelling concordance of observation and theory, it would be foolhardy to completely disregard a collisionless Maxwell–Boltzmann model. However, one cannot neglect the concerns raised in the previous discussion. The standard galactic halo model [Bax+21; DFS86] is therefore a truncated Maxwellian distribution:

$$f(\vec{v}) = \begin{cases} \frac{1}{N_{esc}} \left(\frac{3}{2\pi\sigma^2} \right)^{3/2} \exp\left(-\frac{3\vec{v}^2}{2\sigma^2} \right) & |\vec{v}| < v_{esc} \\ 0, & |\vec{v}| > v_{esc} \end{cases} \tag{2.147}$$

where, for our own Milky Way galaxy,

$$
\begin{aligned}
v_{esc} &\approx 544\,\text{km/s} \\
\sigma_v &\approx 288\,\text{km/s} = \text{RMS velocity dispersion} \\
v_0 &= \sqrt{2/3}\sigma_v \approx 235\,\text{km/s} = \text{most probable speed} \\
z &\equiv v_{esc}/v_0 \\
N_{esc} &= \text{erf}(z) - 2\pi^{1/2} z e^{-z^2}
\end{aligned}
$$

Note that $f(v)$ describes the *probability to find a dark matter particles with a speed v per unit velocity-cubed*. The distribution must be integrated over velocities to obtain the probability, $p(v) = \int f(v) d^3v$. The probability distribution is illustrated in Fig. 2.13 for this model. The standard halo with a Maxwell–Boltzmann velocity distribution is widely used by experimentalists to interpret their results. We will explore this later in the text.

Numerical simulations are used by astrophysicists to develop a wide array of possible profiles that diverge from the isothermal model. An example of this is the Navarro–Frenk–White (NFW) model proposed in 1995, which cannot be derived from first principles (Fig. 2.14). It is an example of a "cusp" profile; one with a steep inner slope

$$\rho_{NFW}(r) = \frac{\rho_0}{\frac{r}{r_s} \left(1 + \frac{r}{r_s} \right)^2} \tag{2.148}$$

where r_s is a scale radius. These parameters have to be determined from computer models, astronomical measurements of galaxy populations, or both.

Another example is the Burkert profile, where instead of a cusp, this model describes a "cored" profile with a flatter slope,

$$\rho_{Burk}(r) = \frac{\rho}{(1 + r/r_0)(1 + (r/r_0)^2)} \tag{2.149}$$

where r_0 is the core radius. Again, these are parameters whose values have to be constrained.

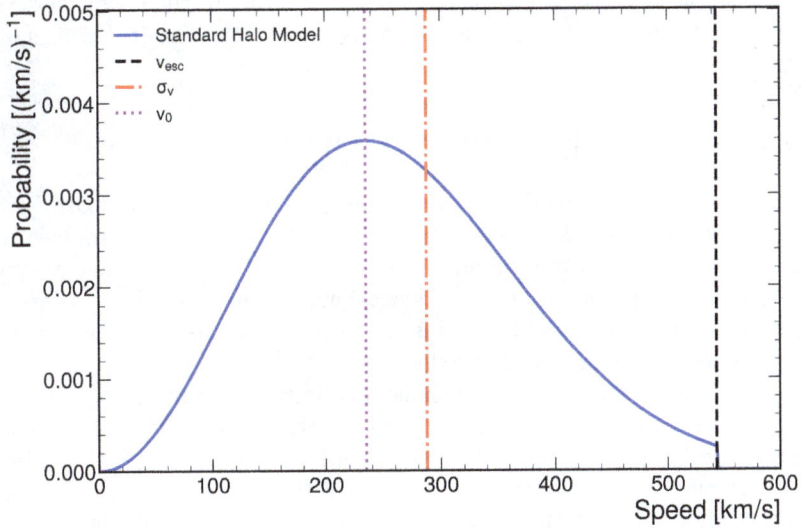

Figure 2.13: The probability distribution vs. dark matter speed for our region of the Milky Way, as described in the standard halo model.

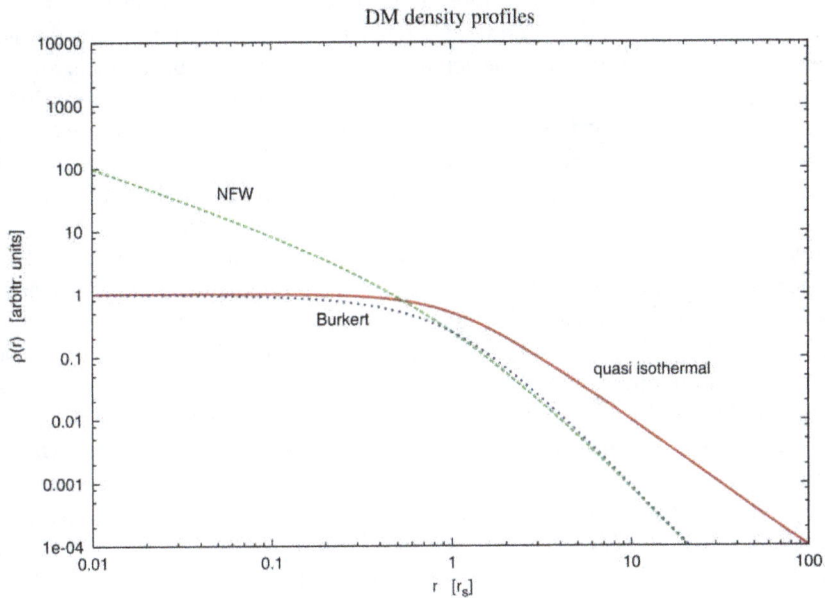

Figure 2.14: A comparison of the density profiles obtained from a quasi-isothermal, NFW, or Burkert model. Figure from Ref. [Del14].

Other common density profile models as the *quasi-isothermal* or *pseudo-isothermal* model as well as the *Einasto* profile [Ein65]. We do not detail the latter model here but its functional forms are widely available. A comparison of the density profiles of some of the models discussed here is shown in Fig. 2.14.

2.8   Concluding Thoughts on Astrophysical Probes

In this chapter we have explored the various key astrophysical probes of mass distributions. Dynamics provides us with information about the kinetic energy of astrophysical objects. Using conservation of energy and momentum, as well as the law of gravity, we can infer from the motion information about the gravitational potential of the system—and thus infer something about the mass that holds the system together. This approach can be applied to galaxies or to clusters of galaxies.

Similarly, we can use established relationships between luminosity and mass to infer something about the stellar contribution to mass distributions. By probing wavelengths other than visible light, we are able to learn about dust and gas, which is also present. Since it is possible that a halo of dust and gas, glowing only in x-rays (for example), may explain the nonluminous (non-visible) matter component of galaxies and clusters of galaxies, it is crucial that we employ a multi-wavelength strategy in assessing light and matter.

Finally, we have explored a deeper relationship between mass and gravity through the general relativistic prediction of light deflection by energy densities. This allows us to map mass densities independent of whether the matter being mapped is, itself, responsive to light. By combining the lensing and light techniques, we have seen that we can separate the gas, stars, and dark matter from one another using colliding galaxy clusters as laboratories. We are left with the conclusion that whatever composes the dark matter, it is relatively collisionless (at least, compared to hot baryonic gas) and follows along with the galaxies it envelopes, rather than to hang behind in its collision with the gas halos.

A recent analysis of galaxy groups. using x-ray information, suggests that the degree of interaction of dark matter with itself (its *self-interaction cross section*) is constrained to be $\sigma/m < 0.16$–6.61 cm^2/g at the 95% confidence level for dark matter velocity dispersions between 200–500 km/s [KD23]. To put this in perspective, we can use unit conversions from Chapter 1. The lowest end of the limit range corresponds, in barns and GeV, to 0.16 cm^2/g $= 0.29$ b/GeV. If dark matter particles have a mass scale around that of the proton (1 GeV), then the self-interaction cross-section would have to be less than 0.29 b. This is consistent with the scale of nuclear interactions at its largest, so dark matter could still have a strong interaction with itself. However, that self-interaction strength could also be much smaller (down to the gravitational interaction scale for two subatomic particles). Astrophysical probes put limits on self-interaction by dark matter and, compared to the hot gas surrounding galaxies in clusters, is far less likely to experience a collision (with itself or the gas particles).

We appear to be left with a large quantity of near-collisionless matter that envelopes galaxies and clusters and noticeably affects their dynamics. A key question at this point is: has this dark matter always been there? One would assume that it has been around for a long time, shaping the structure of the cosmos in the process. It should be possible to combine astronomy and cosmology and rewind the clock to a much earlier era of the cosmos. Do we see the hand of nonluminous, near-collisionless matter in the early universe?

CHAPTER 3

Intersections of Astrophysics and Cosmology

3.1 The History of the Cosmos

The universe is 13.78 billion years old. Since it came into existence (Fig. 3.1), it has been expanding. There is evidence for an extremely rapid period of expansion ("inflation") in just the first moment of time, when it went from nearly zero size to about the size of a grain of sand in the first 10^{-23} s (or so). This was followed by a period of much slower

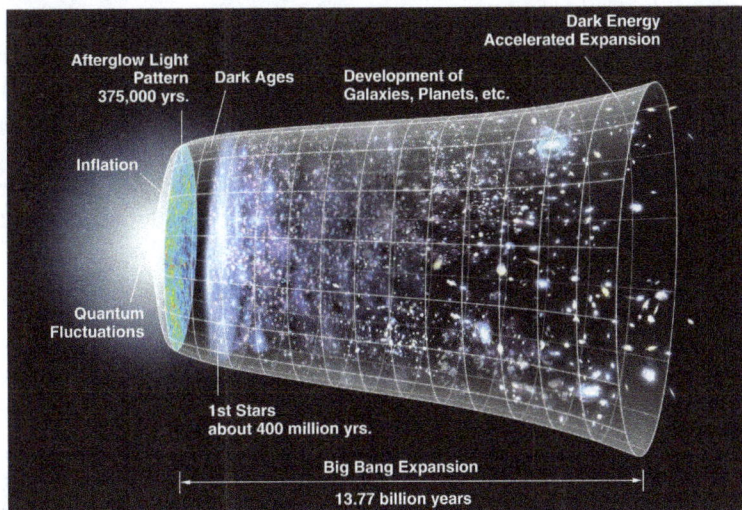

Figure 3.1: A visual timeline of the beginning, rapid expansion, and subsequent expansion of the universe. The formation of light nuclei, then-neutral light elements, followed by star formation and the re-ionization of the universe, as well as galaxy formation, and large scale modern cosmic structure formation, are all indicated. Figure from Ref. [Wik24c].

expansion, such that the universe is now only about a thousand times larger in linear dimension (diameter) than it was a hundred thousand years after its birth.

Presently, the visible universe (the parts of the universe for which there has been sufficient time for light to reach us) is about 93 billion ly across. Why is the visible universe larger than 13.78 billion ly? This is due to that expansion. The farthest parts of the universe from which light has been able to reach us have since moved *even farther* from us due to cosmic expansion. Taking into account current knowledge of this expansion rate, this puts the *visible* universe—the universe that is hypothetically observable—at the stated size. We cannot presently know what is beyond the visible universe.

At its inception, the universe was very small, very dense, and very high-energy. What we know about the earliest moments of the universe, within the first few minutes of time, is inferred from recreating those conditions in particle colliders. By understanding physics in high-energy and energy-dense environments, we can model that earliest phase of the universe. The baryonic (e.g., protons and neutrons) and leptonic (e.g. electrons, muons, and neutrinos) components of the universe are thus established by this approach.

However, we now understand, after nearly a century of researching, that there appears to be more to the universe than baryonic and leptonic matter. There appears also to be a substantial nonluminous and massive component, whose composition is unknown.

Nearly all the hydrogen and helium in the universe, along with other very light elements, were forged in the first few minutes after the beginning of time. This period is known as "recombination," although this is an historical misnomer; it marks the first moment during which electrons and protons combined into atoms, as opposed to they having been together earlier and then later recombined [1]. Predictions arising from the knowledge of particle and nuclear physics allow us to estimate the degree and rate of nucleosynthesis, the formation of those elements, during the minutes after the beginning of time. These estimates conform with the ratio of light elements observed in the universe today.

After the formation of the light nuclei, it would take a few hundred thousand more years for the expanding and cooling universe to reach a temperature such that the kinetic energy of nuclei and electrons was sufficiently lessened that they could form neutral elements (called the "era of recombination"). Once this concluded, the universe was essentially electrically neutral (called the "dark ages") until the first stars burst into existence, re-ionizing large parts of the cosmos. The light that escaped when hydrogen and helium formed for the first time is known as *cosmic background radiation*. These photons, loosed a few hundred thousand years after the beginning of time, free-streamed (for the most part) through the newly neutral universe. As the universe expanded, the wavelength of this light was stretched, shifting it more into the red part of the spectrum.

---

[1]The word "recombination," to apply to a period of the formation of the light elements, predates the Big Bang hypothesis. Certainly, had hydrogen and helium been present at the beginning of time, then ionized (separating protons from electrons), and then cooled, recombination would be the correct idea. However, the Big Bang hypothesis predicts that protons and electrons emerged first in energy states so high that they could not bind to one another through the electromagnetic interaction. In that era, the electromagnetic force was too weak to overcome the large kinetic energies of protons, neutrons, and electrons. However, once the primordial plasma cooled, an initial combination was possible. The historical term, "recombination," unfortunately, has remained.

Currently, it is visible in the microwave band, with frequencies similar to those used by television signals in the era of analog broadcasting (from the 1940s to the 2000s).

If dark matter was present during those earliest moments of the cosmos, we would expect to observe its fingerprints in relics such as the Cosmic Microwave Background (CMB). How might these appear? The CMB is the light released from matter as it formed neutral light elements. One would then expect the density of light across the sky to correspond to a pattern of structure of over- and under-densities of matter and also from the local curvature of space-time. Whatever structure and dynamics existed prior to the release of the CMB would then imprint, at the moment of recombination, on that light. This would provide a snapshot of the universe about 380,000 years after the beginning of time.

Was dark matter already present then, and was it forming structure? Or, are the patterns in the CMB compatible with those possible composed only of the kinds of light baryons and leptons we know (from the ratio of light elements) to have been present at that time?

3.2 The Cosmic Microwave Background

When characterizing an electromagnetic radiation field like the CMB, it is typical to speak of its *spectral radiance*, its radiance per unit wavelength, λ, or frequency, f. The latter choice depends on whether it is useful to parameterize the radiance as a function of one or the other property of the radiation. *Radiance* is the energy, per unit time, per steradian (the unit of solid angle), per unit area ($W \cdot sr^{-1} \cdot m^{-2}$) that is absorbed, emitted, or reflected by a surface. The spectral radiance of the CMB peaks around 160 GHz when parameterized in terms of frequency[2]. The CMB has a near-perfect blackbody spectrum, a key prediction of the Big Bang hypothesis.

Exercise 13

Knowing that the most prominent frequency of the CMB today about 160 GHz when considering the frequency spectral radiance function ($dE_f/dtdfd\Omega dA$), estimate the CMB's current temperature.

These observations allow us to conclude that the CMB presently has a mean temperature of 2.725 K. While a near-perfect blackbody spectrum, the CMB does contain thermal anisotropies at the level of 1 part in 100,000. These are evidence of corresponding anisotropies in the structure of the early universe, a structure from which the CMB photons were loosed early in its history. In turn, this pattern of anisotropies can tell us about the players in the universe at the time of recombination. The origin of the fluctuations is even more subtle, as any large-scale anisotropies present so long after the beginning of time would have been seeded by quantum fluctuations in the fleeting

---

[2]Refer to Eqn. 2.70, the surrounding discussion, and Exercise 11 to be refreshed on the spectral radiance of blackbody radiation.

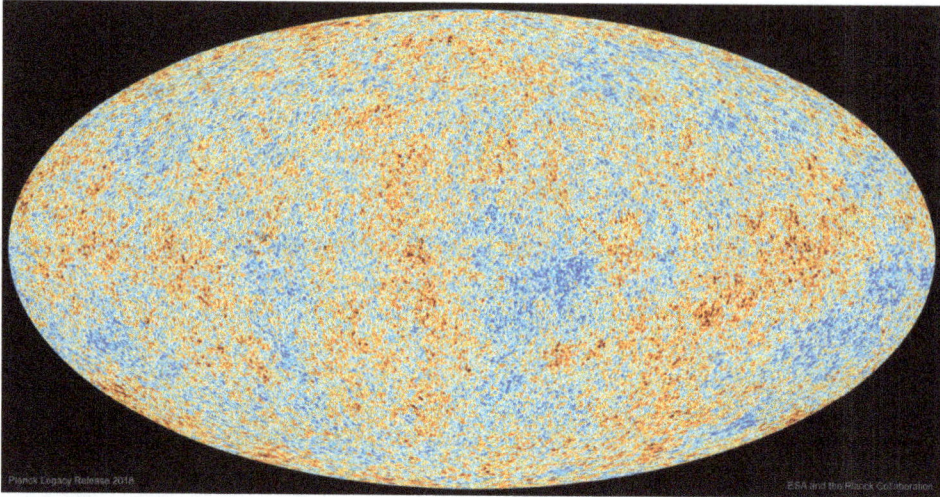

Figure 3.2: An all-sky map of the CMB produced in 2018 by the Planck Collaboration. The colors represent fluctuations from the mean temperature of the CMB (2.725 K). These anisotropies in the temperature of the CMB correspond to anisotropies in the early universe and are at the level of 1 part in 100,000 above (red) or below (blue) the average.

fractions of a second after the beginning. Over time, these fluctuations were gravitationally enhanced and became the structure we see in the CMB, as well as the overall structure we see in the cosmos today.

The patterns in the CMB can tell us many things, from "big picture" stuff to subtle quantum mechanical things. For example, an overall dipole shape in the CMB is evidence of the motion of the solar system about the galaxy, causing a red shift of the CMB light behind us, and a blue shift ahead of us. This dipole effect is purposely removed from most maps, like the one in Fig. 3.2. The pattern of anisotropies in the CMB has provided information about the space curvature of the universe, the baryon and dark matter content of the cosmos, the vacuum energy content of the universe, the value of the Hubble constant H_0 (the present rate of the expansion of space),[3] and has provided information about the period of inflation. The information derived from the CMB is an unfolding scientific story that has not yet been fully told. Most of the rich information found in this radiation is, however, beyond the scope of this book.

An analysis of the temperature of the CMB provides information about the kinds of matter present in the early universe. The key to analyzing the CMB is to unfold its *power spectrum*. "Unfolding" is the act of decomposing the spectrum into its key parts and

---

[3]We will come back to this more formally in Section 3.3, but the notation H and H_0 will appear in discussions of cosmic history. $H \equiv H(t)$ refers to the general *Hubble rate*, the expansion rate of the universe, at any time t. H_0 specifies the Hubble rate now, in the current era, and is also referred to as the *Hubble constant*.

using a model to infer the origin and degree of each of those parts. We will see that the aspects of the model that can influence the CMB each affect the spectrum in key ways.

The Power Spectrum of the CMB

All complex wave forms (for example, a note from a musical instrument) can be represented as a sum of amplitudes ("powers") of harmonics ("frequencies") of simple sine or cosine waves. The *fundamental* can be thought of as a sine wave of a particular frequency, f_0. All *harmonics* are derived from the fundamental by a simple relationship: $f_i = n_i f_0$, where n_i is any integer above zero. A single musical note can be so decomposed. Generally, any complicated waveform can be similarly composed into a sum of cosine waves, each with a distinct frequency f_i and a power (amplitude) of A_i. Finding the best sum of such waves that reproduces the observed complex wave form is the key mathematical and computational trick. This is the foundation of signals analysis, supported, at its heart, by the Fourier transform.

A cosine wave with a single period (T) and thus single frequency $f = 1/T$, corresponding to a single angular frequency $\omega = 2\pi f$, can be written

$$y(t) = A\cos(\omega t + \phi) \tag{3.1}$$

where A is the amplitude of the wave and ϕ is the phase angle of the wave (the degree to which the maximum amplitude is displaced from $t = 0$). A complex wave form like that shown in the top panel of Fig 3.3 can be formed by summing a number of such waves, e.g.

$$Y(t) = \sum_i A_i \cos(\omega_i t + \phi_i). \tag{3.2}$$

For example, the complex wave form in the left of Fig. 3.3 can be realized with a sum of about twenty cosine waves of distinct and specific amplitudes and frequencies (bottom of Fig. 3.3), each new frequency being an integer multiple of a lowest fundamental frequency. More complex wave forms may require a much larger sum of more varied sine waves. In principle, Fourier decomposition allows one to write any arbitrary wave form as an infinite sum of sine waves.

This same approach can be used to analyze the CMB. Instead of treating it as a temporal wave form, it shall be treated as a *spatial* wave form with a wave number (spatial frequency) instead of a time-domain frequency. A graph of the amplitude of the contributing spatial harmonics can be matched to predictions from models that influence that spectrum.

Consider a one-dimensional function, $f(x)$, that is in a finite region of dimension D. Applying the Fourier transform from a spatial distribution to a frequency distribution will allow us to determine what spatial frequencies are present, and at what proportion,

$$f(x) = \sum_{n=0}^{\infty} b_n \cos(2\pi n x/D + \phi_n). \tag{3.3}$$

Information about the system is characterized by b_n and ϕ_n. If we want to characterize the temperature fluctuation over the entire visible sky, we need a two-dimensional function

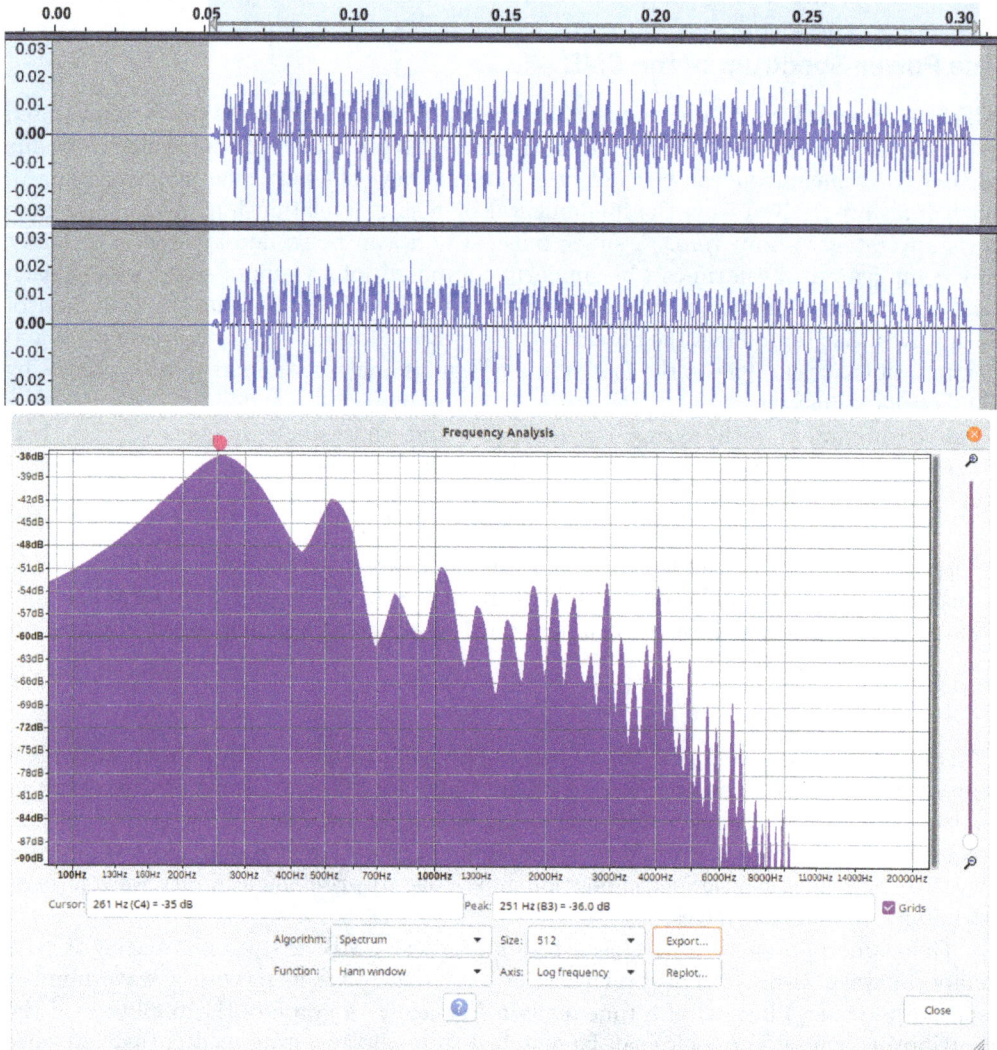

Figure 3.3: A complex wave form (top two charts, wave forms showing stereophonic audio) decomposed into its harmonics (bottom-most graph) showing how these are combined to achieve the singular complex wave form. The wave is the result of a middle C key being struck on a keyboard, ringing for a duration of a quarter-second. This rich sound wave is composed of a fundamental frequency (the largest peak in the bottom graph, marked by a red dot) at about 251 Hz, plus harmonics such that each is an integer multiple of the fundamental.

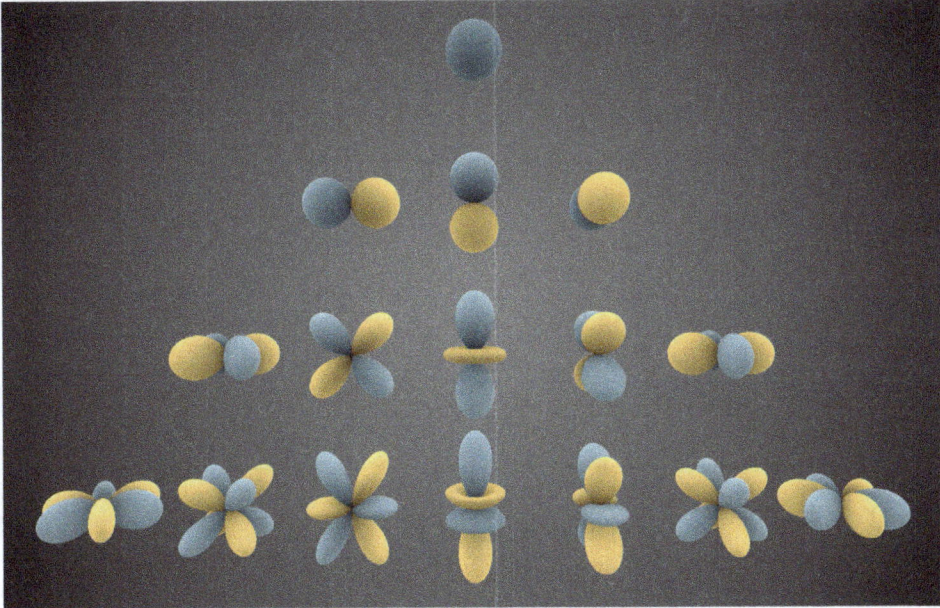

Figure 3.4: The first few real-valued spherical harmonics. The case $\ell = 0$ is shown at the top, with increasing ℓ as one descends down each row of the figure. Figure from Ref. [Wik23h].

$T(\theta, \phi)$. This is more complex. To decompose this requires the application of *spherical harmonics* rather than basic one-dimensional cosine waves. Spherical harmonics are the building blocks of repetitive structure in two dimensions. Thus

$$T(\theta, \phi) = \sum_{\ell=0}^{\infty} \sum_{m=-\ell}^{\ell} a_{\ell m} Y_{\ell m}(\theta, \phi) \tag{3.4}$$

where the functions $Y_{\ell m}(\theta, \phi)$ are the spherical harmonics defined by

$$Y_{\ell m}(\theta, \phi) = e^{im\phi} P_\ell^m(\cos \theta) \tag{3.5}$$

and P_ℓ^m are the Legendre functions. The index ℓ is known as *the multipole moment* of the harmonic. The first few real-valued spherical harmonics are visualized in Fig. 3.4.

The lowest-order ($\ell = 0$) spherical harmonic is a single sphere that covers the entire two-dimensional surface. You can think of this as representing a feature that spans the entire sky. Thus, we can say that such a low-order contribution tells us something about a feature with a large spatial scale. In the case of the CMB, this component represents a single scalar that covers half the entire sky and thus probes an angular scale of 180°. This relates to the average temperature of the CMB across the sky.

As we increase the size of ℓ, selecting higher and higher harmonics, we are probing the structure of the sky at increasingly finer and finer distance scales. To capture very fine

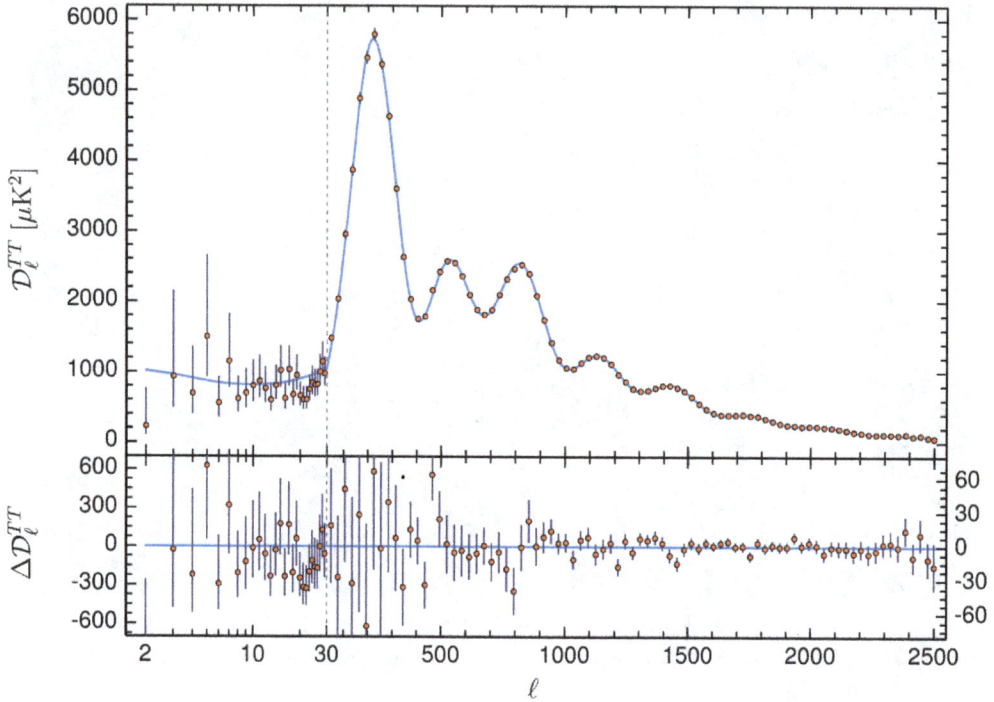

Figure 3.5: Planck 2018 temperature power spectrum. From Ref. [Agh+20b].

structure on the sky—subtle correlations between temperature variations, for example, that may be scattered across the CMB map—requires the application of higher and higher orders of spherical harmonics.

Each coefficient in the Fourier expansion, $a_{\ell m}$, will have an *average* that depends only on ℓ but not m. The distribution of the values should be Gaussian. Thus, we can define an average and characterize the power spectrum:

$$C_\ell = \langle |a_{\ell m}|^2 \rangle. \tag{3.6}$$

Let's review some features of these harmonics. There are $2\ell + 1$ values of m for each ℓ. Consider $\ell = 3$. There would be only seven independent measurements to be averaged; this means that the sample average could be significantly different from the *true* average obtained from the "true" cosmological parameters that determine $a_{\ell m}$. Even with no measurement error at all, we might happen to have fluctuations that give smaller or larger values of C_ℓ. This effect is called "cosmic variance." At large ℓ, the number of independent measurements of $a_{\ell m}$ becomes large enough that cosmic variance is not a limiting factor.

The power spectrum of the CMB, as determined by the Planck Collaboration, is shown in Fig. 3.5. The data points are the observed power spectrum, while the smooth curve is an overlaid model of the power spectrum determined from cosmological considerations. It is worth noting that the analogy between analyzing a musical note by breaking it down into its constituent time domain information (the fundamental and harmonic frequencies) and analyzing the CMB by decomposing its temperature features into spatial harmonics is strikingly good. For example, it is remarkable how the power spectrum of Fig. 3.5 resembles that of middle C in Fig. 3.3. The musical fundamental frequency is a long-wavelength (low frequency) wave upon which harmonics are piled in a mathematical relation. Similarly, the CMB is composed of a large-scale fundamental structure (the low-ℓ peak) and a series of peaks at smaller distance scales (higher ℓ). While not following the mathematical regularity of musical fundamentals and harmonics for a single note, the similarities imply that a Fourier analysis will lend itself to exposing the underlying relationships that yield the observed CMB.

Understanding the Structure of the CMB Power Spectrum

The structure of the power spectrum is related to cosmology by deep physical considerations of the universe just after it burst into existence. The size of the universe in those first moments was infinitesimal, and, as such, would have been ruled by quantum field theory and general quantum mechanical considerations. Quantum theory predicts fluctuations in the vacuum of the early universe which, in turn, has the ability to introduce disturbances in matter and energy density at all scales. What happened in those first moments would have, by expansion of the universe, been stretched across vast distance scales.

These quantum fluctuations would have been places where matter could clump. This clumping activity generates gravitational-potential wells which, in turn, further attract baryons, leptons, and photons ("radiation"). These two quantum roommates, however, act somewhat in opposition to one another. Gravitational baryon clustering is opposed by photon pressure. While the quantum fluctuations can act to gather matter, which, through enhanced gravitation, can gather additional matter and radiation, the radiation brought into these gravitational wells acts to push apart the baryons and leptons. In cosmology, it is the density of a component that generally plays a significant role in the evolution of the cosmos at any given moment. These counteracting processes are illustrated in Fig. 3.6.

Focusing on just the baryons and leptons, the lightest and most stable of these particles have very different mass scales. Even if leptons and baryons were co-equal in population number, neutrons and protons are almost 2,000 times heavier than electrons and at least 10^{12} times heavier than neutrinos. Leptons and their role in population mass densities are generally neglected in discussions of the SM matter contributions to cosmic evolution. The focus in these dynamics is entirely on the baryonic component.

The alternating compression/expansion of baryons generates an oscillation in the baryon density; an acoustic wave. These are known as *baryon acoustic oscillations*.

The quantum fluctuations gather all forms of matter, not just baryonic or leptonic matter. Should nonluminous, low-scattering-rate matter have been present as well, it would have also collected in the quantum fluctuations. However, unlike its baryonic counterpart, this form of matter would have become more dense due to its inward gravitational self-attraction, but would not feel the expansion driven by radiation. So far as we know, a kind of matter matching the description of dark matter would not emit photons nor

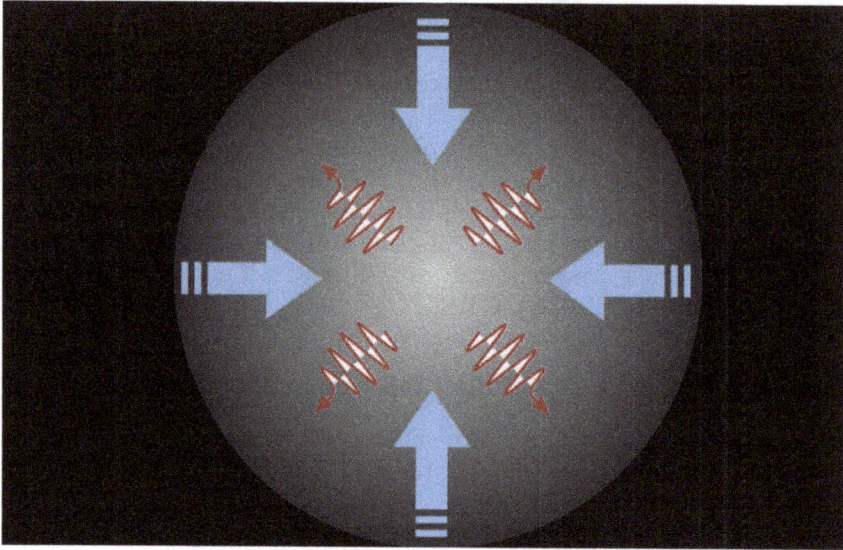

Figure 3.6: An example of a tiny region of the primordial universe with a slight over-density (circular grey gradient) resulting from a random quantum fluctuation. Matter collecting in this over-density would tend to compress under its own gravity (blue arrows), but, in doing so, would heat and emit radiation (red wiggles) that tends to drive it to expand. This creates an oscillatory phenomenon of the two competing tendencies.

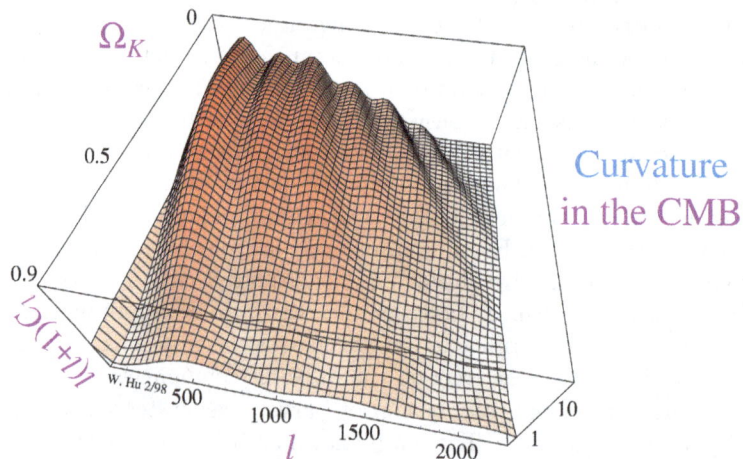

Figure 3.7: The CMB power spectrum and its change in shape as one alters the curvature of space from flat. The position of the first peak at $\ell = 200$ is indicative of a spatially flat universe. Other kinds of universes would have this peak shifted substantially away from $\ell = 200$. Modified from an image by W. Hu. (University of Chicago)

interact with them, experiencing the kind of radiation pressure that baryons routinely experience. In addition, through its own gravitation, dark matter would have aided in further compressing baryons into these gravitational wells. If that matter outnumbered baryons, there is an excellent chance that it is the original cause of the gravitational wells into which baryons fell, nurturing those original baryon acoustic oscillations.

Each peak in the CMB power spectrum (Fig. 3.2) can tell us a part of the story of this early universe. As in the analysis of a musical note from a piano, the most prominent component of the frequency spectrum gives us large-scale core information about the overall pattern. For example, the CMB's tallest peak (containing most of the power of the spectrum) can tell us about the overall geometry of the universe.

Remember that lower multiple moments, ℓ, correspond to probes of larger distance scales. This first peak at $\ell = 200$ tells us about larger-scale temperature structure in the universe. How does this relate to the overall geometry of space (Fig. 3.7)? If space is flat, and there is a "hot spot" in the universe (where a quantum fluctuation, stretched by inflation, led to a matter under-density[4]) then any light passing by that hot spot is not bent toward or away from the hot spot by the curvature of space (since space is not curved at all). In that case, we would observe in the CMB the *true* size of that hot spot. If, on the other hand, space has positive curvature (is "closed"), then light rays passing such a region will bend toward it, making the hot spot appear larger and hotter than it actually is. If the universe has negative spatial curvature, then light rays will appear to bend away from the hot spot, making it appear smaller and cooler than it actually is.

Each of these cases shifts around the location, or changes the shape, of the first peak of the power spectrum, since that peak probes the largest scales over which temperature is distributed in the CMB. In other words, the overall geometric shape of the universe's space would necessarily have the largest impact on what photons do when traversing the entire universe to get to us here on Earth (Fig. 3.7). The fact that the peak is located in the observed position tells us that space is extremely flat, overall, in the universe. Space, overall, appears to possess no strong inherent curvature. If positively curved, the peak would be farther to the left; if negatively curved, the peak would be farther to the right. While stars, galaxies, and other massive gravitating bodies would *locally* distort the curvature of space, the universe itself, on average and across the largest visible distance scales, is flat.

The second peak ($\ell \approx 600$) tells us the amount of overall matter in the universe. Initial quantum fluctuations in the early universe would cause all matter to gravitationally clump toward fluctuations that, later, correspond to higher temperature regions in the CMB. Ordinary matter suffers self-interactions through electromagnetism, converting the potential energy of electromagnetism into the kinetic energy of baryonic matter. This

---

[4]In the CMB temperature anomaly map in Fig. 3.2, there are red ("hot") and blue ("cold") spots. The hot (cold) spots correspond to places where the temperature is a little higher (lower) than the average. An over-density will tend to trap light in the early universe, forcing it to escape a slightly higher mass region and gravitationally red-shifting the photon in the process. This results in a "cooler" photon. An under-density will red-shift the same photon by a lesser degree, leading to a "hotter" photon. Thus it is that hot spots correspond to matter under-densities and cold spots to matter over-densities. This is known as the Sachs-Wolfe effect [SW67]. We are grateful to Professor Joel Meyers (Southern Methodist University) for a discussion of this point.

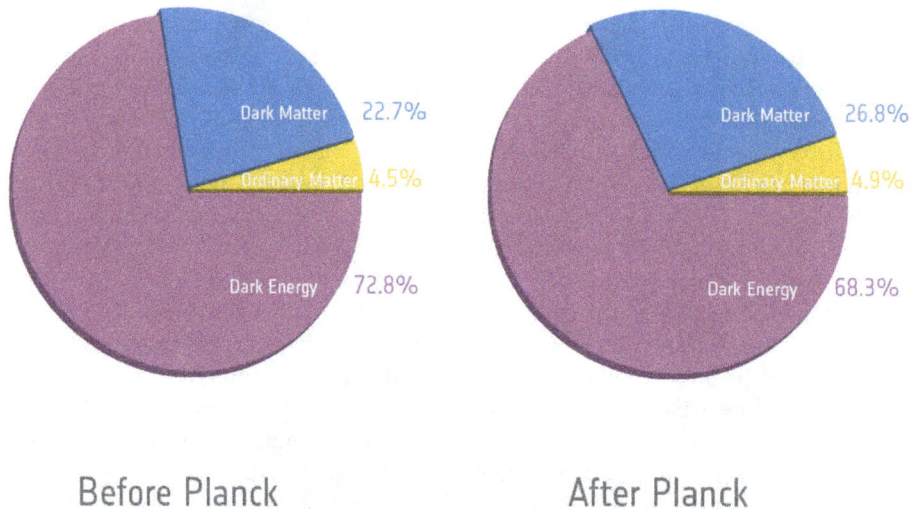

Before Planck   After Planck

Figure 3.8: The "Cosmic Pie Chart" as it follows from the work of the Planck Collaboration to understand the CMB. The situation before Planck's results (including the WMAP satellite and other experiments) is shown at the left, improved in precision by the Planck data shown at the right.

heating results in pressure that opposes the clumping matter. More matter leads to a larger response and has the net effect of damping out the second peak. The prominence of that peak, therefore, is a measure of how much matter was present in the early universe as a fraction of all energy density in the universe.

The third peak ($\ell \approx 850$) tells us the amount of dark (nonluminous, but gravitationally active) matter that was present in the universe. This peak results from oscillations that have had time to reach their extremal (maximum and minimum) densities at least once. Once minimum density is reached, pressure in the over-dense region pushes the ordinary matter away. Dark matter does not interact with photons and isn't strongly affected by the baryon pressure, remaining clumped in the quantum fluctuation that seeded its over-density in the first place. The more dark matter left in a region, the denser it becomes, as gravity takes over again.

More dark matter leads to a higher third peak in the CMB power spectrum. Essentially, dark matter clumping more densely due to a higher original population seeds small-scale temperature hot spots across the sky. As we observed earlier, smaller spatial features correspond to large multipole moments in the CMB.

By using cosmology to determine the amount of curvature, total energy density, inherent vacuum energy, radiation, dark matter density, and baryonic density, one can build a model of the power spectrum of the CMB. This is represented by the pie chart in Fig. 3.8. Based on data from the WMAP and Planck instruments, themselves built on previous balloon-borne and satellite-based experiments, modern understanding of the CMB is that it was imprinted by a universe dominated by a significant and non-zero vacuum energy density (the energy of empty space, if you will) followed by a large dark matter component and then the baryonic component. Baryonic matter is described extremely well overall by the Standard Model of Particle Physics, a framework that combines quantum field theory and mathematical symmetry in order to describe forces and particles. We can understand the behavior of baryonic matter using this theory. How do we measure the density of non-baryonic matter and how do we then describe the behavior that could have led to what is observed?

We begin by measuring the baryon density and then the total matter density. If we then subtract the baryon density from the total matter density, we arrive at the isolated contribution from non-baryonic matter. This is typically referred to, by shorthand, as the dark matter density. If we observe the baryonic density and the total matter density to be equal, this would tell us that all matter whose effects are imprinted on the CMB are explainable entirely by baryons.

The baryon density at any time, t, is defined relative to a quantity known as the *critical density* and can be written

$$\Omega_b = \frac{\rho_b(t)}{\rho_{\text{critical}}}. \tag{3.7}$$

We define the critical density later in Section 3.3, specifically in Eqn. 3.13.

The relationship between the actual density of the universe and the critical density, $\rho_{critical}$, is what helps us to predict the shape of the universe. As we discussed in considering the origin of the largest CMB acoustic peak, space itself can be, naturally and inherently, curved or flat. However, adding matter on top of that natural, or inherent, geometry will then further bend space. It is possible, therefore, to take an inherently perfectly flat universe (which, naively, would allow for unlimited expansion) and cause it to curve, so that it closes—so that the universe eventually re-collapses. Prior to recognizing that the energy density of the vacuum is actually non-zero, the critical density was taken as the threshold for the overall geometry of the universe and defined whether or not the universe would expand forever or would one day collapse. A universe with a matter density precisely equal to the critical density is truly flat, taking into account both its inherent geometric shape and the shape deformations added by the presence of matter.

3.3   Structure Formation and the Evolution of the Universe

The anisotropies of matter and radiation in the early universe should become the seeds of large-scale structure as the universe expands and continues to cool. The CMB temperature map tells us that the scale of the fluctuations in the matter density at early times was $\delta\rho/\rho \approx 10^{-4}$. The universe expands as space expands. These perturbations grow and collect more matter, including dark and baryonic matter, over subsequent hundreds of millions of years. At the scale of hundreds of Mpc, the anisotropies are at the level of $\delta\rho/\rho \approx 10^2$. Galaxies are seeded in these matter over-densities, gravitational potential wells that promote the condensation of gas into stars, stars into galaxies, and galaxies into clusters and superclusters. At the level of galaxies, the anisotropies are at the level of $\delta\rho/\rho \approx 10^6$.

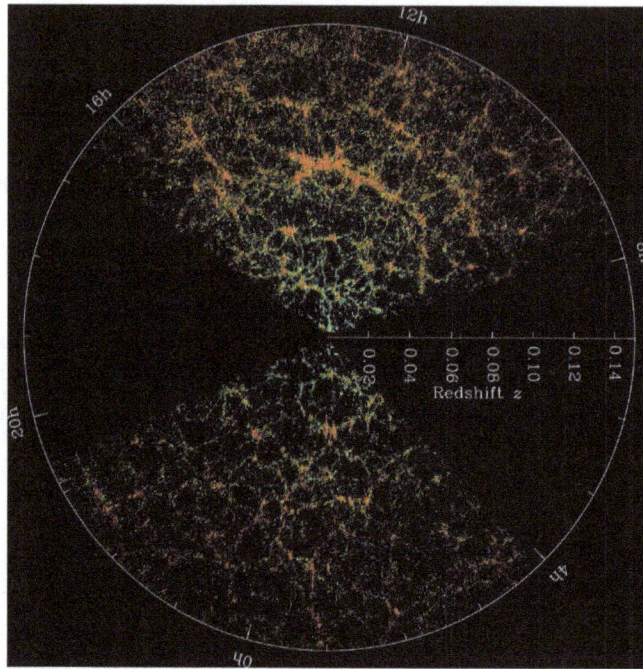

Figure 3.9: The Sloan Digital Sky Survey map of the universe. Each dot is a galaxy, the color bar showing the local density. The anisotropy of the universe on the scale of MPc is observable here, and cosmology would suggest that this was seeded from quantum fluctuations near the beginning of time. Image courtesy of SDSS.

The origin and evolution of structure (galaxies and large scale structure) in the universe is a central area of research in cosmology. Structure is generally thought to arise from the growth of density perturbations which originated in the early universe. Astronomical surveys of galaxies at low redshift ($z < 0.5$, reaching a few billion years into the past) reveal a web-like structure of anisotropies (Fig. 3.9) that may once have been small but, through the expansion of space, have been stretched over vast distances. A key question here is: what might have been the specific origin of those perturbations?

There is not one definitive answer to this, and no strong evidence for any of the proposed ideas. One attractive idea is that these perturbations arise specifically from quantum fluctuations of a scalar quantum field. By invoking the existence of such a field, it is possible to also explain the rapid inflation of the universe just after the beginning of time. Any random fluctuations in the field would then have been amplified by the exponential growth of the universe at that same time.

Since there is no definitive theory of the universe that extends back this early in cosmic history, it may be safest to adhere to what we can know, or infer, from observations of the cosmos. A reliable theory of nature will ultimately be drawn from those observations.

What do we know about early structure formation? We observe fluctuations of size $\delta T/T \approx 10^{-4} \approx \delta\rho/\rho$ in the detailed map and analysis of the CMB. General relativity then dictates that inhomogeneities grow in various ways depending on the component of the universe (matter, radiation, or a cosmological constant) that dominates during a given period. During the radiation-dominated era (when relativistic components like photons and neutrinos dominated the density parameter of the universe) until about 47,000 years after the big bang, $\delta\rho/\rho \approx \ln(t)$. In the matter-dominated era, from about 47,000 years until about 9.8 billion years after the big bang, $\delta\rho/\rho \approx t^{2/3}$. In the dark energy or cosmological constant-dominated present era, $\delta\rho/\rho \approx$ constant [Ryd03]. The fluctuations in the CMB were primarily set in the form we see them now during the matter-dominated era as recombination occurred during this era.

In each period, the inhomogeneities will depend on a different function of a, the cosmic *scale factor*. The scale factor is that time-dependent coefficient that, in a classical picture of space-time provided by the theory of general relativity, relates the distance between two objects at a reference time, t_0, to the distance at a later time, t (e.g. $d(t) = a(t)d_0$). The famous *Hubble parameter*, originally inferred from the observed motion of distant galaxies, is, by definition, $H(t) \equiv \dot{a}(t)/a(t)$. That is to say, the parameter tells us the ratio of the rate of change of the scale factor at any given time, t, to the size of the scale factor at the same time. Thus, by observing the recession velocity (a measure of the change of scale between points in space with time) of distant galaxies as a function of their distance (an effective measure of the scale itself) from us on Earth, one infers the Hubble parameter. Modern measurements constrain H to be about 70 km/s/Mpc (c.f. Ref. [Lee+24] for an example of a recent measurement).

The velocity of distant objects along the line of sight connecting the observer and the object is determined by measuring the Doppler shift of the light from the object (the "radial velocity"). Objects moving away from us have their light waves "stretched" (lengthened), creating a shift of their nominal atomic spectra toward the red end of visible light (longer wavelengths). Objects moving toward us are blue-shifted. The relationship between the expected wavelength (λ), the observed wavelength (λ_{obs}), and the radial velocity relative to the speed of light ($\beta = v/c$) is

$$\lambda_{obs} = \lambda\sqrt{\frac{1+\beta}{1-\beta}}. \tag{3.8}$$

Note that the above argument is on solid ground when discussing a source and observer separated by distances that are well below the cosmological scale, where not only will light also be affected by space-time expansion but also by whatever it experiences along the journey from source to observer. Nevertheless, without taking the full expansion history of the cosmos into account in the equation we can observe that the degree of red shift can be quantified by a fraction,

$$z \equiv \frac{\lambda_{obs} - \lambda}{\lambda} \tag{3.9}$$

which merely denotes the fractional change in the wavelength relative to its original value at the source. Thus we can write $1 + z = \lambda_{obs}/\lambda = \sqrt{(1+\beta)/(1-\beta)}$.

Since, on large cosmic scales (distances greater than Mpc) the radial velocity of objects is determined by the expansion of space, which is in turn tied to the distance to those objects (the farther away they are, the older the observed objects are, due to the finite travel time of light), astronomers tend to speak interchangeably of *red shift* and time (or distance) in the cosmos. An overly simplistic derivation will demonstrate the relationship between the expansion of the universe and red shift (we remind the reader that again, here, we are neglecting the effect of the expansion history on the details of these photon wavelength shifts).

Imagine that, at some reference time, t_1, the distance between Earth and a very distant co-moving galaxy is d_1. Here, "co-moving" means that if space were not expanding at all, at this moment the two objects are tracking each other identically and would, even if moving relative to some third reference point, remain at a fixed distance forever. The expansion of space spoils that, altering the distance between the two objects over time in spite of their identical velocities. At some later time, t_2, owing to the expansion of space, that distance will shift by an amount δd to $d_2 = d_1 + \Delta d$. We can rewrite all of this in terms of the scale factors at the two times, where, again, the scale factor is itself defined with respect to some other reference time in the cosmos. Then $d_2 = a_2 d_0 = a_1 d_0 + \Delta d$. This allows us to write $\Delta d = (a_2 - a_1)d_0$.

Any light wave emitted from the galaxy toward Earth at t_1 will then subsequently experience a lengthening that is proportional to the expansion of space in between its origin and its destination (Earth). In other words, we would expect that, if space expands by 1%, then the wavelength will stretch by 1%. We can write this mathematical relationship simply as $\Delta \lambda / \lambda = \Delta d / d_1$. We see immediately that we can substitute z, the red shift, into this equation:

$$z = \frac{\Delta d}{d_1} = \frac{(a_2 - a_1)d_0}{a_1 d_0} = \frac{a_2 - a_1}{a_1} \tag{3.10}$$

$$= \frac{a_2}{a_1} - 1 \tag{3.11}$$

Thus we arrive at a relationship between the *observed degree of red shift of a light wave* and the *ratio in the cosmic scale factors at the time of observation vs. the time the light departed the object*.

Consider galaxies that are observed to have red shifts $z > 6$, which means mathematically that the observed wavelengths of their light are 600% longer than when emitted at rest relative to the observer. A good example of this would be the shift of the hydrogen "Balmer alpha" line, a prominent wavelength of light emitted by hydrogen at 656.46 nm. This kind of red shift means that we would observe that prominent feature to be at 3938.8 nm ... far into the infrared. If we observe galaxies with red shifts in excess of 6, it means that galaxies must have existed at that time in the universe. Using Eqn. 3.11, this would correspond to a time when the scale factor of the universe was 7 times smaller than it is now. Such cosmological red shifts correspond to co-moving radial distances in excess of 8 Gpc.

Although we observe such galaxies now, they existed at the time that their light left those galaxies to then travel to us over such large distances. For a galaxy to form, the universe must contain matter inhomogeneities at the level of $\delta\rho/\rho \approx 10^6$ relative to the mean density of the universe. The question then becomes: how did the conditions in the universe develop rapidly enough (over just a few million years, at most), from inhomogeneities at the level of 10^{-4}, to be home to those at that level of 10^6?

We can use a naive and simplistic approach to explore answers to this question. Consider a flat, matter-dominated universe, a fairly good approximation for the early universe before the energy density of the vacuum began to dominate over matter a few billion years ago. In this case, owing to the flat metric associated with space, we can apply the Friedmann equation:

$$H^2 - \frac{8}{3}\pi G\rho = -\frac{\kappa}{a^2} \longrightarrow 0 \qquad (3.12)$$

since, for a flat universe, $\kappa = 0$. Here, H is the Hubble parameter and a is the scale factor at the time when this equation was true. This equation applies at the largest size scale of the universe at that time. We wish to emphasize that a number of terms and components in this equation are (or may be) time-dependent and this equation applies for the time at which it is specified. For example, the Hubble parameter would have its value *at this specific time in cosmic history* (which may not, and is not, the value it has today). The scale factor would also be specific to this time. We also note that it is from the Friedmann equation with a flat geometry that the *critical density*, ρ_c, is derived:

$$\rho_{\text{critical}} = 3H^2/8\pi G. \qquad (3.13)$$

In addition to Eqn. 3.12, a second equation must also be true. While the overall universe may be geometrically spatially flat, a *local* region of the universe with an over-density of matter will contain local, non-zero curvature:

$$H^2 - \frac{8}{3}\pi G\rho' = -\frac{\kappa'}{a^2} \qquad (3.14)$$

where the primed quantities indicate they are local, not global, measures of density and curvature. The Hubble parameter is assumed to be the same at this time in both scales: the cosmological scale and the local (e.g., "galactic" or "galaxy cluster") scale. We can then subtract the global equation from the local equation, cancelling any common terms (the Hubble parameter), and isolate the local curvature:

$$\left(H^2 - \frac{8}{3}\pi G\rho'\right) - \left(H^2 - \frac{8}{3}\pi G\rho\right) = -\frac{\kappa'}{a^2} \qquad (3.15)$$

$$\frac{8}{3}\pi G\left(\rho' - \rho\right) = -\frac{\kappa'}{a^2} \qquad (3.16)$$

$$\rho' - \rho = -\frac{3\kappa'}{8\pi Ga^2}. \qquad (3.17)$$

We can then define the *fractional over-density*:

$$\delta \equiv \frac{\rho' - \rho}{\rho} = -\frac{3\kappa'}{8\pi Ga^2\rho} \qquad (3.18)$$

The density of matter in a local region of the universe will appear[5] as $\rho \propto a^{-3}$, so the fractional over-density becomes directly proportional to a in this simple model. There is a linear relationship between the size of the fluctuations and the scale factor of the universe. Since cosmologists prefer to think in terms of red shift, one can employ Eqn. 3.11. The red shift tells us how much light was stretched by expansion between a reference time and another time (the time of large-scale structure formation, for example). Thus we see that in this model as $\delta \propto (z + 1)^{-1}$. The constant of proportionality between red shift and relative over-density in this model depends on Newton's gravitational constant, some numbers, and the scale factor of the universe at some unspecified reference time.

Since we wish to use the model to draw some conclusions, we need to eliminate that unspecified reference scale. Ideally, we would have measured, at some distant time in the past, the size of the universe so that we can now establish the relative scaling with respect to some absolute reference. But that is not possible. Instead, we can use information about the universe at one time and infer the evolution in the model to some other time. We must choose two times.

The time at which the CMB was released from the early hot plasma is a good reference scale, since we can observe the CMB now, and thus, any observations about it will tell us about the universe at that reference time (its release into the cosmos). We know, from the CMB, the level of inhomogeneity in the universe at that time. We can then project forward the relative baryonic over-density of the universe (local density compared to average cosmic density) and see if this explains the universe we actually observe. Recall that inhomogeneities in the universe at the time the CMB formed were at the level of 10^{-4}. At the time of large-scale structure formation, inhomogeneities would need to be at the level of 10^2 (this is the level at which galaxies exist).

Let's now use our model to compare the over-densities at the time of the CMB's release (recombination) and the time when large scale structure is known to have formed. If we see agreement between model and observation, the model survives to make more predictions. If not, the model is incomplete (at best). Recombination occurred at $z = 1100$ while large-scale structure formation occurred at a red shift in the neighborhood of $z = 6$. Thus

$$\frac{\delta_f}{\delta_i} = \frac{(1+z)_i}{(1+z)_f} \tag{3.19}$$

$$\frac{\delta_{LSS}}{\delta_{CMB}} = \frac{(1+z)_{CMB}}{(1+z)_{LSS}} \tag{3.20}$$

$$\delta_{LSS} = \delta_{CMB}\frac{(1+z)_{CMB}}{(1+z)_{LSS}} \tag{3.21}$$

$$= 10^{-4}\left(\frac{1+1100}{1+6}\right) \tag{3.22}$$

$$\approx 0.02 \tag{3.23}$$

---

[5]The relationship $\rho \propto a^{-3}$ owes to the fact that a local matter density will depend on the volume of space it occupies. That volume will have scaled by a in each dimension from some reference time, so the density scales merely as $1/a^3$.

This would tells us that, in this simple model of the universe, in the time between recombination and the observed time of large-scale structure formation, the over-densities would have only increased by a factor of about 200 from what they were at recombination. But this isn't right, in the sense that it is not sufficient to result in the observed universe. Based on observations, we would expect this to be more at the level of 10^2 (a million-fold increase since recombination), not 10^{-2}.

Why did our model fail? In the context of the present exploration, we can posit that it is because our simplified model neglected *dark matter*. This non-baryonic and non-electromagnetic matter would have collapsed to form larger anisotropies much earlier than when recombination happened at $z = 1100$. Baryonic matter was prevented from doing this earlier by radiation pressure. It was only when recombination occurred, making the universe transparent to radiation, that baryons could be ruled primarily by gravitational interactions and pulled into the gravitational wells of any existing anisotropies in the universe. We see here again that, if we are to understand the relationship between early and current anisotropies, we cannot neglect the role played by a dark matter component.

We will define some useful standard notation before we proceed to the discussion of dark matter in the early universe. First, we note that it is often easier to discuss a quantity called the *density parameter*, Ω, than it is to discuss actual densities of the components of the universe (SM matter, dark matter, radiation, etc.). This is defined as in Eqn. 3.7, which we write generically as $\Omega \equiv \rho/\rho_{\text{critical}}$. Adopting this concept, we can rewrite the Friedmann equation as

$$H^2 = \frac{8}{3}\pi G \rho - \frac{\kappa}{a^2} \tag{3.24}$$

$$= \frac{8}{3}\pi G \Omega \rho_{\text{critical}} - \frac{\kappa}{a^2} \tag{3.25}$$

$$= H^2 \Omega - \frac{\kappa}{a^2} \tag{3.26}$$

$$H^2(1 - \Omega) = -\frac{\kappa}{a^2} \tag{3.27}$$

$$\Omega - 1 = \frac{\kappa}{H^2 a^2}. \tag{3.28}$$

Since the scale factor and the Hubble parameter can vary with time, it is often useful to define $h \equiv Ha$ and write

$$(\Omega - 1)h^2 = \kappa. \tag{3.29}$$

In a flat universe, $\kappa = 0 \longrightarrow \Omega = 1$ and the components of Ω must satisfy $\Omega \equiv \sum_i \Omega_i = 1$.

Interpretations of the cosmology of the universe, obtained from the Planck satellite and many other astronomical observations, result in constraints on the components of Ωh^2. The current best model that explains the concordance of data, the Λ Cold Dark Matter (ΛCDM) model, requires the cosmos consist of a (a) baryonic matter (SM matter) component, (b) a radiation component, (c), a non-baryonic ("dark") matter component, and a scalar energy density (Λ) associated with space itself. This results in the following constraints from the Planck satellite power spectrum-only data [Agh+20b]:

$$\Omega_b h^2 = 0.02233 \pm 0.00015 \text{ (Baryonic Matter)},\tag{3.30}$$

$$\Omega_m h^2 = 0.1428 \pm 0.0011 \text{ (All Matter)}.\tag{3.31}$$

The above imply that the non-baryonic component of the matter is $\Omega_{\text{non-bary}} h^2 \approx 0.12$, which accounts for $\Omega_{\text{non-bary}}/\Omega_m \approx 84\%$ of the matter density parameter.

3.4   A Brief Look at the Synthesis of the Light Elements

The CMB is a relic from 380,000 years after the big bang. It is valuable to verify that the information from the CMB regarding baryonic matter, dark matter, and their relative roles in cosmic structure, are confirmed by at least one independent source. The source we glimpse at here is the abundance of the light elements, focusing on hydrogen and helium (and their isotopes). *Big bang nucleosynthesis* (BBN) happened in the first minutes of time and essentially dictated the overall relative abundances of hydrogen, deuterium, and helium in the observable universe[6]. In a sense, the populations of light elements (and their relative values) is a relic from even earlier in the universe and should provide information about baryonic matter (protons and neutrons). Recent reviews of BBN are available in Refs. [Cyb+16; GF23].

We summarize the physics that leads to the expectation of specific relative abundances of the lightest elements. At very high temperatures (which we will better define in a moment), the average kinetic energy of the matter and radiation particles in the universe is high enough that the lowest-mass baryonic states—the neutron and the proton—cannot form. Constant collisions of quarks and gluons, as well as collisions with high-energy photons, will disrupt the ability to create stable bound states. Destruction of bound states by high-energy photons is called *photodissociation* and is important in the next stage.

Once the temperature of the universe falls below $m_n \approx 939.6$ MeV/c^2, and then $m_p \approx 938.3$ MeV/c^2, it is possible for neutrons and protons to form and, in principle, persist. The time at which this happens is around 0.1 s after the big bang. Free neutrons are not stable and have a lifetime of about 878 s (15 minutes). Free protons are stable for lifetimes beyond that of the universe, and experiments have only been able to determine that, if the proton *can* decay, its lifetime is longer than 10^{29} years. In general, once $k_B T \ll m_p$ the lightest baryons will be describable as an isotropic non-relativistic gas, with each population following the Maxwell-Boltzmann distribution for such a gas,

$$N_i = g_i \left(\frac{m_i k_B T}{2\pi\hbar^2} \right)^{3/2} e^{-mc^2/2k_B T},\tag{3.32}$$

where g_i is the "degrees of freedom" of species i, counting the number of internal states available to the particle. For protons and neutrons, both spin-1/2 fermions, there are two

---

[6]We know that the birth of stars altered the primordial populations of light elements. Stars burn hydrogen, fusing it into helium. Heavier stars can later fuse helium and other elements, up to iron. These effects can be measured by studying stars and other regions of the cosmos. From these observations, corrections can be applied to current elemental populations to infer the primordial populations.

ways to arrange the spins of the particles. Thus $g_p = g_n = 2$. The relative numbers of two populations (protons and neutrons) is given at a specific speed (or kinetic energy) and temperature by

$$\frac{N_p}{N_n} = \left(\frac{m_p}{m_n}\right)^{3/2} \exp\left(-\frac{(m_p - m_n)c^2}{k_B T}\right) = 1.002 \exp(1.300/k_B T). \quad (3.33)$$

The above ratio is thus extremely close to one, until the temperature drops below the proton-neutron mass difference, at which time the ratio begins to increase to favor protons, the lighter of the two species.

While the protons and neutrons are in thermal equilibrium, their populations are maintained by a set of reactions that allow protons to become neutrons and vice versa. These reactions are $n^0 + \nu_e \leftrightarrow p^+ + e^-$ and $n^0 + e^+ \leftrightarrow p^+ + \overline{\nu}_e$. While protons and neutrons are easily able to encounter neutrinos and charged leptons, the reactions will proceed readily in both directions. However, as the universe expands, it becomes harder for these encounters to happen. Thus, the rate at which these interactions occur will fall. This effect is further enhanced by the weakness of the neutrino interaction cross-section, a purely weak interaction phenomenon. Once $k_B T < 0.8$ MeV these reactions are essentially stopped by the combination of the universe's expansion and the smallness of the weak interaction probability. The universe then "freezes out"' a population of protons. This concept of freeze-out will return later in a similar discussion about dark matter populations in the early universe.

The freeze-out population from this process is $N_p/N_n = 1.002 \exp(1.300/0.8) \approx 5$. After freeze-out, and before free neutrons decay, we see that we expect 5 protons for every 1 neutron.

Though the universe cools to a temperature below the proton mass ($k_B T < m_p$), the kinetic energies of protons and neutrons remain high enough to induce nuclear fusion. Fusion occurs when protons overcome their Coulomb repulsion and bind by the strong interaction, and when neutrons bind to either protons or neutrons via the same interaction. Some possible reactions are $p^+ + p^+ \rightarrow d + e^+ + \nu_e$, where d is a bound state of a proton and neutron (a *deuteron*) and $p^+ + n^0 \rightarrow d + \gamma$. The fusion of two neutrons can happen and will result in a di-neutron state, but this state is extremely short-lived and was only first observed in 2012 [Spy+12]. The di-neutron process can be neglected in this discussion.

The proton–proton reaction proceeds through the weak interaction while the proton–neutron proceeds through the strong interaction, so the second interaction is expected to dominate in the early universe. Thus we expect a fraction of the remaining neutrons to fuse with some of the protons and produce deuterons, which comprise the first isotope of hydrogen, *deuterium*. Deuterium is stable over the lifetime of the universe (it is not known to decay at all).

If a proton collides with a deuteron, the reaction produces helium-3, $p^+ + d \rightarrow {}^3\text{He} + \gamma$. If two deuterons collide and fuse via $d + d \rightarrow \text{He} + \gamma$, they produce helium-4. Both isotopes of helium are stable over the lifetime of the universe and are not known to decay.

The radiation in the universe during this period can work against the fusion process. For a time, the above reactions are reversible. The thermal bath of photons in the early universe can have acted to dissociate these heavier nuclei if their typical energy, $k_B T$, lies above the Q *value* of the reaction, which is defined as $Q = K_f - K_i$. If we assume that the proton and deuteron in the helium-3 reaction are just barely at rest when they fuse, then $Q_{^3{\rm He}} = (m_p + m_d - m_{^3{\rm He}})c^2 = (938.27208943 + 1875.61293 - 2808.39151)$ MeV ≈ 5.5 MeV. This will be the energy of the photon released in the fusion reaction. Conversely, a photon with at least 5.5 MeV will dissociate helium-3 into its constituents, forcing them to wait for another fusion reaction to occur.

Let us take account of the Q values of other reaction products. The Q value for the helium-4 process is higher, at 23.8 MeV, while for deuterium formation the value is 2.2 MeV. A pattern emerges. As the universe cools, the average energy of the photons declines. When that energy falls below the Q value for any of these reactions, photodissociation will cease. This helps to freeze-out the population of that isotope, since the reaction will no longer be in equilibrium (the reverse of the reaction, $\gamma + X \rightarrow Y + Z$ will be unavailable). The other side of the equation, fusion, will eventually also stop. This will occur when the expansion rate of the universe outpaces the interaction rates of the initial-state particles. Photodissociation ends before production ends; first we see the cessation of photodissociation and second, the fusion production stops. For example, the fusion process producing deuterons proceeds above $k_B T \approx 0.06$ MeV, when the universe is about 340 seconds old.

To make helium, you have to first make deuterons. Based on the physics above, we expect the story of the early cosmos to look something like the following. First, the universe produced protons and neutrons. When their reactions fell out of equilibrium, the initial population of neutrons and protons had a 5:1 proton-to-neutron ratio. Deuterium fusion was possible at this point because the proton–neutron reaction had sufficient kinetic energy to proceed. It ended after about 340 seconds, but in that time some of the free neutrons also decayed and added more protons to the overall population. This depleted the neutrons available to make deuterons. Once deuterons were formed, helium isotopes formed through fusion. We can anticipate that there will have been be more helium-4 than helium-3, in part owing to the easier photodissociation of helium-3 and the more favorable reaction rate to helium-4.

So we expect a progression over time, first with a population of deuterons appearing and growing, but then being depleted as we run out of neutrons and deuterons get fused into helium nuclei. The fusion of heavier low-mass elements, such as lithium, is also expected to occur. This will act to deplete some of the helium. Since free neutrons decay, any neutrons used in fusion, primarily and ultimately, end up in helium.

As an approximation, we can assume that helium-4 (the most probable surviving helium product) and hydrogen (due to the free proton abundance) are the most prevalent nuclei after BBN. Under this hypothesis, we can predict the expected ratio of hydrogen-to-helium after a few hundred seconds. We have a population of N_n neutrons after 0.1 s (we'll revisit this momentarily as we know neutrons will decay between 0.1 s and 340 s). The total baryonic mass of this early universe would have been determined by $N_n + N_p$. Since each helium-4 nucleus requires two neutrons, we expect the number of helium nuclei to be determined by $N_{He4} = N_n/2$. Thus the ratio of the mass of helium-4 to the total baryonic mass of the universe should be $N_{He4}/N_H = (N_n/2)/(N_n + N + p) = 2/(1 - N_p/N_n)$. For our original estimate, before free neutron decay, this yields 0.33.

We cannot neglect free neutron decay. The proton-to-neutron ratio will evolve over those 340 s based on an exponential law as the neutrons decay to protons. The probability of neutron decay goes as $\exp(-t/\tau_n)$, where τ_n is the lifetime of the free neutron (878 s)[7]. Thus we expect the proton-to-neutron ratio to evolve in that time to $N_p/N_n = 5 \exp(t/878\ s) \xrightarrow{t=340\ s} 7.3$. This alters the helium-4 mass fraction to be 0.24. This is often denoted as Y_p and is called the *primordial mass fraction*. As an example of comparison to experimental measurement, recent results from the Sloan Digital Sky Survey data found that $Y_p = 0.245 \pm 0.007$ [Fer+18], including important corrections from stars. Stars affect the abundances of light elements through both their consumption of hydrogen (and helium) and their production of heavier elements than these. In general, a range of independent measurements are compatible with each other and with this estimate.

The density parameter of the universe affects the rate of reactions in BBN. For example, if $\Omega_b h^2$ is a larger fraction of Ωh^2, or even just $\Omega_m h^2$ (Eqns. 3.30 and 3.31), then the overall density of matter will be greater. Populations with greater densities of neutrons and protons will result in more fusion in the same amount of time (340 s). We would expect this to affect the relic abundances. For example, if the fusion of deuterium with deuterium is enhanced, this will result in more helium-4 production. However, this will deplete the primordial abundance of deuterium in the process and reduce the probability of helium-3 formation. So, the trend we expect as $\Omega_b h^2$ increases is for there to be (a) more primordial ^{4}He (larger Y_p) and (b) less primordial deuterium and ^{3}He. This is illustrated clearly in a full calculation whose results are shown in Fig. 3.10.

Nucleosynthesis stops once the temperature of the universe drops low enough that fusion is no longer possible. This is determined by when deuteron formation ends. Thus, BBN will begin after 0.1 s and end after 340 s. The present relative abundances of these elements in the modern cosmos is thus a powerful probe of the microscopic physics of the universe during its first several minutes of existence. The baryon density in the early universe will be a key driver of the nucleosynthesis outcome. The ratios of light elements (hydrogen, deuterium, helium-3, helium-4, etc.) from cosmic surveys can tell us something about the baryon density in the early universe, and vice versa (Fig. 3.10).

The CMB can tells us about the matter density overall. If these two phenomena, BBN and CMB, yield the same answer, then the dark matter hypothesis is reliable. Measurements of the relative abundances of deuterium (D) and helium-3 provide useful constraints on the baryon density. A review of BBN after the measurements of the Planck satellite [Fie+20] averaged 11 measurements for deuterium abundance, reporting that

$$\frac{D}{H} = (2.55 \pm 0.03) \times 10^{-5}. \tag{3.34}$$

Surveys of helium are made challenging by the fact that processes after BBN can result in helium production. It is often typical to survey the relative abundance of helium-3 and helium-4, corrected for known effects that can alter the primordial populations. Surveys

---

[7]We will discuss unstable particles and radioactive decay in more detail in Section 4.4.

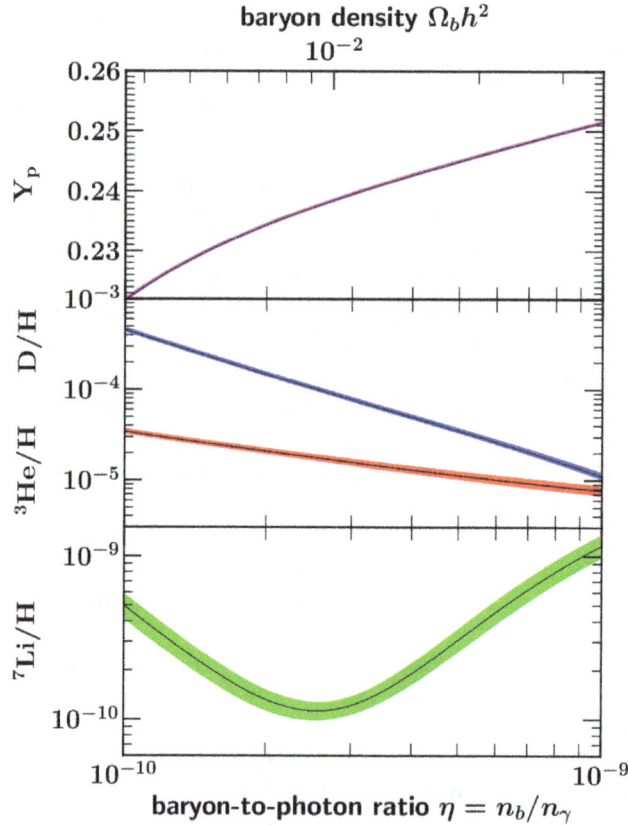

Figure 3.10: The relative abundances of the light elements (Y_p, the primordial helium-4 fraction; the relative abundances of deuterium (D), ^{3}He, and ^{7}Li to hydrogen) vs. the baryon-to-photon ratio and the baryon density, $\Omega_b h^2$. Figure from Ref. [Fie+20].

of helium-3 over many years and, in particular, gaseous regions of the Milky Way [Bal+99] have determined that

$$\frac{^3\text{He}}{H} = (1.6 \pm 0.5) \times 10^{-5}. \tag{3.35}$$

and more recent surveys have used to helium-3 and helium-4 ratio to infer

$$\frac{^3\text{He}}{H} \leq (1.09 \times 0.18) \times 10^{-5}. \tag{3.36}$$

Imposing the above measurements on Fig. 3.10 would lead to the conclusion that $\Omega_b h^2 \approx 0.02$. The CMB results from Planck [Agh+20b] constrain this baryon density independently to be $\Omega_b h^2 = 0.02233 \pm 0.00015$. This is a remarkable concordance of two lines of evidence, and both are consistent with a universe whose matter density is dominated by something other than SM particles.

3.5 Combining Astrophysics and Cosmology to Infer Dark Matter's Properties

If we are to incorporate dark matter into a Friedmann-like equation for the universe, we must infer its particle properties. Various observations about dark matter's bulk properties—that it appears to form halos—allow us to put some broad constraints on those properties. For example, we can set a *lower bound on the mass of dark matter particles* by applying Bose and Fermi spin statistics to a hypothetical dark matter particle. Be warned that, for this exercise, we make a highly simplifying assumption: dark matter is composed of only one type of particle. This may not actually be the case. Nevertheless, the exercise can be instructive.

The lower-mass bound is set by the number of particles that can be confined to a given region of phase space. Let us further assume that we are dealing with scalar (spin-0) dark matter. This makes for a bosonic matter particle. Bose statistics tells us that there is then no limit to the number of such particles that can be placed in the same point in phase space. As a result, we can effectively treat dark matter as a consequence of a classical field. What, then, would maintain the stability of dark matter halos? In this case, it would be the Heisenberg uncertainty principle,

$$\Delta x \Delta p \geq \hbar/2. \tag{3.37}$$

In natural units, where $\hbar = c = 1$, and taking the Heisenberg limit to set the boundary condition for our thought experiment,

$$\Delta x \Delta p \approx 1. \tag{3.38}$$

Treating dark matter as a classical, non-relativistic gas imposes the requirements that $\Delta p \approx m_{scalar} v$ and $\Delta x \approx 2R_{halo}$. This is the pair of conditions that would drive the situation to its limit: that we can be no more certain of a given dark matter particle's position than the scale of the halo itself while being no more certain about the momentum than that it lies somewhere between 0 and the average momentum of that particle in the halo. This set of conditions allows us to set the minimum bound on the dark matter mass,

$$\Delta x \Delta p \gtrsim 1$$
$$2R_{halo} m_{scalar} v \gtrsim 1$$
$$m_{scalar} \gtrsim 2R_{halo}/v \tag{3.39}$$

From our earlier review of probable halo sizes (e.g., given galactic rotation curves) and from halo properties (Sections 2.3 and 2.7), we can insert some estimated numbers for the size of a galactic halo and the typical speed of halo particles and find

$$m_{scalar} \gtrsim 10^{-22} \text{ eV}. \tag{3.40}$$

Of course, it is possible that, in our single-component dark matter hypothesis, the particles are fermions, not bosons, and thus obey Fermi statistics. In that case, we know that it is not possible to put more than one fermion at any specific point in phase space. We must apply the Pauli exclusion principle to constrain the properties of our fermionic dark matter particle. The total mass of the halo will be related to the individual fermion mass by multiplying it with N, the total number of particles in the halo. The latter quantity can be derived from the volume of the halo V and the typical number density of particles n (number of particles per unit energy (or momentum) per unit volume),

$$M_{halo} = m_{ferm}N = m_{ferm}Vn = m_{ferm}V \int g^\star f(p)d^3p. \qquad (3.41)$$

Here, $g^\star$ is the degeneracy (the number of particles per state, which is 2 for spin-1/2 fermions, as they can be placed in the same location in phase space if their spins are oriented in opposing directions) and m_{ferm} is the mass of the hypothetical fermion particle. For the volume, we can assume a spherically symmetric halo, so that $v = (4/3)\pi R_{halo}^3$. The key ingredient is $f(p)$, the density of states, the number of available energy (momentum) states per unit volume. For a gas of dark matter that has decoupled from interactions with normal matter at some temperature T_D, the typical density of states for a dark matter particle as a function of momentum p will be

$$f(p) = \left[e^{(\sqrt{p^2c^2 - m^2c^4} - \mu)/k_B T_D} \pm 1 \right]^{-1} \qquad (3.42)$$

where we use the plus (minus) sign for fermions (bosons).

Here, the energy of a particle is given in its full special relativistic form, $E = \sqrt{p^2c^2 + m^2c^4}$. The chemical potential, μ, relates to the amount of energy that can be absorbed or released when the number of particles in the gas changes (for example, if a phase change occurs and some of the gas particles are trapped in a liquid phase, this decreases the number of gas particles while also changing the energy of the gas). The maximum occupancy is achievable when the most particles are placed in the fewest available phase-space states, which occurs when the gas is highly non-relativistic such that $pc \ll mc^2$. This is given by

$$f_{max} = \left[e^{(mc^2 - \mu)/T_D} \pm 1 \right]^{-1}. \qquad (3.43)$$

Let us apply the fermion case, $(+)$. Considering possible values of μ and T_D, we see that this f_{max} can never exceed $1/2$ for such particles. Thus it is that $n_{max} = 2f_{max} = 1$. Thus we can consider a limiting case that yields a bound on the mass of the particles,

$$M_{halo} = m_{ferm}V \int f(p)d^3p \lesssim m_{ferm}V \int n_{max}d^3p \approx m_{ferm}R_{halo}^3(m_{ferm}v)^3. \qquad (3.44)$$

We can use the relationship between the velocity of particles in the halo and the gravitational potential energy to establish:

$$M_{halo} \lesssim m_{ferm}R_{halo}^3(m_{ferm}v)^3 = m_{ferm}R_{halo}^3 \left(m_{ferm}\sqrt{\frac{GM_{halo}}{R_{halo}}} \right)^3. \qquad (3.45)$$

This ultimately leads to

$$m_{ferm} \gtrsim (R_{halo}^3 G^3 M_{halo})^{-1/8} = \mathcal{O}(10) \text{ eV}. \tag{3.46}$$

A more complete treatment of the fermion case was provided by Tremaine and Gunn in 1979 [TG79], but the fundamental approach is the same. At the time, they were focused on using the problem of unseen matter to constrain the neutrino mass. They estimated that $m_{\text{fermion}} \geq (10 \text{ eV})(H/50)^{1/2} g_\nu^{-1/4}$ where H is the Hubble constant (now known to be closer to 70 km/s/Mpc) and g_{fermion} are the degrees of freedom of the fermions. This yields a constraint in the same general range as our estimate above, $\mathcal{O}(10)$ eV. Such estimates depend on the assumptions about the structure of the astrophysical object under consideration (the dark matter halo of a cluster or a galaxy) so such constraints should be considered as merely guidance on reasonable range of mass orders of magnitude.

A bosonic gas of dark matter can have particle masses as low as about 10^{-22} eV, while a fermionic gas can only go as low as about 10 eV. It's important to stress that this is inferred entirely from estimates of galactic halos as well as the assumption of a single-particle explanation for dark matter. Nevertheless, since we have to start somewhere, this helps to set the stage for what will be required to search directly or indirectly for the particle constituent(s) of dark matter. Experiments designed to do this will potentially need to attempt to probe masses at the eV scale or even lower.

Upper limits on masses of dark matter particles can be established from prior work on the hunt for macroscopic explanations for dark matter, collectively denoted by the acronym *MACHOs* (Massive Astrophysical Compact Halo Objects, a term for macroscopic phenomena that populate galactic halos). A popular, but now widely disfavored hypothesis, was that dark matter was a population of littering black holes or brown dwarf stars (or other such low- or no-luminosity objects—referring to their visible-light component). However, gravitational lensing techniques applied to our own galaxy have established upper limits on how much such MACHO candidates could contribute to dark matter halos. These surveys have determined that, to explain the halo of dark matter enveloping our own galaxy, such objects would have to be possessed of masses satisfying

$$m_{ferm} = m_{scalar} \lesssim 10^{59} \text{ eV}. \tag{3.47}$$

For perspective, the mass of our sun is about 10^{30} kg, which, translated into natural units and electron-Volts, is a mass-energy of about 10^{66} eV, so the above limit is comparable to the mass of the Earth.

In principle, the ultimate limit on the mass of any fundamental subatomic particle is the Planck mass. *Planck units* are ways of expressing fundamental quantities — mass, temperature, time, distance — using constants of nature. This is what ultimately defines the *natural units* outlined in Chapter 1: units defined by the properties inherent in empty space (the speed of light, Planck's constant, Newton's gravitational constant, and the Boltzmann constant). The Planck mass is given as $m_P = \sqrt{\hbar c / G}$ and is equal to 2.176434×10^{-8} kg or 1.22089×10^{28} eV/c^2. At this mass-energy (on such a small scale as a fundamental subatomic particle), a quantum theory of gravity becomes necessary. Since present efforts at such theories have unsolved physical and mathematical difficulties it is at this mass-energy that our theoretical understanding of nature currently and completely breaks down. This provides a *purely theoretical* limit on dark matter mass, more stringent than the empirical limit derived above. We believe it is important not to pick one over the other at this stage, as the Planck mass represents a barrier in human understanding (that may or may not be a true physical barrier) while the empirical limit represents a physical knowledge barrier (a number larger than this would be incompatible with the observed universe).

3.6 Cosmology, Particle Physics, and Constraints on Dark Matter Interactions

Cosmology, combined with nuclear and particle physics, have delivered some clear and accurate predictions. One example of these is the abundance of light elements (hydrogen, helium, etc.). The concept of *thermal decoupling* applied to particle physics in the early universe has proved a successful framework for establishing long term properties of the universe from those first moments of time.

What can this framework do to help us make predictions about dark matter? Let us imagine that dark matter may have a range of possible interactions:

- Interactions with itself;

- Interactions with baryonic and/or leptonic matter.

We already know the kinds of interactions that baryonic and leptonic matter can have with itself. These are described in the Standard Model of Particle Physics. We can attempt to apply that same framework—essentially, a quantum field theory framework—to dark matter. Once more, this exercise will depend on the hypothesis that there is only one kind (or one dominant kind) of dark matter particle that participated in interactions in the very early universe.

Two major classes of interaction emerge from this picture: elastic and inelastic collisions between dark matter χ and standard model matter Y (Fig. 3.11). We must first consider a new, stable particle χ that was produced in the early universe. If this particle can interact with standard model matter, then, while the reaction $\chi\chi \to YY$ is in equilibrium,

(a) Inelastic Collision   (b) Elastic Collision

Figure 3.11: Examples of effective-field-theory Feynman diagrams describing modes of scattering between dark matter and baryonic matter.

dark matter particles will be produced and annihilated in equal numbers, maintaining a population of these particles at very early times.

There are effectively two paths along which this equilibrium interaction process can lead to a relic population of dark matter particles. If dark matter is sufficiently heavy such that its mass tends to exceed most or all of the standard model particle masses, then a chief means by which a relic population is achieved is via "freeze-out." For $m_\chi > m_{SM}$, there will come a time when the average temperature of the universe decreases so much that the SM particles (Y) no longer have sufficient kinetic energy to annihilate into dark matter, even if they are able to interact. The reaction $YY \rightarrow \chi\chi$ no longer contributes and the interactions become non-equilibrated. Thus whatever population of dark matter particles was present just before the temperature dropped below $T \approx 2m_\chi$ will be "frozen" into the universe and unable to increase.

The second path is "freeze-in" and it occurs when the dark matter is much lighter than the typical SM particle and/or when the interaction strength between dark and baryonic matter is extremely small. In this case, $\chi\chi \rightarrow YY$ is the process that is cut off by an expanding and cooling universe. Dark matter can no longer annihilate into SM particles, and whatever population was present before the universe reaches the relevant temperature condition is "frozen into" the universe.

In either case we see that, as the universe expands and cools, there is the possibility for the equilibrium condition to end. Nevertheless, something else must also prevent further SM or dark matter annihilations from significantly altering the population of dark matter. What might this be? The answer is the expansion of the cosmos. As the universe continues to expand, that expansion reduces the chance that a pair of particles will get into close enough proximity to have a reaction. "Freeze-out" and "freeze-in" both effectively can occur when the universe is large enough *and* cool enough that the fundamental interactions become asymmetric and the spatial separation of any random pair of particles prohibits a significant degree of interaction.

The two competing factors so far are the probability to annihilate or create. Let's consider an example from freeze-out. The annihilation cross-section can be denoted by σ_A. The average relative velocity of the dark matter decreases commensurate with the decreasing temperature. This reduces the chance that two dark matter particles will reach each other in order to annihilate. What determines the fate of dark matter is the velocity-averaged product of these two factors, $\langle \sigma_A v \rangle$.

We know from Planck Collaboration data that the relic dark matter density is

$$\Omega_\chi h^2 = 0.120 \pm 0.001. \tag{3.48}$$

We would expect the relic abundance of dark matter to have resulted from, first, the annihilation process exceeding the creation process, then the freeze-out of the annihilation process. The greater $\langle \sigma_A v \rangle$, the lower $\Omega_\chi h^2$. Thus we expect:

$$\Omega_\chi h^2 \approx \frac{3 \times 10^{-27} \text{cm}^3 \text{s}^{-1}}{\langle \sigma_A v \rangle} \tag{3.49}$$

where the numerator on the right-hand side is the thermal annihilation rate required to explain the Planck data. This translates into an approximate interaction cross section of

$$\sigma_\chi \approx 10^{-37}\,\mathrm{cm}^2. \tag{3.50}$$

In standard particle physics units (barns), this is between nanobarns and picobarns—right in the "sweet spot" for a weak-interaction-like cross section. This coincidence is what spawned the concept of a *weakly interacting massive particle*, or WIMP, a dark matter particle candidate with a mass in excess of standard model particle masses (e.g. $\mathcal{O}(100-1000)\,\mathrm{GeV}$) and with a weak-interaction-like coupling to standard model matter. WIMPs have been a mainstay of dark matter phenomenology and project planning partly as a result of this cosmic and mathematical coincidence. It is important, however, to stress the drawbacks of this conclusion; it is based on the singular freeze-out of a one-species dark matter population. If dark matter is more complex than this, this picture is altered.

A More-detailed Look at Freeze-out

Let us spend a little more time on heavy dark matter and freeze-out. From particle physics, we know that the rate of interactions Γ (in this case, annihilations) in a population of particles will be given by

$$\Gamma = n \cdot \sigma \cdot v \tag{3.51}$$

where n is the number density of particles (cm^{-3}), σ is the interaction cross section (in units of cm^3, and v is the relative velocity of particles in the population (e.g., the velocity of incident particles relative to the target particles with which they can interact). For a large number of particles, each with their own kinetic energies (even if the population itself possesses typical kinetic energy), it makes little sense to speak of any one cross section (which can vary with momentum) or velocity. Instead, it is better to think of the average across the populations, $\langle \sigma v \rangle$. When we speak of the rate of interactions, we will thus, instead, mean the average rate of such interactions which we shall simply denote by Γ.

In the early universe, as expansion proceeds, the freeze-out of interactions between two populations of particles occurs at the moment when the annihilation rate is on the order of the *Hubble rate* (expansion rate of the volume of the universe),

$$\Gamma = n\langle \sigma v \rangle \approx H. \tag{3.52}$$

where now, since we are talking about "typical" moments in the time evolution of the population, we employ the velocity-averaged product σv. The Hubble rate H comes from Friedmann's equation,

$$H^2 = \frac{8\pi G}{3}\rho \tag{3.53}$$

where ρ is the matter density.

What does this mean for the kinds of single-type dark matter candidates we are considering in our population? We can imagine three kinetic energy scenarios: cold, warm, and hot dark matter. Each temperature reference corresponds to a sense of the product of the mass and the speed of the particles. A cold gas of dark matter is one that is expected to be moving slowly compared to light, and would sensibly consist of extremely heavy particles given the kinetic energies otherwise available in the hot, dense early universe.

Our earlier exploration of astrophysics and mass limits on fermionic dark matter utilized the number density of a Maxwell-Boltzmann gas of particles. This is most generally written in natural units as

$$n = g \int d^3p \, f(p) \tag{3.54}$$

where g is the number of internal degrees of freedom of the gas particles (the degeneracy of each state in phase space) and $f(p)$ is the occupancy of states (labelled by momentum, p) as defined in Eqn. 3.42. For a spin-zero (scalar) particle, $g = 1$. A relativistic gas will be one where $k_b T \gg \mu$ and $k_b T \gg m_\chi c^2$, whereas, for a slow-moving (non-relativistic) gas, the opposite will be true. If the gas is dense, then $f \approx 1$ whereas, for a sparse gas (where most available states are not populated), then $f \ll 1$. A sparse gas will tend to be the one that suffers fewer collisions, and so would be a good model of a dark matter population at or near freeze-out.

In the case of such a sparse gas, $f \ll 1$ implies that the denominator of f is very large, which in turn tells us that the ± 1 that appears in the denominator of Eqn. 3.43 contributes little or nothing to the total and can be ignored. Thus, in this limit

$$f \approx e^{-[E(p)-\mu]/(k_B T)}. \tag{3.55}$$

If cold dark matter exists and freezes out annihilations, this means that its number density will be

$$n_\chi^{CDM} \propto T^{3/2} e^{-m_\chi/T}. \tag{3.56}$$

Alternatively, if dark matter is very light, it will still be relativistic (referred to as "hot dark matter") at the time of freeze-out and we would expect that

$$n_\chi^{HDM} \propto T^3. \tag{3.57}$$

An intermediate case, so-called "warm dark matter," lies somewhere between these two extrema at the time of freeze-out.

As in the case of thinking about halos of dark matter in galaxies (Section 2.5), we begin with the Boltzmann equation

$$\hat{L}[f] = \hat{C}[f]. \tag{3.58}$$

With some effort, it can be shown that this yields

$$\frac{dn}{dt} + 3Hn = -\langle \sigma v \rangle (n^2 - n_{eq}^2). \tag{3.59}$$

Here, n_{eq} is the number density of dark matter particles at thermal equilibrium (when the annihilation and creation rates are equal). For cold dark matter,

$$n_{eq} = g \left(\frac{mT}{2\pi} \right)^{3/2} e^{-m/T} \tag{3.60}$$

where m is the dark matter particle mass.

The above equations are complex because, as the universe expands, the density of particles will be decreased. We want instead to convert to co-moving quantities, factoring out the expansion of the universe. We define the *Yield Y* as

$$Y \equiv \frac{n}{s} \tag{3.61}$$

$$Y_{eq} \equiv \frac{n_{eq}}{s} \tag{3.62}$$

where s is the entropy density, $s \equiv S/V$ where V is a volume.

$$s = \frac{2\pi^2 g_{*s} T^3}{45}. \tag{3.63}$$

This entropy applies to a universe that is dominated by radiation (photons). Here, g_{*s} refers to the degeneracy of particles in the entropy in the universe that are in thermal equilibrium, satisfying $m \leq T$. This quantity takes into account that the universe, at any time, would be a mix of bosons (e.g., photons and, possibly, dark bosons) and fermions (e.g., standard model particles and, possibly, dark fermions). We can write this degeneracy as

$$g_{*s} = \sum_{i \in \text{bosons}} g_i \left(\frac{T_i}{T}\right)^3 + \frac{7}{8} \sum_{j \in \text{fermions}} g_j \left(\frac{T_j}{T}\right)^3 \tag{3.64}$$

where T_i is the temperature of particle population i.

While the temperature of the universe exceeds about 1 MeV (above the QCD scale, where quarks continue to be asymptotically free in the early universe), all T_i are simply equal to the photon temperature, T. Essentially, all SM particles are in thermal equilibrium at very high temperatures due to their interactions, as they are constantly exchanging energy. We don't know how the dark sector (dark fermions and bosons) might contribute to this.

In Eqn. 3.64, g_i is the multiplicity of a given energy level for a particle [Hus16]. This multiplicity is related to the available internal states specific to a particle. For example, at high temperatures ($T \gg 100$ GeV) the Z, W^+, W^-, and photon are all effectively massless vector bosons. Such particles have only two available polarization states (transverse to their direction of motion), so $g_i = 2$. When $T \ll 100$ GeV, the $W^\pm$ and Z^0 acquire mass and gain an additional longitudinal polarization state so that then $g_i = 3$. Gluons are massless and have two polarization states (like the photon), but they also have 8 available color states, so $g_g = 16$. The Higgs boson is a scalar and only ever has $g_i = 1$, as we indicated earlier for scalar particles. For each quark, there are 3 color states, a particle and an antiparticle state, and two spin states (for a total of 12 degrees of freedom per quark). For each charged lepton there are particle and antiparticle states and two spin states, for a total of four degrees of freedom each. Neutrinos have particle and antiparticle states, but due to the fact that all particles are left-handed and all antiparticles are right-handed, there is only one spin state available (for a total of two degrees of freedom for each neutrino). Particles in the dark sector have unknown spin properties, so one must hypothesize how (if at all) they contribute to this degeneracy.

Constructing the above definitions allows us to do something clever:

$$\frac{dY}{dt} = \frac{d}{dt}\left(\frac{n}{s}\right) \tag{3.65}$$

$$= \frac{d}{dt}\left(\frac{a^3 n}{a^3 s}\right) \tag{3.66}$$

By introducing the scale factor of the universe, we construct something ... $a^3 s$ ... which is *constant* even in an expanding universe that is iso-entropic—that is, one in which the total entropy of the universe remains fixed. Since $S = a^3 s$, by definition, then $\dot{S} = 0$. This allows us to write

$$\frac{dY}{dt} = \frac{d}{dt}\left(\frac{a^3 n}{a^3 s}\right) \tag{3.67}$$

$$= \frac{1}{a^3 s}\left(3a^2 \dot{a} n + a^3 \dot{n}\right) \tag{3.68}$$

$$= \frac{1}{s}\left(3Hn + \dot{n}\right). \tag{3.69}$$

Remember that $H \equiv \dot{a}/a$, which has been utilized above. Thus we are able to write

$$\frac{dY}{dt} = -s\langle \sigma v \rangle (Y^2 - Y_{eq}^2). \tag{3.70}$$

It is convenient and useful to introduce an independent variable $x \equiv m/T$, the relative mass of the dark matter particle to the temperature of the system. Since the temperature of the system is the inverse of the scale factor of the system, $T \propto 1/a$ (e.g. the photon gas temperature is the inverse of its wavelength, which is, in turn, determined by the scale factor of the universe), then

$$\frac{dx}{dt} = Hx. \tag{3.71}$$

If we assume that the universe is still very radiation-dominated at the time of dark matter freeze-out (which would have been true up to recombination), then

$$H(m) = 1.66g_\star^{1/2}\frac{m^2}{M_{Planck}} = Hx^2. \tag{3.72}$$

where $H = 1.66g_\star^{1/2}T^2/M_{Planck}$ and here we encounter $g_\star$, the effective degrees of freedom associated with the energy density of the universe (instead of the entropic effective degrees of freedom, $g_{\star s}$). The energy density effective degrees of freedom is defined

$$g_\star \equiv \sum_{i \in \text{bosons}} g_i \left(\frac{T_i}{T}\right)^4 + \frac{7}{8}\sum_{j \in \text{fermions}} g_j \left(\frac{T_j}{T}\right)^4. \tag{3.73}$$

Using the chain rule, we can write

$$\frac{dY}{dt} = \frac{dY}{dx}\frac{dx}{dt} = \frac{dY}{dx}Hx \longrightarrow \frac{dY}{dx} = \frac{dY}{dt}(Hx)^{-1}. \tag{3.74}$$

If we then substitute this into Eqn. 3.70, we obtain

$$\frac{dY}{dx} = -\frac{s\langle\sigma v\rangle}{H(m)}(Y^2 - Y_{eq}^2). \tag{3.75}$$

If we then employ the definition for entropy density (Eqn. 3.63) and define $\lambda = 1.66(2\pi^2/45)(g_{\star s}/g_\star^{1/2})$ then

$$\frac{dY}{dx} = -\frac{\lambda\langle\sigma v\rangle}{M_{Planck}m_\chi x^2}(Y^2 - Y_{eq}^2). \tag{3.76}$$

This is a *Ricciati equation* and has no closed analytical form. Solutions to this equation can only be developed numerically, and those solutions can then be used to see how interacting dark matter evolves as the universe expands.

Let's estimate the yield ($Y = n/s$) needed to produce the correct DM relic abundance $\Omega_\chi h^2 \approx 0.1$. We begin with the definition of the relic abundance,

$$\Omega_\chi h^2 = \frac{\rho_\chi}{\rho_{critical}}h^2 = \frac{m_\chi n_\chi h^2}{\rho_{critical}} = \frac{m_\chi Y_\infty s_0 h^2}{\rho_{critical}} \tag{3.77}$$

where s_0 is the current entropy density (recall that we are assuming an iso-entropic universe wherein $S = s/a$ remains constant); Y_∞ is the dark matter yield today at $t \approx \infty$ after the beginning of the universe. The yield should remain constant (assuming no great moment of additional annihilation) after freeze-out (when $Y = Y_f$), so

$$\Omega_\chi h^2 = \frac{m_\chi Y_f s_0 h^2}{\rho_{critical}}. \tag{3.78}$$

This tells us that

$$Y_f = \frac{\Omega_\chi h^2 \rho_{critical}}{m_\chi s_0 h^2}. \tag{3.79}$$

The entropy density of the present universe is $s_0 = 2970\,\text{cm}^{-3}$, and the critical density is $1.054 \times 10^{-5}\,\text{GeV}\cdot\text{cm}^{-3}$, so normalizing the dark matter mass to the proton mass ($\approx 1\,\text{GeV}$) we find

$$Y_f \approx 3.55 \times 10^{-10}\left(\frac{1\,\text{GeV}}{m_\chi}\right). \tag{3.80}$$

This corresponds to an $x = m_\chi/T \approx 25$. This convergence of information is illustrated in Fig. 3.12.

Details of the Yield Estimate

We can begin by defining

$$\Delta Y = Y - Y_{eq} \tag{3.81}$$

and noting that

$$(Y^2 - Y_{eq}^2) = (2Y_{eq} + \Delta Y)\Delta Y. \tag{3.82}$$

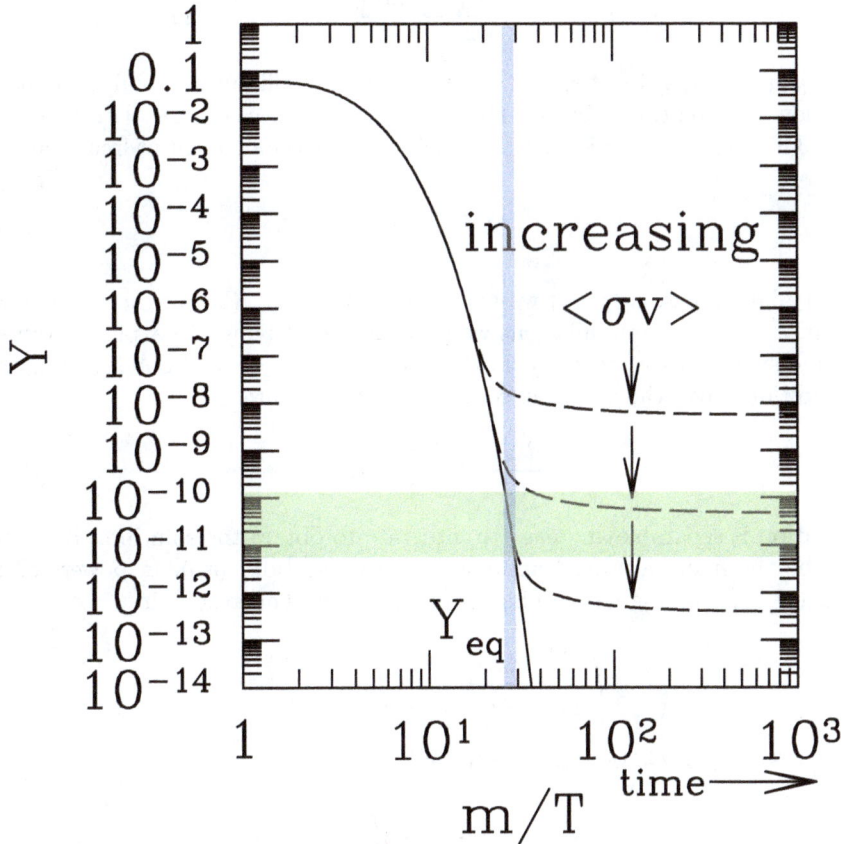

Figure 3.12: The freeze-out of dark matter annihilations in the early universe, represented as Yield (co-moving number density) vs. x (the mass-to-temperature ratio, a proxy for time after the beginning of time). The shaded rectangle indicates the decade in Y about the value estimate in Eqn. 3.80, while the blue box indicates the corresponding region of x permitted by that estimate. Original figure from Ref. [Gel15].

The departure from equilibrium will be given by

$$\frac{d}{dx}\Delta Y = \frac{dY}{dx} - \frac{dY_{eq}}{dx} \tag{3.83}$$

$$= \frac{-\lambda\langle\sigma v\rangle}{x^2}(Y^2 - Y_{eq}^2) - \frac{dY_{eq}}{dx} \tag{3.84}$$

$$\frac{d}{dx}\Delta Y = -\frac{dY_{eq}}{dx} - \frac{\lambda\langle\sigma v\rangle}{x^2}(2Y_{eq} + \Delta Y)\Delta Y. \tag{3.85}$$

At very early times, $1 < x < x_f$, $Y \approx Y_{eq}$ as we are still in the equilibrium phase when annihilation and creation rates are about equal. In that case, $\Delta Y \approx d\Delta Y/dx$. Solving the Eqn. 3.85 for this condition gives the following at the moment of freeze-out:

$$\Delta Y \approx \frac{x_f^2}{2\lambda\langle\sigma v\rangle}. \tag{3.86}$$

At very late times, $x \gg x_f$, we expect that $Y \gg Y_{eq}$ (keep in mind that without freeze-out, as $x \to x_\infty$, annihilations would continue, driving the expected value of Y_{eq} smaller and smaller, closer to zero). In that case, $\Delta Y \approx Y \gg Y_{eq}$. Solving Eqn. 3.85 with these conditions gives the following condition at freeze-out:

$$\frac{d\Delta Y}{dx} = \frac{\lambda\langle\sigma v\rangle}{x^2}(\Delta Y)^2. \tag{3.87}$$

This equation is separable and easy to integrate to obtain the solution. To do this, we expand the thermally averaged annihilation cross section, $\langle\sigma v\rangle$, in powers of x^{-1}, as $\langle\sigma v\rangle = a + \frac{b}{x} + \ldots$ and ignore the higher-order terms. This results in

$$\int_{\Delta Y_f}^{\Delta Y_\infty} \frac{d\Delta Y}{(\Delta Y)^2} = -\int_{x_f}^{x_\infty} \frac{\lambda\langle\sigma v\rangle}{x^2} dx. \tag{3.88}$$

Recalling that $x_\infty \gg x_f$, we integrate to find:

$$\frac{1}{Y_\infty} = \frac{1}{\Delta Y_f} + \frac{\lambda}{x_f}\left(a + \frac{b}{2x_f}\right) \tag{3.89}$$

which results in

$$\Delta Y_\infty = \frac{x_f}{\lambda\left(a + \frac{b}{2x_f}\right)}. \tag{3.90}$$

We can now express the relic density as

$$\Omega_\chi h^2 = \frac{m_\chi Y_\infty s_0 h^2}{\rho_{critical}} \tag{3.91}$$

$$\approx \frac{10^{-10}\,\text{GeV}^{-2}}{a + \frac{b}{50}} \tag{3.92}$$

$$\approx \frac{3 \times 10^{-27} \, \text{cm}^2\text{s}^{-1}}{a + \frac{b}{50}} \tag{3.93}$$

where we have used the numbers obtained, or Y_f and x_f, from the calculations in the previous section. The above implies that, in order to have the correct relic abundance of ≈ 0.1, the thermally averaged annihilation cross section must be $\langle \sigma v \rangle \approx 3 \times 10^{-26} \, \text{cm}^3\text{s}^{-1}$.

If we assume a simple leading-order massive gauge boson, exchange is sufficient to describe the annihilation process depicted in Fig. 3.11, we can infer the coupling constant G_X associated with that interaction:

$$\langle \sigma v \rangle \sim G_X^2 m_\chi^2 \longrightarrow G_X = 1.16 \times 10^{-5} \, \text{GeV}^2 = G_F. \tag{3.94}$$

The assumption of a weak-scale interaction consistent with the Fermi coupling constant, G_F, leads to the expectation that the dark matter particles must have a mass at the level of $\mathcal{O}(GeV)$.

These calculations are rather basic and can be upset by some simple factors (not the least of which would be multi-component dark matter). For example, if there are new annihilation channels that open up, such as $\chi + \chi \to A + B$ (when $2m_\chi \sim m_A + m_B$), due to significant dark matter kinetic energy, this alters the picture. We have assumed, as is common in the field, that dark matter is heavy and non-relativistic (cold) even at early times in the universe before the freeze-out occurs. If this is not the case, as due to lighter dark matter species that nonetheless have such annihilation channels open to them, then this additional kind of process may be present.

Another factor would be the creation of an intermediate resonance during the annihilation process. For example, if there is a dark gauge-boson sector where one or more of those bosons have low masses, collisions leading to dark matter annihilation can lead to an intermediate resonance. Annihilation cross sections could be suppressed or enhanced when intermediate resonances are available.

A multi-component dark matter population can lead to co-annihilations where one subspecies interacts with another; e.g., $\chi_1 + \chi_2 \to \dots$. As noted above, this kind of additional physical process alters the picture of freeze-out.

A More Detailed Look at Freeze-In

Freeze-in[8] (c.f., Refs. [BPZ20; Hal+10]) involves the decay of SM particles to lighter (and/or very weakly coupled) dark matter particles (or vice versa), a process that then ceases as the universe expands and cools. The process is no longer reversible, leaving a population of cold dark matter particles unable to decay to SM partners. The key modification to freeze-out occurs in Eqn. 3.59, where instead of a population of light particles at $t = 0$ (when the scale factor of the universe was also $a = 0$) we introduce that population first through SM particle decay. Eventually the decay process ceases as the universe cools, and the light relic population is frozen into the cosmos.

---

[8]We are particularly grateful to a private communication with Professor Aaron Vincent (Queen's University) regarding the matter of calculating the dark matter yield due to the freeze-in mechanism.

We begin with a thermal bath not of DM particles but of some other particle, which we label Q. The population of χ will be governed by a version of the Friedmann equation where we incorporate Q's decay into χ on the right-hand side (the part of the equation that encapsulates the dynamical process that affects the number density of DM particles, n),

$$\frac{dn}{dt} + 3Hn = g_Q \int d^3p \frac{f_Q \Gamma_Q m_Q}{(2\pi)^2 E_Q}. \tag{3.95}$$

Here, Γ_Q is the partial width of the particle, Q, describing the probability of its decay to χ (the ratio of the partial width to the natural, or full, width (Γ) of the particle is the "branching fraction" of that particle decaying to χ, the probability that it decays to a DM final state).

We can assume the very early universe is nearly isotropic, such that integrating $d^3p = p^2 \sin\theta d\theta d\phi dp$ yields $4\pi dp$ with no angular dependence in any of the quantities in the integrand. Use of the relativistic mass-energy-momentum relationship (in natural units) yields

$$p^2 = E^2 - m^2 \longrightarrow 2pdp = 2EdE \tag{3.96}$$

where m is assumed to be constant. Thus

$$\frac{dn}{dt} + 3Hn = g_Q \int d^3p \frac{f_Q \Gamma_Q m_Q}{(2\pi)^2 E_Q} \tag{3.97}$$

$$= g_Q \int 4\pi p_Q^2 dp_Q \frac{f_Q \Gamma_Q m_Q}{(2\pi)^2 E_Q} \tag{3.98}$$

$$= g_Q \int p_Q^2 \frac{E_Q dE_Q}{p_Q} \frac{f_Q \Gamma_Q m_Q}{\pi E_Q} \tag{3.99}$$

$$= g_Q \int_{m_Q}^{\infty} \sqrt{E_Q^2 - m_Q^2} \frac{f_Q \Gamma_Q m_Q}{\pi} dE_Q \tag{3.100}$$

$$= \frac{g_Q \Gamma_Q m_Q}{\pi} \int_{m_Q}^{\infty} \sqrt{E_Q^2 - m_Q^2} f_Q dE_Q. \tag{3.101}$$

We now need to sort out the occupancy of states, $f(p)$. This is a system where the species Q is declining with time as it decays (e.g., into dark matter), but the state Q can be repopulated by annihilation of dark matter back into Q. However, the latter process will cease when one (or both) of two conditions are met: the dark matter can no longer easily encounter other dark matter to annihilate (due to expansion of the universe) or, in the case that $m_\chi < m_Q$, the average kinetic energy of the dark matter (due to the average kinetic energy of the parent particle, Q) declines below the energy required to (pair) produce Q from dark matter annihilation. We can then approximately describe this as $f(p) \approx e^{-E_Q/T}$ where T is the temperature of the universe at a given time, t. Thus we find that

$$\frac{dn}{dt} + 3Hn \approx \frac{g_Q \Gamma_Q m_Q}{\pi} \int_{m_Q}^{\infty} \sqrt{E_Q^2 - m_Q^2} e^{-E_Q/T} dE_Q. \tag{3.102}$$

Note also that this equation encodes the case that the coupling of Q and χ is very small, which can also act to cut off the production of dark matter. If $m_\chi > m_Q$ but $\Gamma_Q/\Gamma \to 0$ then it becomes extremely improbable for the decay to occur later in time, freezing in a relic population of dark matter (albeit one with the possibility of decay, though rare) once annihilation is no longer common in the expanded universe.

We can then estimate how the yield of dark matter varies with the parameters of the universe. From Eqn. 3.69 we can write, for freeze-in, that

$$s\frac{dY}{dt} \approx \frac{g_Q \Gamma_Q m_Q}{\pi} \int_{m_Q}^{\infty} \sqrt{E_Q^2 - m_Q^2}\, e^{-E_Q/T}\, dE_Q. \tag{3.103}$$

Recall that $\frac{dY}{dx} = (Hx)^{-1}\frac{dY}{dt}$, where for a radiation-dominated universe $H = 1.66 g_\star^{1/2} T^2/M_{planck}$. Employing once more that the entropy density of the dark matter is $s = (2/45)\pi^2 g_{\star s} T^3$ and borrowing from our earlier algebra and calculus, we can now define $x = m_Q/T$ and $y = E_Q/T$ to write

$$\frac{dY}{dx} = (Hx)^{-1}\frac{dY}{dt} \tag{3.104}$$

$$= \frac{M_{planck}}{x 1.66 g_\star^{1/2} T^2} \frac{s}{s} \frac{dY}{dt} \tag{3.105}$$

$$\approx \frac{M_{planck}}{1.66 g_\star^{1/2} x s T^2} \frac{g_Q \Gamma_Q m_Q}{\pi} \int_{m_Q}^{\infty} \sqrt{E_Q^2 - m_Q^2}\, e^{-E_Q/T}\, dE_Q \tag{3.106}$$

$$\approx \frac{M_{planck}}{1.66 g_\star^{1/2} (2/45)\pi^2 g_{\star s} T^5} \frac{1}{x} \frac{g_Q \Gamma_Q m_Q}{\pi} \int_{m_Q}^{\infty} \sqrt{E_Q^2 - m_Q^2}\, e^{-E_Q/T}\, dE_Q. \tag{3.107}$$

Let us define $\lambda' = 1.66(2\pi^2/45)g_\star^{1/2} g_{\star s}$. Then

$$\frac{dY}{dx} \approx \frac{M_{planck}}{\lambda' T^5} \frac{1}{x} \frac{g_Q \Gamma_Q m_Q}{\pi} g_Q \int_{m_Q}^{\infty} T\sqrt{\frac{E_Q^2}{T^2} - \frac{m_Q^2}{T^2}}\, e^{-E_B/T} T\, d(E_Q/T) \tag{3.108}$$

$$\approx \frac{M_{planck} \Gamma_Q m_Q g_Q}{\pi \lambda'} \frac{1}{x T^5} \int_{m_Q}^{\infty} T\sqrt{y^2 - x^2}\, e^{-y} T\, dy \tag{3.109}$$

$$\approx \frac{M_{planck} \Gamma_Q m_Q g_Q}{\pi \lambda'} \frac{1}{x T^3} \int_{m_Q}^{\infty} \sqrt{y^2 - x^2}\, e^{-y}\, dy \tag{3.110}$$

$$\approx \frac{M_{planck} \Gamma_Q m_Q g_Q}{\pi \lambda'} \frac{x^3}{x m_Q^3} \int_{m_Q}^{\infty} \sqrt{y^2 - x^2}\, e^{-y}\, dy \tag{3.111}$$

$$\approx \frac{M_{planck} \Gamma_Q g_Q}{\pi \lambda'} \frac{x^2}{m_Q^2} \int_{m_Q}^{\infty} \sqrt{y^2 - x^2}\, e^{-y}\, dy \tag{3.112}$$

For a renormalizable coupling between Q and χ, the partial width Γ_χ should have the standard Yukawa coupling form, $\Gamma_\chi = (1/8\pi)\lambda_\chi^2 m_\chi$. Here we finally introduce the dark matter mass itself.

The integral is a modified Bessel function of the second kind, $K_1(x)$, making this a challenging equation to simplify. We emphasize the value of numerical methods by noting that the equation can be explored using modern computational tool kits that allow for visualization of the function under varying conditions. For example, we can select values of Γ_χ, M_Q, and g_Q and graph the resulting dY/dx and $Y(x)$. This allows us to explore the dependence of yield on factors such as the mass of the dark matter candidate.

Using the Python programming language and the NumPy and SciPy libraries of numerical functions, we can define our function for dY/dx,

```python
from scipy.integrate import quad # numerical integration
from scipy.special import kn # Modified Bessel Function of 2nd Kind
import numpy as np

g_star = 100 # c.f. arXiv:2001.02142 [hep-ph]
g_chi = 2 # assume a fermion, 2 spin degrees of freedom
lambda_prime = 1.66*(2*np.pi**2/45)*np.sqrt(g_star)*g_chi
M_planck = 1.22e19 # GeV

M_Q = 200 # GeV, weak-scale particle
g_Q=1 # Assume a scalar parent

lambda_chi = 1e-8 # small coupling to Q
m_chi = 1e-3 # GeV, dark matter mass

def B1x(x):
    (result, result_err) = \
        quad(lambda y: np.sqrt(y**2 - x**2) * np.exp(-y), x, np.inf)
    return result

def dYdx_FI(x, lambda_chi, m_chi):
    Gamma_chi = (1/(8*np.pi)) * lambda_chi**2 * m_chi
    vB1x = np.vectorize( B1x )
    return (M_planck * Gamma_chi * g_Q)/(np.pi) * x**2/(lambda_chi*M_Q**2) * vB1x(x)
```

Defining the function dY/dx numerically then permits numerical integration to obtain $Y(x)$.

```python
def Y_FI(x, lambda_chi, m_chi):
    (result, result_err) = quad(dYdx_FI, 0, x, args=(lambda_chi, m_chi))
    return result
```

Standard graphing tools, such as MatPlotLib, can then be used to study how these functions change as one alters parameters such as the dark matter mass. An example of this is presented in Fig. 3.13. The yield itself is a strong function of the dark matter mass, with larger masses yielding a higher yield. In the freeze-out mechanism, the final yield is inversely proportional to the dark matter mass; in the freeze-in mechanism, it is directly proportional to that mass. We see the key features expected of this model: the rate of change of dark matter yield is large at early times and high temperatures, falling rapidly as the universe expands and cools (large x). This freezes in the population of dark matter, whose final value depends on the model parameters and which ultimately defines the relic abundance in the universe, $\Omega_\chi h^2$.

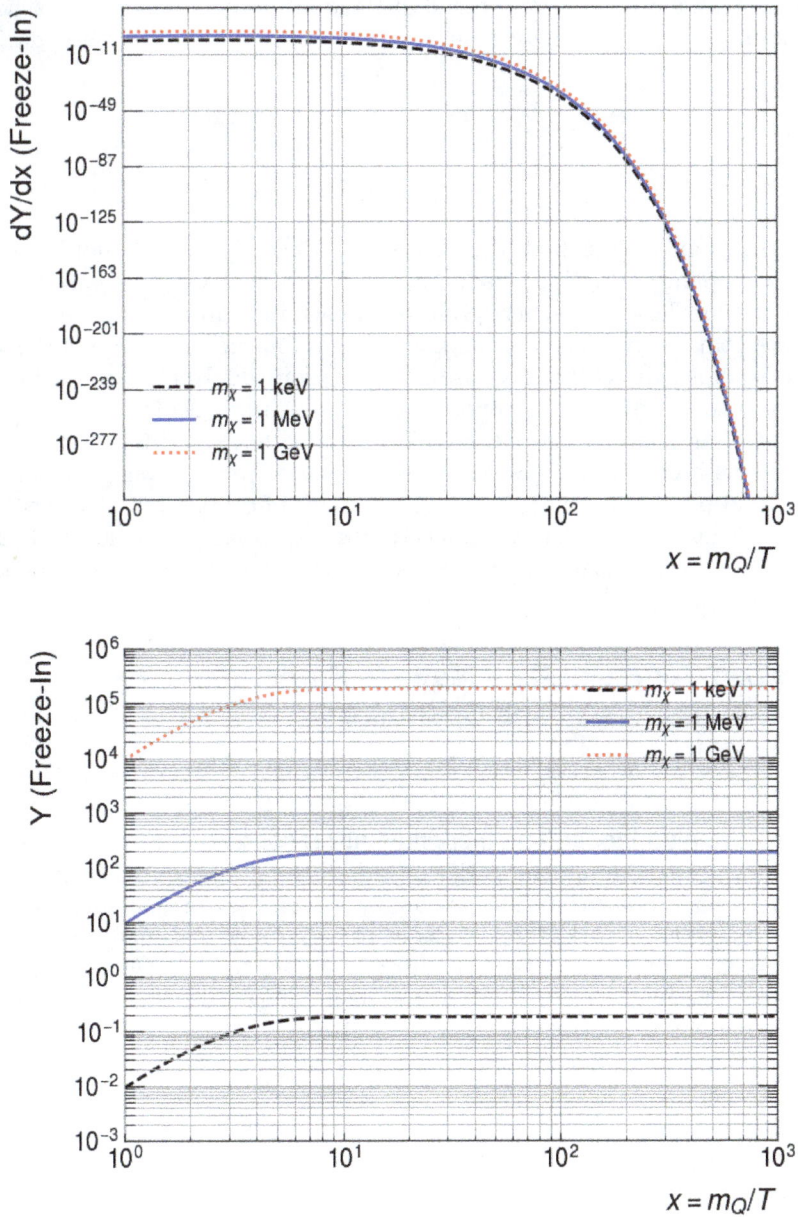

Figure 3.13: The rate of change of yield (left) and the yield itself (right) vs. the ratio of the mass of a particle, Q, and the temperature of the universe, denoted x. These graphs are generated using a weak-scale mass for Q and a Yukawa coupling strength $\lambda_\chi = 10^{-8}$.

From these more detailed investigations of the dark matter yield from either of the two mechanisms, we can draw from broad distinctions between these approaches. From Eqns. 3.79 and 3.112, one can conclude that (for a generic coupling λ and dark matter mass m),

$$Y_{FO} \propto \frac{1}{\lambda^2 m} \tag{3.113}$$

and

$$Y_{FI} \propto \lambda^2 m. \tag{3.114}$$

There are two complications from this discussion. First, since we do not know well the mass or the coupling strength to standard model particles, we get a very broad set of possible values for these parameters given the known relic abundance. Second, there is no guarantee that the particles interacting with dark matter are those in the standard model; for example, the Q in the freeze-in calculation could be a new particle from a previously undiscovered sector of interactions. Third, these calculations tend to assume only one kind of dark matter particle, a problem related to the second point. There could be an entirely new sector of physics yet-to-be-discovered. Fourth, even if we did know the precise sector of physics in which this occurs, the lack of constraints from the first point means we don't know for certain which mechanism led to the production of dark matter.

Progress on these and many other questions depend on additional observations of dark matter's interactions.

3.7 A Broad Survey of Dark Matter's Properties from Astrophysics and Cosmology

So far in this text, we have established a few key features of dark matter.

Dark Matter is nonluminous: it neither emits nor responds to light   This observation translates into constraints on the dark matter's electric charge and electric or dipole moment. However, those constraints are highly model- and/or mass-dependent. Since dark matter is optically dark, the process of losing energy via photon radiation must be inefficient—dark matter is effectively dissipationless. In addition, dark matter does not accrete onto black holes or collapse at the center of galaxies as efficiently as baryons (because dark matter is a low-friction gas, unlike baryonic matter which readily interacts with other baryonic matter, leading to all kinds of drag-like effects on its motion).

It is worth noting that even this previous statement is subject to model-dependence. For example, there are extended effective field theory models of a "dark sector" that include features like "dark electromagnetism" under which dark matter might be highly charged. Such a hypothetical "dark photon" might be readily exchanged between dark fermions or dark scalars and lead to energy dissipation.

Dark Matter is essentially collisionless   What does it mean to be collisionless? The dark matter self-interaction-to-mass ratio is constrained by observations of galaxy cluster mergers and the ellipticity of galactic halos. Sufficient interactions between dark matter particles would lead to significant ellipticity of halos, as it does in baryonic matter, or in the dragging of one dark matter halo upon another during galaxy cluster collisions (which would cause the halos to slow down and lag behind the galaxies after the collision). The figure of merit for this claim is the mean free path being smaller than galaxy cluster scales (Mpc) and dark matter densities ($\approx 1\,\mathrm{GeV/cm^3}$).

This means, if the dark matter mass is that of a proton, the interaction cross section can be as large as a barn (consistent with the strength of nuclear forces and the strong interaction). To provide a sense of scale, one barn is the cross section of a heavy atomic nucleus, like uranium. This is NOT a small cross section, since Uranium is one of the largest nuclei and phenomena associated with such a scale therefore correspond to very high probability subatomic events. For comparison, the interaction cross sections of electrons and positrons, which are electro-weak in nature, are at the level of picobarns or nanobarns.

This means that dark matter can have interactions with itself as large as those permitted in the strong nuclear interaction and yet still be consistent with observations. In addition, the interactions of dark matter with ordinary baryonic or leptonic matter also has weak constraints (except in certain mass ranges, which we will discuss later in the text).

Dark Matter is non-baryonic   The nature of dark matter as a non-standard-model (non-baryonic) form of matter present in the early universe is evidenced by observations from the CMB and BBN. First of all, a baryonic component (one that may interact with radiation) and a non-baryonic component seem equally important to enable understanding the power spectrum of the CMB. Remove these, or change their relationship in the model, and the CMB power spectrum defies explanation. In addition, the size of the density fluctuations now appears explicable only if a non-baryonic matter component, liberated from radiation pressure, was able to collapse in quantum fluctuations much earlier than recombination. Finally, if dark matter were explicable by baryons, then the careful balancing act of primordial nucleosynthesis, yielding the observed ratios of hydrogen, helium, and other light elements, would have been upset.

One key issue we will return to later is whether dark matter is explicable solely by existence of neutrinos. In past discussions, we have excluded leptons from consideration. but the astute reader will note that, as we are seeking a form of matter with mass (neutrinos have small but undetermined masses), and ones that experience the weak, but not the electromagnetic interactions, neutrinos would seem to fit the bill. Neutrinos, however, are relativistic given their very low masses. (The sum of their individual particle-level masses is constrained by the CMB, as well as by other considerations.) They would constitute hot dark matter. This would upset the freeze-out picture presented in Section 3.6. That may be permissible to some degree because of the caveats on that picture that we presented earlier.

Standard model neutrinos as dark matter have been ruled out by a growing body of evidence. However, since the origin of neutrino mass is, itself, an open question, and may be resolved by the existence of heavy neutrinos, it is possible that non-Standard Model neutrinos and dark matter may yet share some relationship.

The mass of dark matter particles is not well-constrained by astrophysics or cosmology   Our exercises using astrophysics to place bounds on single-particle dark matter mass has only confined such particles to a 90 (70) order-of-magnitude range for bosons (fermions). The considered mechanisms for producing dark matter in the very early, high-temperature universe also do not provide strong constraints. While the relic abundance is well-measured, the lack of information on dark matter mass or interaction coupling(s), and the lack of information about what population(s) interact most readily with dark matter (outside of gravity), lead to poor constraints. Dark matter models have tremendous freedom to imagine the mass range of such particles.

We have good information on large-scale dark matter abundance, but not small-scale abundance   We have a reasonably precise measure of the total of dark matter density on cosmological scales as well as the density and velocity distribution on galactic scales. Information at smaller scales is not yet as precise. This will pose challenges to terrestrial experiments making assumptions about local densities, etc.

What we do know, broadly speaking, about dark matter's abundance is that it was there in large quantities near the beginning of time and it is still abundant today. Dark matter is primordial and extant.

PART II

MECHANISMS FOR NEW DETECTION

CHAPTER 4

Dark Matter Detection by Scattering

4.1 Introduction

Detecting dark matter by looking for it to scatter off of a target is commonly referred to as *direct detection*. The use of direct here means observing the effect of dark matter as a particle in a controlled experiment that isolates its interactions. This has proven to be a great challenge. Given what we know about it from astrophysics and cosmology—that it interacts with certainty only through gravitational influence—poses, perhaps, the greatest challenge to experimentalists. We are not assured of its interactions by any other known force, and should it interact by forces *unknown*, we do not know what character those forces might possess. As such, there is a vast phase space of theoretical consideration that lies open to experiment.

To design a dedicated *direct* dark matter detection experiment (one that aims to be sensitive to interactions with dark matter particles), one needs to factor in a number of considerations:

- What may be the rate of dark matter passing through the experimental apparatus?

- What may be the rate of interaction between dark matter and matter described in the SM?

- What known interactions and particles might confound the observations?

- What will be the potential totality of experimental signatures of dark matter interacting with the instrumentation?

It is possible, of course, that dark matter interacts all the time with experiments that have been operated in the last century. That does not mean it would have been noticed, or even be noticeable as an effect. If a non-SM particle was causing obvious signals, it should have been noticed (even in the context of searches that were not originally intended for dark matter, an example of which we give in Section 5.5). If dark matter is interacting with instrumentation at all, it is likely not doing so with high frequency.

Experiments capable of observing the scattering of a weakly interacting, electrically neutral particle have existed for decades. The scattering process might happen off

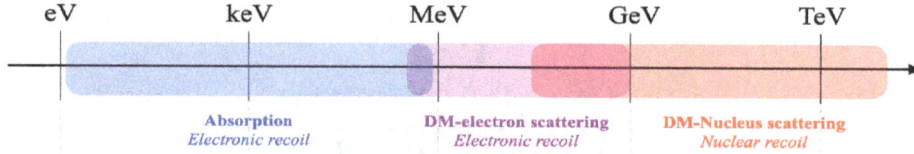

Figure 4.1: A logarithmic plot of possible masses for dark matter particles, as well as general physical interactions in various ranges that might be applied to constrain dark matter in that region.

an atomic nucleus or electrons, or both. In principle, however, the challenge to the experimentalist lies in the lack of constraints on just what kind of particle we are hunting, perhaps most well-represented by the wide range of potential masses for such a particle. It could be so low in mass that its effective de Broglie wavelength spans crystal lattices, lab spaces, or whole planets (see Chapter 6). On the other hand, it could be so heavy that, in a single collision, it disturbs or displaces heavy nuclei that make up atoms in solids, liquids, or gases. An example of the challenge, and the categories of experiments that have, presently do, or will rise to meet that challenge, is shown in Fig. 4.1.

4.2   Dark Matter Distributions

To search for dark matter's particle constituents is for naught if the distribution of dark matter in our local region of the Milky Way is non-existent. Since naturally occurring dark matter populations must be the source of incident particles for these experiments, we will first focus on the distribution of relic WIMP(-like) dark matter in the Milky Way halo. Understanding the distribution is a crucial ingredient in considerations for the construction of experiments and interpretation of their (non-)observations.

A Simplified Model of WIMP Distribution in the Milky Way

WIMPs in the Milky Way are assumed to be organized in an isothermal spherical halo. This is confirmed by computational N-body simulations (c.f. Ref [Boz+16])[1]. They are also assumed to have settled into a circumstance where the halo consists of particles whose mean velocity is $\vec{v}$ (speed v) with some spread in that velocity distribution, σ_v. Those particles are assumed to be very weakly interacting with one another and with SM matter and dispersed enough that interactions do not strongly affect the overall kinematics of the halo. This allows the halo to be described using a Maxwell-Boltzmann model with velocity distribution,

$$f(\vec{v}) = \frac{1}{\sqrt{2\pi}\sigma}e^{-|\vec{v}|^2/(2\sigma_v^2)}. \tag{4.1}$$

[1]Recent work has led to important corrections to the baseline assumptions in this part of the book. For example, Milky Way stellar surveys suggest a local anisotropy in the halo. We explore this effect briefly in Section 4.7. In addition, effects from the Large Magellenic Cloud may have an impact on halo assumptions in our region of the galaxy [Smi+23].

The dark matter velocity is taken to be that as observed from Earth, the relevant perspective for an experiment.

The halo itself is assumed to be non-rotating. This is justified by previous insights into galactic halos and the shapes of galaxies and by measurement. Numerical calculations at least as early as 1973 estimated that galaxies under certain conditions are more likely to evolve from smooth uniform discs into long, thick, rotating bar-shaped distributions [OP73]. This was inferred from the virial theorem, which can be written for a virialized system in term of two kinetic energy components and one potential energy component,

$$2T_{mean} + 2T_{rand} + V = 0 \longrightarrow \frac{T_{mean}}{|V|} + \frac{T_{rand}}{|V|} = \frac{1}{2} \longrightarrow t + u = \frac{1}{2}, \qquad (4.2)$$

where T_{mean} is the rotational kinetic energy for a simple system with steady rotational motion while T_{rand} contains the contribution of the kinetic energy due to the dispersion of velocities in the system, related to the pressure in the gas of rotating particles. The t term, related to the ratio of the steady rotational kinetic energy to the magnitude of the total potential energy, is bounded by $0 \leq t \leq 1/2$. For a fluid-like system, it was determined that, if $t \geq 0.14$, the rotating system becomes unstable. Rapid rotational motion can drive the system to a new equilibrium state in the shape of a bar-like formation that maintains the angular momentum, mass, and the central density of the system.

A static dark matter halo will increase $|V|$ with no contribution to the kinetic energy, and thus overall decrease t. This can drive a galaxy below the instability threshold. However, if the halo also rotates this alters the kinetic energy terms. Studies of co-rotating dark matter halos suggest that they can drive the visible matter to the barred state much faster than the visible matter alone achieve, implying that a galaxy with a rotating halo will become bar-like much earlier in its evolution [SN13]. A counter-rotating halo can slow bar formation.

Observational evidence suggests that the Milky Way halo has a rotational velocity generally consistent with zero, and which at larger radii (20 kpc) could be slightly counter-rotating (with a velocity at the level of 10 km/s). A recent example of such a measurement was done using 91 globular star clusters in the Milky Way to probe the dynamics of the gravitational potential [PH19]. Given the expectation and the evidence, it is reasonable to assume that our galaxy's halo is approximately at rest and the only net motion contribution to the dark matter, relative to Earth, will be due to our Sun's motion around the center of the galaxy.

The Sun's orbital speed tangent to the line connecting the galactic center to our solar system is $v_c \approx 220$ km/s. The speed dispersion, σ_v, is related to the local circular speed by

$$\sigma_v = \sqrt{\frac{3}{2}} v_c. \qquad (4.3)$$

The halo density profile is assumed to be $\rho(r) \propto r^{-2}$ and $\rho_0 = 0.3$ GeV/$c^2 \cdot$ cm^3, where the local density of dark matter near Earth is taken to be the average of a small volume within a few hundred parsecs of the sun. There are two approaches to measuring the local dark matter density [Rea14]. The first involves the use of the vertical kinematics of stars near the sun, [Kap22; Oor32; Oor60; Zha+13] and the second involves extrapolation from galactic rotation curves [Wd10]. Recent measurements suggest that the halo profile trends more like $r^{-3.3}$ [PH19], but for the purposes of *interpreting* scattering experiments it is sufficient to choose a halo model that all experiments then use for design and interpretation. As new measurements refine our understanding of the halo density profile, experiments can re-evaluate their results in light of updated information.

Any material body with a velocity in excess of the local escape velocity in a galaxy (at some distance R from the center of the galaxy) will not be gravitationally bound to that region of the galaxy. Therefore, if one precisely measures the distribution of velocities at a given R, one should find that the population is truncated; there will be no objects with $v > v_{esc}$. In our region of the galaxy that speed is $v_{esc} = 650$ km/s.

It is interesting to think about what will be the population of WIMPs in the area where you are sitting reading this text. We have inferred the local dark matter density in our region of the Milky Way. To complete the estimate, you need to next select your "favorite" mass for a WIMP candidate. We will choose two options: one lighter species, $m_{\chi 1} = 5$ GeV/c^2, and one heavier species, $m_{\chi 2} = 100$ GeV/c^2. What then will be their respective number densities? These are,

$$n_1 = \rho_0/m_{\chi 1} = 6 \times 10^{-2} \text{ cm}^{-3}, \tag{4.4}$$
$$n_2 = \rho_0/m_{\chi 2} = 3 \times 10^{-3} \text{ cm}^{-3}. \tag{4.5}$$

Using these number densities, we can consider a room with a certain volume and determine at any moment in time what are the number of particles in that room. Let's choose a more familiar volume—that of a large (2-liter) water or soda bottle. In that case,

$$N_1 = n_1 V_{bottle} = 120, \tag{4.6}$$
$$N_2 = n_2 V_{bottle} = 10. \tag{4.7}$$

This is not just a convenient human-scale exercise. Real detector volumes will range from compact (liters in volume) to large-scale (room-sized or larger in volume). It is clear that, given what we know about the local dark matter halo, there is a good chance for dark matter to be passing through such volumes. What will control the rate of actual detection are two factors: the fundamental strength of the interaction between SM and dark matter (which may be weak-scale, but since that was determined from a very simplistic hypothesis, perhaps not) and the acceptance and efficiency of an active detection medium for the energy deposited by these interactions. We will consider possible interactions with nuclear matter (protons, neutrons, quarks, etc.) and electrons.

4.3 Nuclear Scattering of Dark Matter

It is generally assumed, owing to the extreme weakness of the gravitational interaction between two atomic or subatomic objects, that the detection of an interaction between those two objects cannot rely only on gravity. Designing an experiment to directly detect dark matter generally means assuming that *dark matter has an interaction with SM*

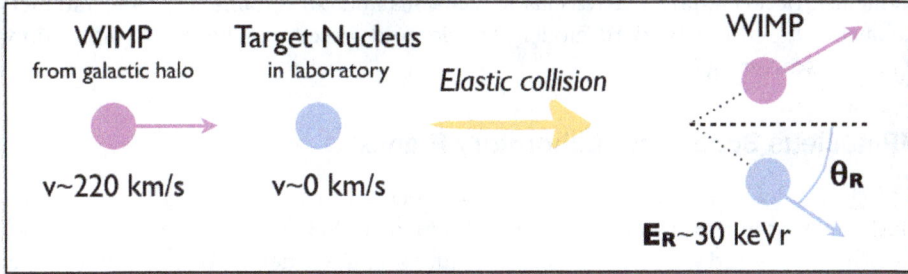

Figure 4.2: A schematic of dark matter (WIMP) elastic scattering off a target nucleus.

matter that is not only through gravitation[2]. A popular assumption is to consider some kind of weak-interaction-like framework (even to include the actual SM weak interaction involving an exchange of weak bosons or the Higgs boson). This is couched in the language of "WIMP searches," referring to the weakly interacting massive particles inspired by freeze-out considerations discussed elsewhere [KT90]. The term "WIMPs" refers to one candidate-class of dark matter particles that can be heavy and weakly interacting.

The specific nature of the interaction, broadly speaking, doesn't matter here. All that matters is that it be stronger than the gravitational interaction alone which would be so feeble in a subatomic terrestrial experiment as to be totally undetectable.

Our consideration of galaxy dynamics and ideal gas models has led us to conclude that a typical dark matter particle velocity in the Milky Way's dark matter halo at our radius from the galactic center would be about 220 km/s, or $\beta = v/c = 7.34 \times 10^{-4}$. We can begin there—that a single dark matter particle has an average speed at this level. This certainly qualifies as non-relativistic, so we can treat a hypothetical scattering process using non-relativistic quantum mechanics and kinematics. Let us hypothesize a dark matter particle mass of 100 GeV/c^2 (about 100 times the proton mass). What kind of dark matter kinetic energy would we expect in this case?

$$K = \frac{1}{2}mv^2 = \frac{1}{2}m\beta^2 c^2 = 27 \text{ keV} \qquad (4.8)$$

Compare that to the typical binding-energy-per-nucleon in heavier nuclei, which is 6–8 MeV. This is nowhere near enough to fission nuclei. Therefore, we can anticipate that interactions with the nucleus will be *elastic* in nature. Even for dark matter with mass of $\mathcal{O}(1TeV)$, we still expect elastic interactions. For this discussion we will focus on elastic scattering. It should be noted that inelastic scattering is also possible [SW01].

The consequences of this kind of nuclear interaction are depicted in Fig. 4.2. This scattering process will impart a small amount of energy to the nucleus. While elastic, spin-dependent interactions may nevertheless play a role in the degree and kind of scatter and will need to be factored into scattering interpretations. In addition, it will be important

---

[2]We will revisit this claim in Chapter 7. In fact, gravitational-only dark matter might yet yield to laboratory experimental tests.

to distinguish between dark matter elastic collisions and *background scatters*—interactions in the nuclei caused by standard model particles and forces that have nothing to do with dark matter scattering.

WIMP-nucleus Scattering: Laboratory Frame

Using Fig. 4.2, let us consider the scattering process in the *laboratory frame*, the frame in which the medium containing the nuclei is at rest. While the overall system may be at rest, the atoms and nuclei may not. We can estimate whether or not atomic/nuclear motion is going to play a significant role in any scattering process in comparison to the average expected dark matter velocity. A typical nuclear mass (which dominates the mass of any atomic system) is going to be $m_N \sim \mathcal{O}(10 \text{ GeV}/c^2)$. Experiments in this field are carried out at a range of temperatures, from essentially room temperature ($T \sim 293$ K) down to $T \sim \mathcal{O}(1$ K). The range of velocities for nuclei/atoms in thermal motion is going to be

$$K = \frac{1}{2} m_N v_N^2 \approx k_B T \longrightarrow v_N \approx \sqrt{\frac{2 k_B T}{m_N}} \approx (40\text{--}700) \text{ m/s} \tag{4.9}$$

or a $\beta_N = v_N/c \approx 10^{-6}$ in the higher-temperature case (room temperature). Since the dark matter is expected to be traveling at speeds consistent with the solar system's speed in orbit around the center of the galaxy (220 km/s $\longrightarrow \beta_\chi \approx 10^{-3} \gg \beta_N$), we can safely approximate the nuclei to be at rest within the laboratory frame during a collision process.

We can now develop the mathematics for this scattering. Given the low speeds involved, we will treat this as a *non-relativistic collision primarily subject to classical mechanics*. In an elastic collision, we conserve both momentum (always conserved in closed and isolated systems) and kinetic energy (only conserved in this case). We can denote initial-state quantities without primes and final-state quantities with primes. Let $\vec{p}$ ($\vec{k}$) denote the dark matter (nucleus) momentum. Our conservation equations are thus,

$$\vec{p} + \vec{k} = \vec{p'} + \vec{k'} \tag{4.10}$$

$$\frac{p^2}{2m_\chi} + \frac{k^2}{2m_N} = \frac{(p')^2}{2m_\chi} + \frac{(k')^2}{2m_N}. \tag{4.11}$$

For the nucleus, $k = 0$ (the at-rest approximation), these simplify to:

$$\vec{p} = \vec{p'} + \vec{k'} \tag{4.12}$$

$$\frac{p^2}{2m_\chi} = \frac{(p')^2}{2m_\chi} + \frac{(k')^2}{2m_N}. \tag{4.13}$$

Let us further simplify this by placing the original dark matter particle entirely along the horizontal axis in Fig. 4.2. The nucleus will then be scattered relative to that axis at the indicated angle, θ_R, so that $k'_{||} = k' \cos \theta_R$ and $k'_\perp = k' \sin \theta_R$, where the parallel ($||$) and perpendicular ($\perp$) symbols indicate direction relative to horizontal. Since there was no net momentum in the transverse direction in the initial state, and since momentum is conserved in each direction separately, it must be true in the final state that

$$\vec{p'}_\perp + \vec{k'}_\perp = 0 \longrightarrow \vec{p'}_\perp = -\vec{k'}_\perp, \tag{4.14}$$

which, of course, tells us

$$p' \sin \phi = k' \sin \theta_R \longrightarrow \sin \theta_R = \frac{p'}{k'} \sin \phi \tag{4.15}$$

and relates the scattering angles to one another through the ratio of the final-state momentum magnitudes. The angle ϕ is that between the final-state dark matter particle and the initial-state dark matter momentum.

It is useful to define the *momentum transfer*, $\vec{q} = \vec{p'} - \vec{p}$, which is the amount of dark matter momentum that is transferred to the nucleus in the collision (or vice versa). We can then identify that

$$q^2 = p^2 - 2\vec{p'} \cdot \vec{p} + (p')^2. \tag{4.16}$$

and, for our conditions,

$$\vec{q} = -\vec{k'}. \tag{4.17}$$

This lets us write the kinetic energy of the recoiling nuclear (also known as the *recoil energy*) as

$$E_R \equiv \frac{(k')^2}{2m_N} = \frac{q^2}{2m_N}. \tag{4.18}$$

Let us try to determine a relationship between the initial dark matter speed and the final recoiling nucleus speed. To do this, we will need to eliminate the final dark matter particle speed (v') from our equations. From the conservation of kinetic energy,

$$\begin{aligned}
\frac{1}{2}m_\chi v^2 &= \frac{1}{2}m_\chi (v')^2 + \frac{1}{2}m_N (u')^2 \\
m_\chi v^2 &= m_\chi (v')^2 + m_N (u')^2 \\
m_\chi^2 v^2 &= m_\chi^2 (v')^2 + m_\chi m_N (u')^2 \\
m_\chi^2 (v')^2 &= m_\chi^2 v^2 - m_\chi m_N (u')^2
\end{aligned}$$

where u' is the speed of the recoiling nucleus. We can then employ the conservation of momentum,

$$\begin{aligned}
m_\chi \vec{v} &= m_\chi \vec{v'} + m_N \vec{u'} \\
m_\chi \vec{v'} &= m_\chi \vec{v} - m_N \vec{u'} \\
m_\chi^2 (v')^2 &= m_\chi^2 v^2 - 2m_\chi m_N \vec{v} \cdot \vec{u'} + m_N^2 (u')^2.
\end{aligned} \tag{4.19}$$

Now relate these two initial dark matter speed terms between the conservation of kinetic energy and momentum,

$$\begin{aligned}
m_\chi^2 v^2 - m_\chi m_N (u')^2 &= m_\chi^2 v^2 - 2m_\chi m_N \vec{v} \cdot \vec{u'} + m_N^2 (u')^2 \\
-m_\chi m_N (u')^2 &= -2m_\chi m_N \vec{v} \cdot \vec{u'} + m_N^2 (u')^2 \\
2m_\chi m_N \vec{v} \cdot \vec{u'} &= m_\chi m_N (u')^2 + m_N^2 (u')^2.
\end{aligned} \tag{4.20}$$

Let us now consider the case of *maximum energy transfer*, which will happen when the initial dark matter velocity is entirely collinear with the final nucleus velocity, so that $\theta_R = 0$. In this case, $\vec{v} \cdot \vec{u'} = vu'$ and

$$
\begin{aligned}
2m_\chi m_N \vec{v} \cdot \vec{u'} &= m_\chi m_N (u')^2 + m_N^2 (u')^2 \\
2m_\chi m_N vu' &= m_\chi m_N (u')^2 + m_N^2 (u')^2 \\
2m_\chi m_N vu' &= (m_\chi m_N + m_N^2)(u')^2 \\
2m_\chi m_N v &= (m_\chi m_N + m_N^2)u' \\
\frac{v}{u'} &= \frac{m_\chi m_N + m_N^2}{2m_\chi m_N}.
\end{aligned}
\tag{4.21}
$$

This can be transformed into a statement about the ratios of the kinetic energies of the initial dark matter particle and the final nucleus. This is a more relevant physical quantity since most detectors we will consider are *calorimeters*, meaning their primary sensitivity is to the amount and kind of energy deposited in the medium:

$$
\begin{aligned}
\frac{v}{u'} &= \frac{m_\chi m_N + m_N}{2m_\chi} \\
\left(\frac{v}{u'}\right)^2 &= \frac{(m_\chi + m_N)^2}{4m_\chi^2} \\
\frac{\frac{1}{2}m_\chi v^2}{\frac{1}{2}m_N (u')^2} &= \frac{m_\chi (m_\chi + m_N)^2}{4m_\chi^2 m_N} \\
\frac{\frac{1}{2}m_\chi v^2}{\frac{1}{2}m_N (u')^2} &= \frac{(m_\chi + m_N)^2}{4m_\chi m_N}.
\end{aligned}
$$

Since $E_R \equiv (k')^2 / 2m_N = \frac{1}{2}m_N (u')^2$,

$$
E_R^{max} = \frac{1}{2}m_\chi v^2 \frac{4m_\chi m_N}{(m_\chi + m_N)^2}.
\tag{4.22}
$$

The quantity $\frac{m_\chi m_N}{(m_\chi + m_N)}$ is also known as the *reduced mass* of the system. If we put ourselves in the center-of-mass frame of the colliding system (a special point that, by conservation of momentum, is fixed in space), the reduced mass appears quite naturally. It can be thought of as if a system of that singular mass μ was experiencing the collision, taking into account the relative motion of the original two bodies before the collision. Thus we can write:

$$
E_R^{max} = 2m_\chi v^2 \frac{\mu^2}{m_\chi m_N} = 2\frac{\mu^2 v^2}{m_N}.
\tag{4.23}
$$

WIMP-nucleus Scattering: Center-of-Mass Frame

Let's explore the kinematics of this kind of scattering in more detail, but moving to the center-of-mass frame. While much of the notation in this section will look similar to

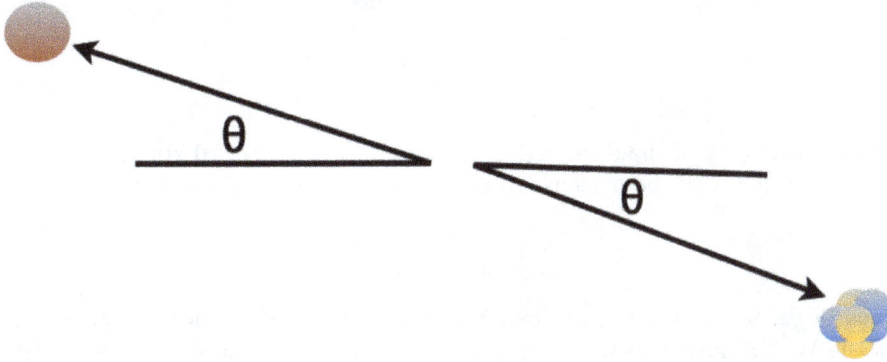

Figure 4.3: WIMP-nucleus scattering in the center-of-mass frame.

the previous section, it is more of convenience than invariance. For example, the initial speed of the dark matter in the lab frame is given by the earth's motion through the dark matter halo of the Milky Way; the nucleus's velocity is taken to be zero. In the frame of an observer at rest with respect to the center of mass of the dark matter-nucleus system, the dark matter and the nucleus will appear to be moving with equal but opposite momentum, by construction (Fig. 4.3).

We can derive the relationship between the lab velocity of the dark matter and the center-of-mass frame velocity of the dark matter. In the laboratory frame, the velocity of the dark matter with respect to the nucleus is defined by the first derivative, with respect to the time of displacement between the dark matter particle and the nucleus. Let us denote this as $\Delta(t)$. In the center-of-mass frame, the velocity of the dark matter particle will be defined by the time derivative of the displacement between the dark matter particle and the center of mass position, which we will denote $x(t)$. Let us place the center of mass at the origin of our coordinate system. In terms of these other displacements, the position of the center of mass along the x-axis will be given by

$$0 = \frac{m_\chi x + m_N(\Delta - x)}{m_\chi + m_N}. \tag{4.24}$$

Thus,

$$
\begin{aligned}
0 &= m_\chi x + m_N(\Delta - x) \\
m_N(\Delta - x) &= m_\chi x \\
m_N \Delta &= (m_\chi + m_N)x.
\end{aligned}
$$

We can now relate the velocities by taking the time derivative of this equation:

$$\frac{d}{dt} m_N \Delta = \frac{d}{dt}(m_\chi + m_N)x$$

$$
\begin{aligned}
m_N \dot{\Delta} &= (m_\chi + m_N)\dot{x} \\
m_N v &= (m_\chi + m_N) v_{com} \\
v_{com} &= \frac{m_N}{m_\chi + m_N} v
\end{aligned}
\tag{4.25}
$$

where v is the velocity of the dark matter in the lab frame $v = 220$ km/s. You can also show that the velocity of the nucleus (u) relative to the center of mass will be:

$$
u = \frac{m_\chi}{m_\chi + m_N} v. \tag{4.26}
$$

Let us begin by calculating the recoil energy of the nucleus in the center-of-mass frameof the WIMP-nucleus system. Let p label the three-momentum of the WIMP and k label the three-momentum of the nucleus; now in the center-of-mass frame, not the laboratory frame. Then,

$$
\begin{aligned}
\vec{p} &= -\vec{k} \quad \text{(initial momentum)} \tag{4.27} \\
\vec{p'} &= -\vec{k'} = \vec{q} + \mu \vec{v}_\chi \quad \text{(final momentum)}, \tag{4.28}
\end{aligned}
$$

where the WIMP-nucleus reduced mass, μ, was defined earlier as:

$$
\mu = \frac{m_\chi m_N}{m_\chi + m_N}, \tag{4.29}
$$

v_χ is the mean WIMP velocity *relative* to the target nucleus and q is the momentum transfer between the WIMP and the nucleus, $\vec{q} \equiv \vec{p'} - \vec{p}$. In the center-of-mass frame, and for elastic scattering, $|\vec{p}| = |\vec{p'}|$. We can then write:

$$
\frac{q^2}{2} = \frac{1}{2}(\vec{p'} - \vec{p})^2 = p^2 - \vec{p} \cdot \vec{p'} = p^2(1 - \cos\theta) = \mu^2 v_\chi^2 (1 - \cos\theta). \tag{4.30}
$$

The angle, θ, is as defined in Fig. 4.3. This angle is often referred to as the *recoil angle*.
The nuclear recoil (NR) energy will then be given by

$$
E_R = \frac{q^2}{2m_N} = \frac{\mu^2 v_\chi^2}{m_N}(1 - \cos\theta_R). \tag{4.31}
$$

Using this, we can then calculate the minimum dark matter particle velocity for which we expect a recoil with energy E_R. This will happen under the limiting case $\cos\theta_R = -1$. Then,

$$
\begin{aligned}
E_R &= \frac{\mu^2 v_\chi^2}{m_N}(1 - \cos\theta) \\
&= \frac{\mu^2 v_\chi^2}{m_N}(1 - (-1)) \\
&= 2\frac{\mu^2 v_\chi^2}{m_N}
\end{aligned}
$$

$$\longrightarrow \quad v_{min} = \sqrt{\frac{m_N E_R}{2\mu^2}} = \frac{q}{2\mu}. \tag{4.32}$$

This relationship between the minimum speed that generates a recoil and the other properties of the system has implications:

- Lighter dark matter particles ($m_\chi \ll m_N$) must have larger minimum (threshold) velocities to generate a recoil;

- *Inelastic scattering*, where the nucleus is fissioned, or where additional particles are generated during the scattering process, can further increase the minimal velocity needed to induce the reaction (by the addition of other mass terms in the final state that further alters μ).

Consider the average momentum transfer in an elastic scattering of a WIMP and an atomic nucleus. Assume a 10 GeV/c^2 WIMP whose speed is $\approx$ 100 km/s.

$$p = m_\chi v = (10 \times 10^9 \text{ eVc}^{-2})(100 \times 10^3 \text{ ms}^{-1})\frac{c}{3 \times 10^8 \text{ ms}^{-1}} \approx 3 \text{ MeV}. \tag{4.33}$$

If we increase the mass of the dark matter by a factor of ten, we would then also increase the energy transferred to the nucleus to $\approx$ 30 MeV.

We can also consider the de Broglie wavelength that corresponds to the momentum transfer of $\sim$ 10 MeV. This is given by

$$\lambda = \frac{hc}{pc} = \frac{1239 \times 10^{-6} \text{ eV} \cdot \text{m}}{10 \times 10^6 \text{ eV}} \sim 12 \text{ pm}. \tag{4.34}$$

Typical nuclear size scales are on the order of $r \approx (1.25 \text{ fm})A^{1/3}$, where A is the atomic mass number. A uranium nucleus, for example, would have a radius of about 8 fm. The de Broglie wavelength of such a momentum transfer exceeds the size of even large nuclei, implying that any elastic scattering will be *coherent scattering*—scattering off of the entirety of the nucleus, as if the nucleus were a singular object and not a collection of nucleons; nor, at a deeper level, a collection of quarks and gluons.

Scattering Rates in a Detector—A Simple Picture

Let us now take a simplified picture of elastic scattering rates following the calculations of Lewin and Smith [LS96] using an imagined detector composed of atoms, where the primary target of interest is the nucleus of the atom. In the next section, we will motivate the origin of the assertions made here. Our goal in this short section is to motivate the details that follow.

The differential event rate for simplified WIMP interaction (a detector stationary in the galaxy) is given by:

$$\frac{dR}{dE_R} = \frac{R_0}{E_0 r}e^{-(E_R/E_0)r} \tag{4.35}$$

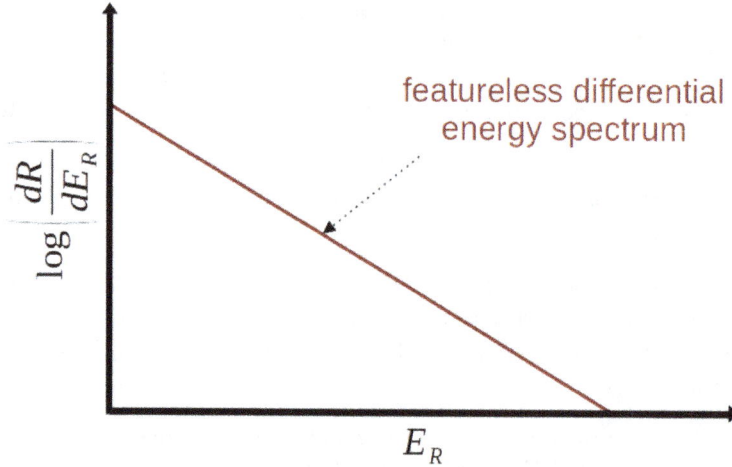

Figure 4.4: An example of the nucleus recoil energy spectrum from elastic WIMP-nuclei scattering.

where R is the rate of elastic scattering events, E_R is the recoil energy of the nucleus, R_0 is the total rate of such events, and E_0 is the most probable incident energy of a WIMP (e.g. given the Maxwell-Boltzmann distribution of galactic dark matter). r is a kinematic factor defined by

$$r = \frac{4m_\chi m_N}{(m_\chi + m_N)^2}. \tag{4.36}$$

The total event rate is recovered by

$$\int_0^\infty \frac{dR}{dE_R} dE_R = R_0 \tag{4.37}$$

and the mean recoiling energy is given by

$$\langle E_R \rangle = \int_0^\infty E_R \frac{dR}{dE_R} dE_R = E_0 r. \tag{4.38}$$

The spectrum of recoil energy for a given WIMP mass and WIMP velocity distribution is shown in Fig. 4.4. Note that this spectrum is featureless. While changing the hypothetical WIMP mass or kinematics, or the target nucleus mass, will alter the spectrum, it will always be straight and featureless (albeit with altered slope, etc.) on a logarithmic scale like this one.

We can now compute the mean nuclear recoil energy deposited in a detector. Let us assume that we have designed the detector such that $m_\chi \approx m_N = 100$ GeV/c^2. Then

$$\langle E_R \rangle = E_0 r = \left(\frac{1}{2} m_\chi v^2 \right) \left(\frac{4m_\chi m_N}{(m_\chi + m_N)^2} \right). \tag{4.39}$$

For our particular case of equivalent or near-equivalent masses, $r = 1$, the typical recoil energy of the nucleus will be the kinetic energy of the WIMP before the collision. If we assume that the WIMP halo is stationary and that the Sun–Earth system is passing through it at our orbital speed about the center of the galaxy, we obtain

$$\langle E_R \rangle \approx 30 \text{ keV}. \tag{4.40}$$

This hurried glance at the implications of scattering theory, applied to dark matter, is meant to provide an encouraging sign before we review the details. An energy at the level of 30 keV is sufficient to cause a range of atomic and nuclear phenomena. While dark matter particles may appear elusive (and they are), an estimate like this suggests that experimental techniques have a strong hope of revealing their existence.

Scattering Rates in a Detector—A More Complex Picture

Scattering will, of course, be more complex than presented in the previous section. We need to take into account the following considerations. First, dark matter will have a certain velocity distribution, $f(v)$. Second, the detector is on Earth, which moves around the Sun, which in turn moves around the galactic center. Additionally, the cross-section will depend on the spin structure of the underlying interaction. The broadest cases can be described as either spin-independent (SI) or spin-dependent (SD) interactions.

In addition, the dark matter will scatter on nuclei that have finite size. As such, we have to consider form-factor corrections which are different for SI and SD interactions. We must also take into account that the recoil energy is not necessarily the observed energy. The detection efficiency in real life is not 100% because real detectors have certain energy resolution and energy thresholds, as well as physical processes that don't necessarily translate into pure conversion of recoil energy of some detectable form.

The total number of dark matter particles, N, to which an experiment is subjected is a product of the dark matter flux (particles per unit area per unit time), the effective area of the target, and the exposure time of the experiment. There are two obvious ways to increase the number of dark matter particles on target: increase the effective target area and/or operate the experiment longer. The formula is

$$N = tnvN_T\sigma, \tag{4.41}$$

where n is the dark matter number density, v the average speed of the particles, N_T the number of target particles, and σ the interaction cross-section of dark matter with the target particles.

Number alone is not sufficient information in planning an experiment. We will need to determine the spectrum of dark matter recoils, which is the energy dependence of the number of detected dark matter particles. This information will help us to design an experiment that meets the challenge of how, in phase space, dark matter particles are distributed. If we are to detect dark matter, we must do so by creating conditions that maximally sample the population of those particles.

We determine this spectrum by simply taking the derivative of Eqn. 4.41 with respect to the recoil energy of the target particles,

$$\frac{dN}{dE_R} = tnvN_T\frac{d\sigma}{dE_R}.$$ (4.42)

We need to consider that the dark matter particles are described by their local velocity distribution, $f(\vec{v})$, where $\vec{v}$ is the velocity of the dark matter particles in the reference frame of the detector (the laboratory frame). We can insert this in the above equation by use of an integral that, performed over all velocities, would yield the average speed of the dark matter particles:

$$\frac{dN}{dE_R} = tnN_T\int_{v_{min}} vf(\vec{v})\frac{d\sigma}{dE_R}d\vec{v},$$ (4.43)

where v_{min} is the minimum speed of a dark matter particle that results in recoil energy in a collision with the nucleus that, in turn, results in a detectable signal. The minimum dark matter velocity will vary by experiment and design. For example, if an experiment using a target material with atomic mass number $A = 40$ was sensitive to recoil energies above a threshold of $E_R = 1$ keV and optimal sensitivity for an experiment is targeted for $m_\chi = 100$ GeV, we can use Eqn. 4.23 to estimate that (assuming maximal energy transfer in the scattering process) the minimum dark matter velocity that can be studied is $v_{min} \approx 50$ km/s, or about one-quarter of the average dark matter speed in our region of the Milky Way. Such an experiment, in this case, should be sensitive to most of the local dark matter population (see Fig. 2.13). The integral is implicitly, then, performed up to some maximum speed that we will consider to be a variable for now (limited, obviously, by the escape velocity and, ultimately, the speed of light).

We have at our disposal some useful relationships based on past considerations of number, volume, density, mass, and time:

$$n = \frac{\rho}{m_\chi}$$ (4.44)

$$N_T = \frac{M_T}{m_N} \text{ (with } M_T \text{ the total target mass)}$$ (4.45)

$$\epsilon \equiv tM_T \text{ (the } exposure\text{).}$$ (4.46)

We can write:

$$\frac{dN}{dE_R} = \epsilon\frac{\rho}{m_\chi m_N}\int_{v_{min}} vf(\vec{v})\frac{d\sigma}{dE_R}d\vec{v}.$$ (4.47)

The *event rate* is defined as $R \equiv N/\epsilon$ (number of observable events per unit target mass per unit time), so the *differential event rate* (observable events per keV of recoil energy, per kilogram, per day) is then

$$\frac{dR}{dE_R} = \frac{\rho_0}{m_\chi m_N}\int_{v_{min}}^{v_{esc}} vf(\vec{v})\frac{d\sigma_{\chi N}}{dE_R}d\vec{v}$$ (4.48)

where we solidify the notation so that ρ_0 is specifically the *local dark matter density* (in our region of the Milky Way), $f(v)$ is the WIMP speed probability distribution in the detector frame, and $\sigma_{\chi N}$ is the WIMP-nucleus scattering cross-section. To convert this latter cross-section to a WIMP-*nucleon* cross-section requires nuclear physics to map overall nuclear structure onto individual nucleons in the nucleus. We pause here to note that a nucleus is more than the sum of its parts; being a quantum mechanical object by definition, there is no clear distinction in a nucleus between neighboring nucleons (there are not marbles packed in a tight ball).

Dark matter detection, then, is a synthesis of astrophysics (to estimate the local dark matter density), particle physics (to establish a framework for the scattering), and nuclear physics (to map the nuclear to the nucleon properties). We summarize the key equations that can guide direct-detection experiments in the box below. We will spend the next sections of this chapter dissecting the components of these equations, as each is complex and rich in and of itself.

Dark Matter Detection Equations

In summary, the dark matter detection equations of note are

$$E_R^{lab} = \left(\frac{m_\chi m_N}{m_\chi + m_N}\right)^2 \frac{v^2}{m_N}(1 - \cos\theta_R) \qquad (4.49)$$

and

$$\frac{dR}{dE_R} = \frac{\rho_0}{m_\chi m_N} \int_{v_{min}}^{v_{esc}} v f(\vec{v}) \frac{d\sigma_{\chi N}}{dE_R} d^3\vec{v} \qquad (4.50)$$

We illustrated earlier in Fig. 4.4 a featureless and exponential shape describing the recoil energy spectrum in these elastic scattering processes. We can motivate, using what has already been shown, that this spectrum must be exponential (linear on a logarithmic graph like the ones we use to illustrate these spectra). The standard halo model $f(\vec{v})$ depends on dark matter speed as $\exp(-v^2)$. The differential scattering rate (Eqn. 4.50) depends on an integral of the product of v, $f(v)$, and $d\sigma/dE_R$. The latter is inversely proportional to v^2. For an isotropic halo, $d^3\vec{v} = v^2 dv d\Omega$, so the integrand's underlying dependence on v is $\int v f(v) v^{-2} v^2 dv = \int v f(v) dv$. It is therefore of the form $\int x \exp(-x^2) dx$, whose solution is $-\exp(-x^2)/2$.

Thus, we are comparing a function (the scattering rate per unit recoil energy) that goes as $\exp(-v^2)$ to a recoil kinetic energy threshold (based on Eqn. 4.23), which itself depends on dark matter speed as v^2. Substituting $u = v^2$, we recognize that we are simply graphing a function $\exp(-u)$ vs. u ... a featureless exponential function (Eqn. 4.35).

The Scattering Cross-Section

A basic "feasibility criterion" for any experiment is to know the rate at which dark matter "events"—interactions between the particles and the experiment that result in energy deposition—are expected to occur. An infeasible experiment is one where fewer than one event (especially including detection efficiency) can plausibly ever be observed given the exposure possible in the instrument. The event rate, R, in a potential detector is determined by summing (integrating) the overall possible WIMP-nucleus interactions that can occur,

$$R = \int_{E_{threshold}}^{\infty} dE_R \frac{\rho_0}{m_\chi m_N} \int_{v_{min}}^{\infty} v f(\vec{v}) \frac{d}{dE_R} \sigma_{\chi N}(v, E_R) \, d\vec{v}. \tag{4.51}$$

A key ingredient in this computation is the WIMP-nucleus cross-section, a quantity with units of area that is a measure of the probability of interactions. In this section, we want to focus on the elements that enter the differential cross-section, $d\sigma_{\chi N}/dE_R$.

First, let us consider that the scattering process takes place within the non-relativistic limit. That is plausible given previous arguments that favor a cold dark matter component whose relative velocity to us is dominated by the sun's motion around the galactic center. We can further approximate the scattering process as an isotropic one (no dependence on direction), such that

$$\frac{d\sigma}{d\cos\theta_R} = \text{constant} = \frac{\sigma}{2}. \tag{4.52}$$

The maximum kinetic energy that can be transferred to the nucleus occurs when $\cos\theta_R = -1$ (100% back-scatter of the nucleus in the center-of-mass frame), such that $E_R^{max} = 2\mu^2 v^2/m_N$. In terms of this quantity, Eqn 4.49 becomes

$$E_R = \frac{E_R^{max}}{2}(1 - \cos\theta_R). \tag{4.53}$$

We can obtain the differential of the recoil energy with respect to scattering angle:

$$\frac{dE_R}{d\cos\theta_R} = \frac{E_R^{max}}{2}. \tag{4.54}$$

From this, we can then write the differential of the cross-section with respect to recoil energy using the chain rule:

$$\frac{d\sigma}{dE_R} = \frac{d\sigma}{d\cos\theta_R} \frac{d\cos\theta_R}{dE_R} = \frac{\sigma}{2} \frac{2}{E_R^{max}} = \frac{\sigma}{E_R^{max}} = \frac{m_N}{2\mu^2} \frac{\sigma}{v^2}. \tag{4.55}$$

The momentum transfer involved in these collisions corresponds to a de Broglie wavelength in excess of the nuclear size. Thus, we are coherently scattering off the entire nucleus. The WIMP doesn't "see" the nucleons at all, especially for lighter nuclei (c.f., Eqn. 4.34). While we emphasized that a whole nucleus is more than just the sum of its nucleons, for light nuclei this is actually a close approximation to observation. For light nuclei, it is therefore possible to translate a scattering rate on the nucleus into a scattering rate on individual nucleons (to infer the WIMP-proton or WIMP-neutron cross-section) without large corrections for making this assumption.

It is typical for heavy nuclei to be used in experiments. Given our previous considerations about maximizing event rates, this is for obvious reasons: presenting greater target cross-sectional area means more anticipated WIMP interactions per unit mass, per unit time. The drawback is that heavy nuclei need significant corrections when making the "sum of nucleons" approximation as a description of the nucleus as a whole. These corrections are summarized by *nuclear form factors, F*. For example, let us consider that we are not certain whether the WIMP-nucleus interactions may have spin-independent or spin-dependent effects. It is natural, therefore, to write the cross-section as a sum of both possibilities with unknown relative contributions, e.g.,

$$\frac{d\sigma}{dE_R} = \left[\left(\frac{d\sigma}{dE_R} \right)_{SI} + \left(\frac{d\sigma}{dE_R} \right)_{SD} \right], \tag{4.56}$$

where "SI" ("SD") denotes the spin-independent (spin-dependent) scattering. Spin-independent effects would arise from a scalar or vector dark matter candidate coupling to quarks inside the nucleons, while spin-dependent effects would arise from an axial-vector coupling to those same quarks. To facilitate basic calculations, it is straightforward to begin by assuming that these effects add coherently with corrections from the parton level (quarks and gluons) up to the nuclear scale:

$$\frac{d\sigma}{dE_R} = \frac{m_N}{2\mu^2 v^2} \left[\sigma_0^{SI} F_{SI}^2 + \sigma_0^{SD} F_{SD}^2 \right]. \tag{4.57}$$

The form factors encode the dependence on the momentum transfer and the nuclear structure without making explicit the details of these dependencies; the cross-sections encode the parton-level (quark-level) particle physics of the WIMP-parton interaction (Feynman diagrams describing the leading interactions possible between fundamental particles). The form factors thus correct the parton-level interactions up to the nuclear scattering effects that will actually be observed in a running experiment.

Form factors are an area of active research. The nature of quantum chromodynamics and the energy domain of the nucleus makes high-precision, direct analytical calculations of such form factors nearly impossible. Instead, a synthesis of nuclear scattering data and computational approaches are typically used to constrain or infer these corrections. It is also possible to make models of these form factors based on theoretical or data-driven considerations.

One example is the Helm form factor [DKG07; Hel56], which applies to spin-independent interactions,

$$F(q) = \left(\frac{3j_1(qR_1)}{qR_1} \right)^2 \exp(-q^2 s^2 / 2). \tag{4.58}$$

Here, j_1 is the spherical Bessel function,

$$j_1(x) = \frac{\sin(x)}{x^2} - \frac{\cos(x)}{x}, \tag{4.59}$$

q is the momentum transfer, s is the nuclear skin thickness ($s \sim 1$ fm), and R_1 is the effective nuclear radius, which should be approximated as $R_1 = 1.25$ fm $\times A^{1/3}$, where A is the atomic mass number. The nuclear skin results from an effect, at the level of nuclear forces, akin to surface tension in a drop of water. The dynamics of the nuclear surface are not the same at those of the nuclear bulk. This approach accounts for that. Here, it's approximately the diameter of a single nucleon.

For spin-dependent interactions, we can write

$$F^2(E_R) = \frac{S(E_R)}{S(0)} \tag{4.60}$$

where

$$S(E_R) = a_0^2 S_{00}(E_R) a_1^2 S_{11}(E_R) + a_0 a_1 2 S_{01}(E_R) \tag{4.61}$$

and

$$a_0 = a_p + a_n, \ a_1 = a_p - a_n. \tag{4.62}$$

Here, S_{ij} are the isoscalar (0), isovector (1), and interference form factors, while a_i are the isoscalar or isovector coupling constants. The spin-dependent form factor is the superposition of form-factor components (isoscalar, etc.) normalized to that superposition at zero recoil energy, the case that no energy is transferred to the nucleus.

The forms of these SI and SD cross-sections that most often appear in the literature are

---

Spin-Independent and Spin-Dependent Cross Sections

$$\sigma_0^{SI} = \frac{4\mu^2}{\pi} \left[Z f_p + (A - Z) f_n\right]^2 \propto A^2 \tag{4.63}$$

$$\sigma_0^{SD} = \frac{32 G_F^2 \mu^2}{\pi} \frac{J + 1}{J} \left[a_p \langle S_p \rangle + a_n \langle S_n \rangle\right]^2 \tag{4.64}$$

Equation 4.63 is valid for spin-independent interactions only. Furthermore, we generally assume a low momentum transfer which corresponds to the small value of q^2. As such, the scattering process cannot resolve the difference between the proton and neutron at the level of their constituent partons. Essentially, a dark matter particle scattering with these kinds of energy transfers is blind to the distinction between a proton and a neutron—they "look the same" in the interaction. In that case, the four-fermion coupling factors satisfy $f_p \sim f_n$ [Gon96] (This is a model-dependent assumption that may need to be revisited if WIMP scattering is observed.) The scattering is, overall, quadratically dependent on the atomic mass number, which implies that heavier nuclei should lead to large increases in sensitivity to this process.

However, in Eqn. 4.64 the nuclear angular momentum, J, explicitly appears, as do the individual couplings to the proton and neutron (a_i) and the expectation values of the nucleon spins $\langle S_i \rangle$. Different nuclei can lead to very different scattering rates, even for nearly identical values of A; in fact, isotopic variations among elements will, potentially significantly, shift the scattering process. A table from Ref [Tov+00] summarizes efforts that have gone into modeling the nuclear physics considerations which influence experimental choices and design. These data are an example of how to factor such

Nucleus	Z	Odd Nucleon	J	$\langle S_p \rangle$	$\langle S_n \rangle$	C_λ^p/C_p	C_λ^n/C_n
^{19}F	9	p	1/2	0.441	-0.109	7.78×10^{-1}	4.75×10^{-2}
^{23}Na	11	p	3/2	0.248	0.020	1.37×10^{-1}	8.89×10^{-4}
^{27}Al	13	p	5/2	-0.343	0.030	2.20×10^{-1}	1.68×10^{-3}
^{29}Si	14	n	1/2	-0.002	0.130	1.60×10^{-5}	6.76×10^{-2}
^{35}Cl	17	p	3/2	-0.083	0.004	1.53×10^{-2}	3.56×10^{-5}
^{39}K	19	p	3/2	-0.180	0.050	7.20×10^{-2}	5.56×10^{-3}
^{73}Ge	32	n	9/2	0.030	0.378	1.47×10^{-3}	2.33×10^{-1}
^{93}Nb	41	p	9/2	0.460	0.080	3.45×10^{-1}	1.04×10^{-2}
^{125}Te	52	n	1/2	0.001	0.287	4.00×10^{-6}	3.29×10^{-1}
^{127}I	53	p	5/2	0.309	0.075	1.78×10^{-1}	1.05×10^{-2}
^{129}Xe	54	n	1/2	0.028	0.359	3.14×10^{-3}	5.16×10^{-1}
^{131}Xe	54	n	3/2	-0.009	-0.227	1.80×10^{-4}	1.15×10^{-1}

Figure 4.5: From Ref. [Tov+00], this table provides values for terms that appear in Eqn. 4.64.

considerations into the choice of a particular target element and isotope. The table is shown in Fig. 4.5.

Effective Field Theory and the Complexity of Dark Matter Scattering

The treatment of scattering, so far, has been simple and generic, devoid of a particular underlying theoretical framework, except to assume basic non-relativistic quantum mechanics. In reality, we know that these more approximate models are rooted in quantum field theories. We do not yet know which field theory framework provides the proper description of dark matter, so it is impossible to be explicit about scattering calculations.

However, effective quantum field theories (EFTs) have proven to be a useful tool for establishing the many possible *kinds* of interactions that can take place between fermions and bosons. As measurements are made with experiments, it may be possible to eliminate certain interaction terms from consideration; some may be ruled out *a priori* as they provide undesirable or incompatible features for dark matter particles. Such effective field theories usually begin by considering leading-order and next-to-leading-order contributions to interactions, where the "order" here refers to the cut-off in a perturbation expansion of the theory that is usually conducted in terms of the coupling strength(s) of the interaction(s) involved in the theory.

Such an EFT would contain fourteen operators that rely on a range of nuclear properties in addition to the SI and SD cases [Fit+12][Fit+13][AFH14][Sch+15a]. They can be combined and grouped such that the WIMP-nucleon cross-section depends on six independent nuclear response functions: (a) one "spin-independent," (b) two "spin-dependent," (c) and three "velocity-dependent" functions. Two pairs of these interfere, resulting in eight independent parameters that can be probed by comparing EFT predictions with real experimental data (e.g. scattering observations or non-observations). Each operator connects an initial state and a final state, and, in effect, determines the nature and outcome of each kind of interaction.

Basic matrix element considerations apply here. Consider some generic operator, $\mathcal{O}$, that describes an interaction that takes some initial state $|i\rangle$ to some final state $|f\rangle$. In scattering theory and using *bra-ket notation*, the basic formalism is to consider an initial state $|i\rangle$ that is transformed into a final state $|f\rangle$ via an operator $\mathcal{O}$ that describes the transformation. This is written mathematically as

$$\mathcal{M} = \langle i|\mathcal{O}|f\rangle. \tag{4.65}$$

This is related to the probability of obtaining a final state from some initial state given the intermediate operation. However, such a product involves the multiplication of complex numbers whose results are also complex and thus not guaranteed to yield only a real-valued result. Real numbers are associated with direct observations of the natural world. Thus it is convention to identify the above matrix element as related to the square root of the probability, while the multiplication of the matrix element with its own complex conjugate,

$$|\mathcal{M}|^2 = |\langle i|\mathcal{O}|f\rangle|^2, \tag{4.66}$$

results always in a real-valued quantity, and is interpreted as being directly related to probability. The fourteen operators in an EFT framework that are used to describe WIMP-nucleon interactions are

$$\mathcal{O}_1 = 1_\chi 1_N \tag{4.67}$$

$$\mathcal{O}_3 = i\vec{S}_N \cdot \left[\frac{\vec{q}}{m_N} \times \vec{v}^\perp\right] \tag{4.68}$$

$$\mathcal{O}_4 = \vec{S}_\chi \cdot \vec{S}_N \tag{4.69}$$

$$\mathcal{O}_5 = i\vec{S}_\chi \cdot \left[\frac{\vec{q}}{m_N} \times \vec{v}^\perp\right] \tag{4.70}$$

$$\mathcal{O}_6 = \left[\vec{S}_\chi \cdot \frac{\vec{q}}{m_N}\right] \cdot \left[\vec{S}_N \cdot \frac{\vec{q}}{m_N}\right] \tag{4.71}$$

$$\mathcal{O}_7 = \vec{S}_N \cdot \vec{v}^\perp \tag{4.72}$$

$$\mathcal{O}_8 = \vec{S}_\chi \cdot \vec{v}^\perp \tag{4.73}$$

$$\mathcal{O}_9 = i\vec{S}_\chi \cdot \left[\vec{S}_N \times \frac{\vec{q}}{m_N}\right] \tag{4.74}$$

$$\mathcal{O}_{10} = i\vec{S}_N \cdot \frac{\vec{q}}{m_N} \tag{4.75}$$

$$\mathcal{O}_{11} = i\vec{S}_\chi \cdot \frac{\vec{q}}{m_N} \tag{4.76}$$

$$\mathcal{O}_{12} = \vec{S}_\chi \cdot \left[\vec{S}_N \times \vec{v}^\perp\right] \tag{4.77}$$

$$\mathcal{O}_{13} = i\left[\vec{S}_\chi \cdot \vec{v}^\perp\right]\left[\vec{S}_N \cdot \frac{\vec{q}}{m_N}\right] \tag{4.78}$$

$$\mathcal{O}_{14} = i\left[\vec{S}_\chi \cdot \frac{\vec{q}}{m_N}\right]\left[\vec{S}_N \cdot \vec{v}^\perp\right] \tag{4.79}$$

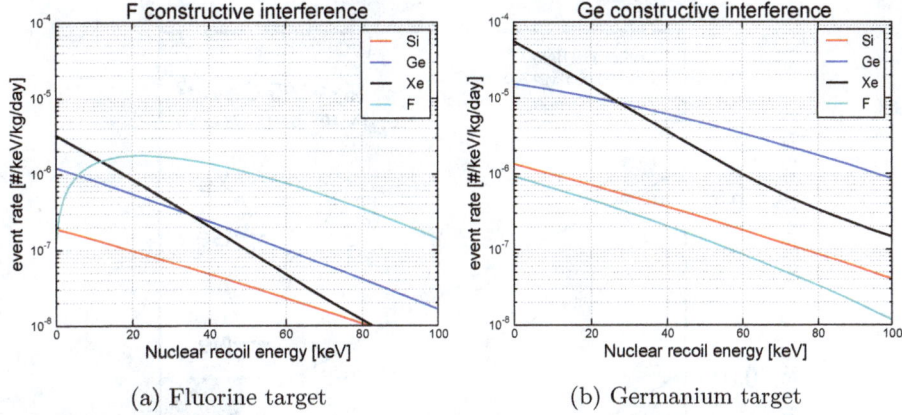

(a) Fluorine target   (b) Germanium target

Figure 4.6: Figures from Ref. [Sch+15a]. Differential event rates are evaluated in fluorine and germanium targets, where the $\mathcal{O}_8$ and $\mathcal{O}_9$ operators contribute with different kinds of interference and alter the scattering rates from the simple spin-independent or spin-dependent expectations.

$$\mathcal{O}_{15} = -\left[\vec{S}_\chi \cdot \frac{\vec{q}}{m_N}\right]\left[\left(\vec{S}_N \times \vec{v}^\perp\right) \cdot \frac{\vec{q}}{m_N}\right] \qquad (4.80)$$

where $\vec{v}^\perp \equiv \vec{v} + \vec{q}/2\mu_N$ a velocity operator related to both the velocity of the dark matter and the momentum transfer to the nucleus, $\vec{q}$; $\vec{S}_\chi$ is the WIMP spin; and $\vec{S}_N$ is the nucleon spin. In addition, each operator can independently couple to protons and neutrons. You will notice that $\mathcal{O}_2$ is missing from the list; this is because it cannot arise in the non-relativistic limit we are considering here.

As a result of this complex web of operators, representing different spin-independent, spin-dependent, and velocity-dependent interactions, WIMP-scattering can look very different across a range of common or plausible target materials. An example of this is a comparison of fluorine and germanium as targets to the same WIMP, but taking into account the fact that the operators $\mathcal{O}_8$ and $\mathcal{O}_9$ can both contribute to the scattering process, but with interference (Fig. 4.6). The interference of these contributions is more constructive in fluorine and more destructive in germanium. This leads to a suppression of the scattering cross-section in germanium relative to fluorine. This kind of pattern suggests an important strategy: operate different targets at the same time, or have the ability to change targets in a single experiment, to facilitate the study of a possible signal and its pattern of interactions.

The WIMP Detection Challenge

The primary challenge that comes from basic or EFT-driven scattering considerations is the overall low rate at which we can expect nuclear recoils to occur. For example, event rates in different materials, as a function of energy threshold for a 100 GeV/c^2 WIMP

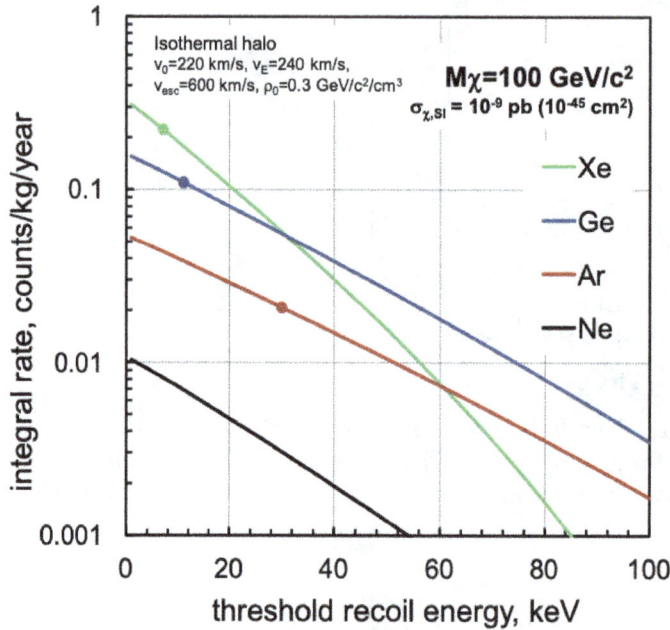

Figure 4.7: Event rates in units of counts per kg of material per year of data-taking vs. energy threshold in keV for different target materials. Graph is from Ref. [CA13]

and a WIMP-nucleon total cross-section of 10^{-45} cm^2 (compatible with upper bounds placed by leading experiments in the field today), are illustrated in Fig. 4.7 [CA13]. You will notice that we only expect a few events per kg-year of exposure.

Elastic scattering of a WIMP deposits small amounts of energy into a recoiling nucleus, amounting to just a few tens of keV. The energy spectrum of the recoil is featureless and falls exponentially as a function of the recoil energy. There are not expected to be any resonances, peaks, knees, or breaks in this spectrum, making it hard to distinguish from other processes having similar featureless energy spectra. The overall event rate is extremely low, requiring large targets and long periods of data-taking. This alone won't be sufficient. Materials are suffused with naturally occurring radioactive backgrounds that are, even under fair conditions, capable of producing nuclear recoils at rates far higher than these.

For lower-mass dark matter candidates, the challenge for detection approaches is even greater. The event rate vs. energy threshold for various WIMP candidate masses is shown in Fig. 4.8a. From the earlier exploration of recoil energy in non-relativistic elastic scattering, we know that, as the WIMP mass declines, the challenge of generating a nuclear recoil grows larger. In addition, different target materials will respond differently to the same kind of low-mass WIMP candidate (Fig. 4.8b). Lower WIMP masses will

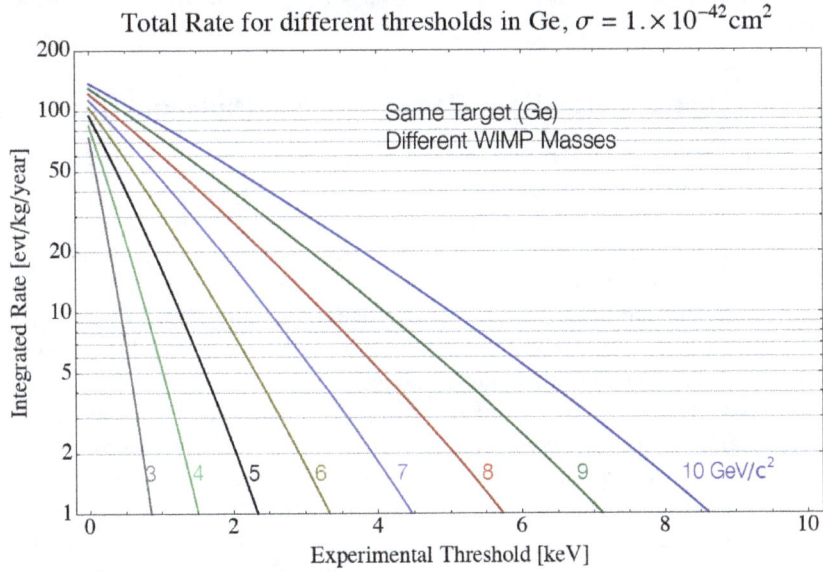

(a) Fixed target material (germanium)

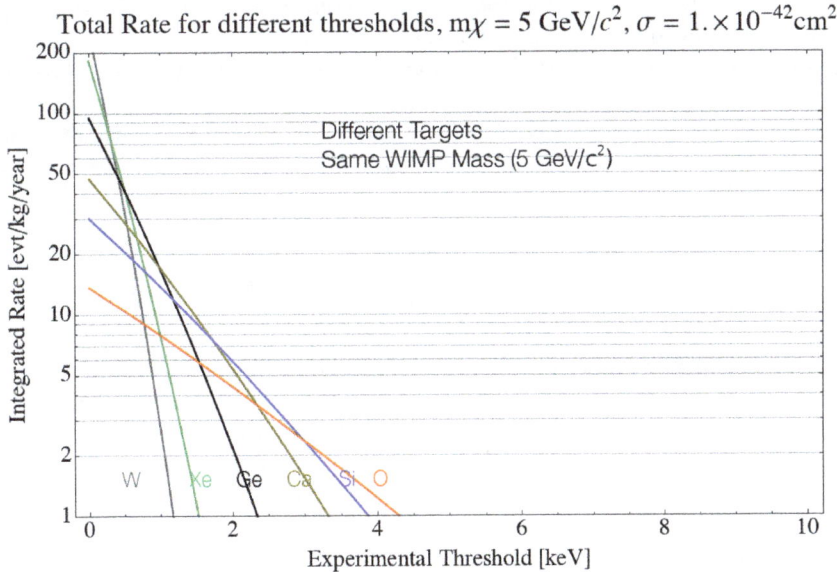

(b) Varying material ($m_\chi \sim 5$ GeV/c^2)

Figure 4.8: Event rates in units of counts per kg of material per year of data-taking vs. energy threshold (in keV) for different low-WIMP masses. Figures courtesy of Enectali Figueroa-Feliciano.

demand more and more materials and approaches that can go to lower and lower recoil energy thresholds.

The Consequence of Large Dark Matter–Nucleus (Nucleon) Cross-Sections

The focus in this text is often on an assumed small dark matter interaction cross-section. That possibility drives the design of ultra-sensitive instrumentation. However, it is important to note that there is an inverse challenge should the dark matter nucleus (nucleon) cross-section be too large. Here, we mean large in the sense that interactions would still be small enough not to upset cosmology and astrophysics (e.g., it would still appear as a nearly collision-less participant in phenomena at or above the galactic scale), but large enough that, when presented with a *dense* target, could suffer interactions that alter the velocity distribution of the dark matter.

Such a cross-section (e.g., 10^{-30} cm^2 or greater) can cause significant energy loss in the population of dark matter that must first pass through the atmosphere and then the rock overburden above an underground-sited experiment. For such large dark matter–nucleon cross-sections, it is possible to shift the velocity distribution $f(v)$ of the dark matter so that recoils have energies below the detectable threshold. This happens as kinetic energy is depleted by scattering off of air, rock, or both. This effect must be accounted for when quoting sensitivity limits, providing a new upper bound on sensitivity (due to having no terrestrially observable dark matter above a certain cross-section) and leaving intact the standard lower bound (due to the limits of the experimental design). The accommodation of this possibility is a more recent development in the field (c.f. Ref. [Alb+23], utilized here as an example).

The effect of overburden material on the dark matter that can ultimately interact with an experiment can be summarized by a *damping factor* that includes the net effect of materials [EK18]. This is given by:

$$\kappa = \frac{\rho \sigma_{\chi-n}}{m_\chi \mu_n^2} \left(\sum_i^{elements} \frac{F_i \mu_i^4 A_i}{m_i^2} \right) d, \tag{4.81}$$

where ρ is the material's density, μ is the reduced mass of the dark matter particle with the nucleon, (μ_n), or a specific element in the material, (μ_i), F_i is the mass fraction of element i in the material, A_i is the atomic number of the element, m_i is the atomic mass of the element, and d is the path length through the shielding material (the atmosphere, a geological structure, etc.).

The damping factor enters into the standard dark matter-scattering framework by altering the velocity distribution of the dark matter population as it passes through each material. For example, after passing through material m, the velocity average of the population would be $v_{f,m} = e^{-\kappa} v_{0,m}$ where $v_{0,m}$ denotes the initial average velocity entering the material. We see that we can build a model of the damping material by breaking it into layers and performing iterative computations of the velocity effects of each layer. Because the damping factors enter into these equations as exponents in the exponential function, it is possible to assemble models by merely adding damping factors for each layer of the model into one overarching damping factor, $\kappa = \sum_m \kappa_m$. The resulting effect on the dark matter distribution is this: if that distribution was originally described by some function, f, of the velocity (e.g., a Maxwell-Boltzmann distribution), then the final distribution will be $f(v_f) = e^{2\kappa} f(v_f e^{\kappa})$.

As an example of this, the SuperCDMS Collaboration utilized a seven-layer model of the atmosphere to incorporate scattering from air in computing the limits of experimental sensitivity. In addition, they built a model of the overburden above the Soudan Mine Underground Laboratory in Soudan, Minnesota, where they had sited an earlier version of the SuperCDMS experiment.

The same approach would need to be adopted by any experiment seeking to assess the effect of a significant overburden cutting off the dark matter distribution for large cross-sections. The collaboration also noted that they assessed the damping effect of their own shielding. This was determined to have the smallest impact of the three possible exterior components: the atmosphere, the earth, and the experiment's shield layers. This is due to the limited amount of high-density material that is used in shielding as compared to the much larger amounts of external natural shielding materials.

The net effect of this approach is to reduce the average WIMP velocity to the point where the product of velocity and cross-section is so small that nuclear interactions in the detector are too infrequent to be observed. This creates a *maximum cross-section sensitivity*, above which a given experiment cannot observe dark matter.[3] This illustrates one of the risks in the assumption of any dark matter experiment: that larger values of numbers, like cross-sections, have been implicitly ruled out by previous effort. In fact, in this scenario, no terrestrial dark matter experiment would observe anything because the dark matter–nucleon cross-section is so high that the dark matter doesn't reach the location on Earth where the sensitive target is located. This opens an interesting question about siting such experiments in space, though we recognize immediately the challenge that is posed by direct exposure to cosmic rays and the solar wind. Recognizing the challenge of the interplanetary radiation environment, we highlight this as a potentially interesting area to consider for future research opportunity.

Time-Dependence of the Scattering Rate

The focus has been on the event rate integrated over time, but not on the potential time structure of the rate itself. There is strong reason to believe that the interaction rate of dark matter on a target will be time-dependent. The most obvious reason for this is that the average speed of the dark matter is determined by the estimated speed with which the sun orbits the center of the Milky Way. If Earth were stationary with regard to the sun, the "WIMP wind" created by the sun's motion would point in the same direction over the span of a typical human lifetime. However, the earth orbits the sun, so, for half of the year, our planet is headed "into the WIMP wind" while, in the other half of the year, we are headed "out of the WIMP wind."

---

[3]All is not necessarily lost in this case. There are very recent suggestions of approaches that could subsequently be used to identify "earth-bound" dark matter particles [EPR24].

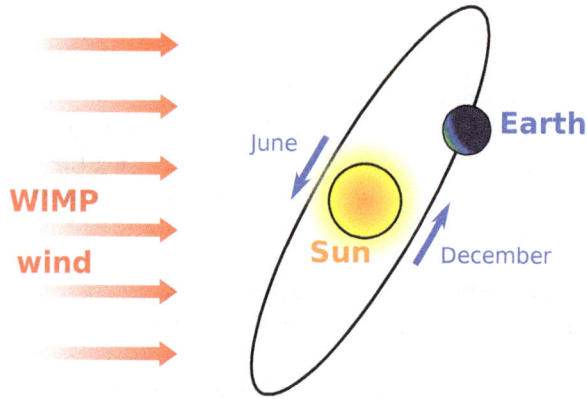

Figure 4.9: A schematic showing the relationship of the earth's orbit around the sun to the direction of travel of the sun through the Milky Way. The WIMP "wind" points opposite that direction of travel. Figure from Ref. [FLS13].

The interaction rate is controlled by many factors; one of them being the speed of the dark matter particles with respect to the target. This is manifest in Eqn. 4.49, as well as in field theory operators that might contribute to dark matter-nucleon interactions (c.f. Eqn. 4.68). Tiny expected variations around the average annual speed of WIMPs could manifest in a detector as an *annual modulation* of the event rate.

The motion of the sun through the WIMP halo is in the direction of the constellation Cygnus and is in the plane of the galaxy. However, the plane of earth's orbit is not aligned with that direction, but tilted about 60° above that line of motion. We should only consider the change in the WIMP wind based on the component of the earth's orbital velocity that is parallel to the sun-Cygnus line (Fig. 4.9).

The earth's speed with respect to the galactic rest frame, taken also to be the WIMP halo rest frame, is largest in the northern (southern) hemisphere's summer (winter) when the components of the earth's orbital velocity in the direction of solar motion is largest. The number of WIMPs with high (low) speeds in the detector rest frame is largest (smallest) in June. As a result, in the northern hemisphere, we would expect any annual modulation of the event rate to peak during the summer months and be minimal in the winter months. The opposite would be true in the southern hemisphere, peaking in the winter and minimal in the summer.

We can estimate the degree of this effect [FLS13]. The earth's speed around the sun is much less than the speed of the sun about the center of the Milky Way. In numbers, $v_{orbit}/v_{sun} \approx 0.07$. The small value of the relative speeds provides an opportunity to Taylor-expand the differential event rate with respect to the small parameter ($v_{orbit}/v_{sun} \ll 1$) to enable a leading order approximation of the effect. This results in

$$\frac{dR}{dE_R}(E_R, t) \approx \frac{dR}{dE_R}\left[1 + \Delta(E_R)\cos\frac{2\pi(t-t_0)}{T}\right], \qquad (4.82)$$

where T is the period of the modulation (expected to be one earth orbital period, or 365.25 days), and t_0 is the phase which, due to the relationship between the plane of orbit and the motion of the sun, is $t_0 = 150$ days. The factor ΔE_R is the *modulation amplitude*. To detect the seasonal variation due to the changing recoil energy spectrum across the year, one needs sufficient observed interactions to tell the difference between the higher number in June and the lower number in December. This requires the observation of at least 1000 such interactions to detect what is expected to be a few percent effect based on the ratio of earth and solar velocities. This effect, applied to an example generic recoil energy spectrum is shown in Fig. 4.10.

In addition to time dependence, if detectors can be made sensitive to the *direction* of the nuclear recoil, this can help to understand a possible modulation signal. Directionality [May+16] can also be used for a variety of other purposes, including the suppression of confounding processes in detector systems.

The earth's motion with respect to the galactic rest frame produces a direction-dependence of the recoil spectrum. The peak WIMP flux will appear to point from the direction of solar motion—from Cygnus toward us. Assuming a smooth and isotropic WIMP halo, the recoil rate will be maximal in the opposite direction of our motion toward Cygnus. In the laboratory frame, this direction varies over the course of a day due to earth's rotation as illustrated in Fig. 4.11. In addition, because the direction to Cygnus varies by hemisphere, detectors placed on different halves of the earth should see slightly different directional effects. This is another example of the importance of the siting of multiple, independent, and complementary dark matter instruments.

As a result of this directional dependence, we should actually expect the number of nuclear recoils along a particular direction in the lab frame to change over the course of a single day. A daily modulation in directionality could be visible with a sensitive enough detector. Assuming the standard halo model, the dependence is given by

$$\frac{dR}{dE_R\cos\gamma} = \frac{\rho_0\sigma_{\chi N}}{\sqrt{\pi}\sigma_v}\frac{m_N}{2m_\chi\mu^2}\exp\left[-\frac{[(v_{orbit}^E v_{sun})\cos\gamma - v_{min}]^2}{\sigma_v^2}\right], \qquad (4.83)$$

where v_{orbit}^E is the component of the earth's velocity around the sun parallel to the sun-Cygnus line, and γ is the angle between the observed nuclear recoil and the sun-Cygnus line. The event rate in the forward direction is expected to be up to an order of magnitude larger than the backward direction [Spe88]. A detector measuring the axis and direction of the recoil with good angular resolution needs only a few tens of events to distinguish dark matter interactions from isotropic backgrounds that have no preferred direction (those events not aligned with the sun-Cygnus line).

4.4 Backgrounds to Direct Detection Methodologies

There are many known and potential sources that confound detector signatures— *background processes* or *backgrounds*, for short. The most prominent of these is

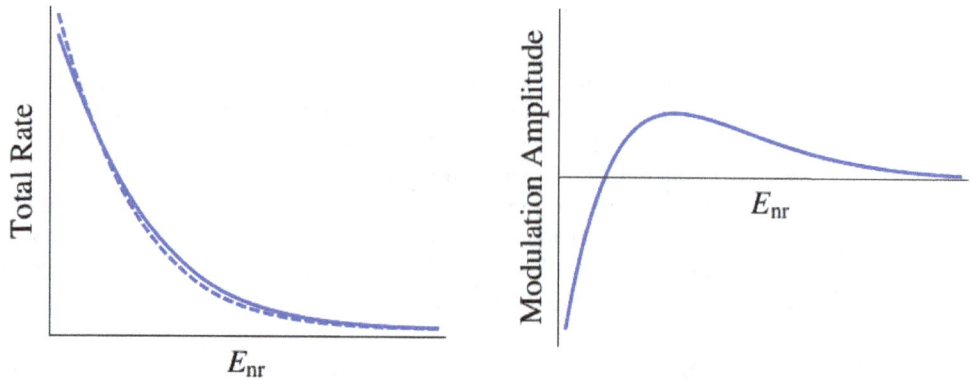

Figure 4.10: The difference in the recoil energy spectrum (left) comparing the maximum (solid line) and minimum (dashed line) times of the year for an annual modulation of WIMP interactions. The right graph shows the ratio of the modulation amplitude ΔE_R for those two cases. Figure excerpted from Ref. [FLS13].

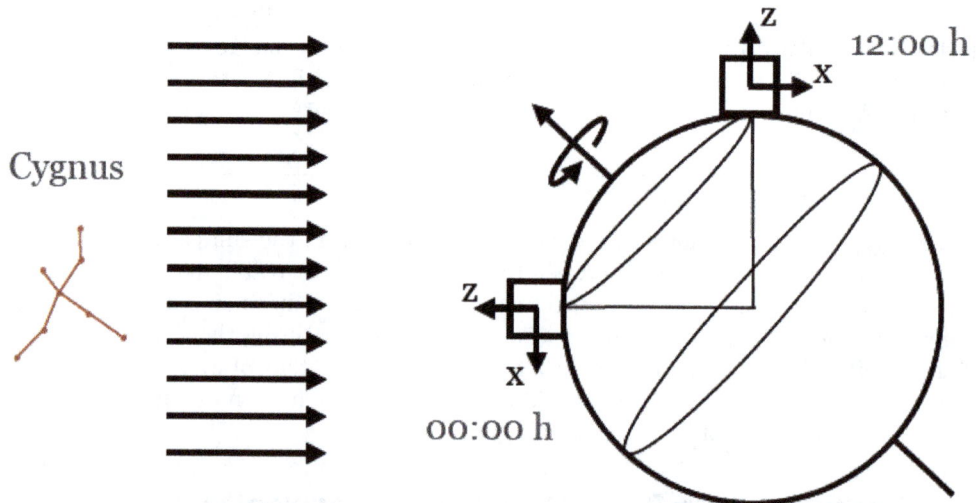

Figure 4.11: Spin axis of Earth with respect to the WIMP "wind" generated by the motion of the earth through the galaxy. Figure excerpted from Ref. [Sch11].

environmental radioactivity. This category includes airborne radon and its progeny, as well as radioactive isotopes or impurities in materials used for the detector and shielding construction. One of the worst potential backgrounds connected to environmental radioactivity is radiogenic neutrons having energies below 10 MeV (neutrons emitted during nuclear decay). These are sourced fission reactions and by alpha particle emission from other nuclei, which can lead to alpha particles interacting with a nucleus and causing the emission of a neutron (referred to as (α, n) *interactions*).

Additionally, there are cosmogenic backgrounds, such as cosmic ray muons passing through the instrumentation or, worse, *spallation* neutrons produced as a secondary by-product of cosmic rays which then enter the detector volume. Some materials, used later in construction, become activated due to natural exposure to cosmic rays or radionuclides near the earth's surface. Neutrinos from a wide variety of sources (ranging from terrestrial origins to solar and cosmic phenomena) offer an explicitly *weakly interacting low-mass background* that can interact with electrons or the nucleus. Of course, it is important to always be aware of the worst background of all: *the one nobody has yet thought of.*

Let's explore the many hierarchies of background and methods by which each might be addressed. Each background is a challenge to be addressed and an opportunity to be seized. After all, if the detector technology cannot leverage background processes to test its own capabilities (e.g., use backgrounds to calibrate or monitor the instrument) what proof can you provide that the project is sensitive to the dark matter interaction process?

Cosmic Rays

Let's explore cosmic rays. The intensity of cosmic rays increases with altitude and decreases as one goes below the surface of the earth. For this reason, direct detection dark matter experiments tend to be set in laboratories situated beneath the surface of the earth, either in specially constructed shallow underground sites or in deep sites (e.g., mines that host laboratory facilities). An overburden of rock will protect an experiment from the leading problem, direct exposure to cosmic rays. However, it will then present a secondary problem: backgrounds *generated* by cosmic ray interactions in the overburden material. A recurring theme of particle physics is that shielding becomes a target which then becomes a new problem (albeit, perhaps, one easier to solve than the original problem).

As shown in Fig. 4.12, such secondary particles can be reduced by further increasing the depth at which the laboratory housing the dark matter experiment is situated. Different sites have different geologies. For example, it can be that at one underground site (2000 m underground) the rock overburden is more dense, on average, than at another even deeper site (which may be at 2500 m depth). It can be that the shallower site, with more dense shielding, is more valuable than the deeper site with less dense overburden. Therefore, to harmonize the way scientists think about depth, it is convention to quote the depth of the overburden in units of *meters–water–equivalent.*

For example, imagine a laboratory located 1000 m underground where the average density of the overburden is 2650 kg/m^3. If we were to replace that overburden with pure water (1000 kg/m^3) at standard temperature and pressure, what would be the equivalent

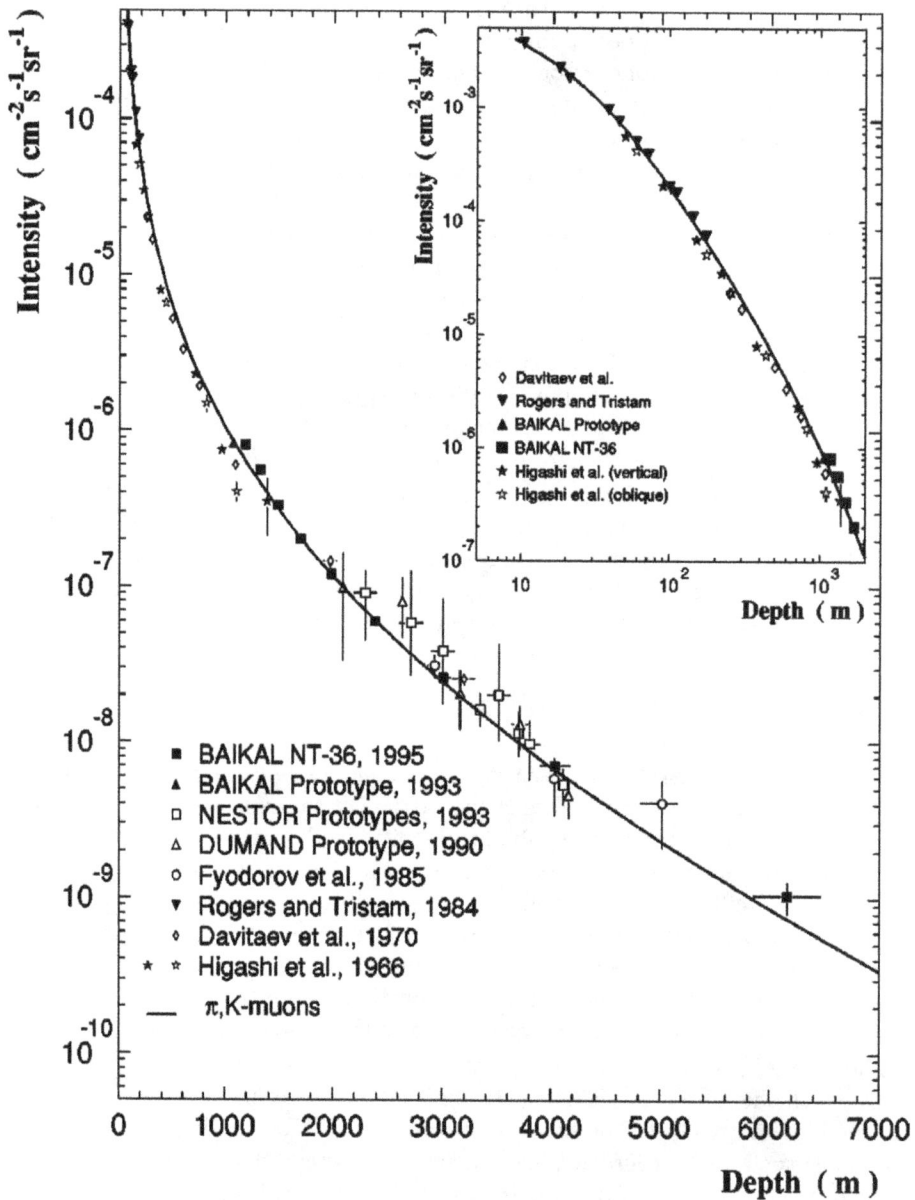

Figure 4.12: Cosmic ray muons detected by multiple experiments and shown as a function of depth in water (in meters). The inset shows the flux for shallow depths. The line shown through the points is a model based on the production of muons from pion and kaon decay, where the progenitors are produced in the original cosmic ray interaction in the atmosphere. Figure from Ref. [Bug+98].

depth of water needed? The relationship is given by this equation:

$$h_{\text{overburden}}\, \rho_{\text{overburden}} = h_{\text{water}}\, \rho_{\text{water}} \qquad (4.84)$$

For our example, the equivalent height of water corresponding to the shielding effect of this rock overburden is 1384 m. The shape of the overburden matters as well. Being 1000 m under the top of a conical mountain, which provides less shielding from cosmic ray muons entering from the sides, is not equivalent to being 1000 m beneath a wide, flat surface (where the amount of protection provided by those sides increases as one conceives of muons entering through the sides).

Many experiments also employ an *active cosmic ray veto system*, a detector subsystem designed to identify moments when a cosmic ray primary or secondary has entered the volume of the main detector system. The time information about these events can then be used to reject from consideration, activity in the primary detector system. An interaction in the surrounding muon veto system that correlates with an event in the primary detector is more likely to be due to cosmic ray activity than to dark matter.

A reasonable approximation of the cosmic ray muon rate, as a function of depth, (h), below the surface of the earth, is given by the empirical formula [MH06]

$$I_\mu(h) = 67.97 \times 10^{-6} e^{-h/285} + 2.071 \times 10^{-6} e^{-h/698}. \qquad (4.85)$$

Radioactivity Primer

Radioactive backgrounds across a range of sources are the primary antagonists to dark matter detectors. It is useful here to remind ourselves of some basic features of radioactive decay. Fundamentally, the mathematics of this process is simple. Imagine a population of N particles that are unstable, so that their number changes with time (they are lost in the process of decay). Radioactive decay is a *spontaneous process*. The rate of change (loss), $-dN/dt$, in the population of particles is observed to be proportional to the size of that population. This is expressed as:

$$-\frac{dN}{dt} = \lambda N. \qquad (4.86)$$

The *activity* of a sample, A, is defined as this rate of change of the population, $A \equiv dN/dt$. Here, λ is referred to as the decay constant (potentially quite different for each radioisotope or new unstable species). One can find a solution to this equation rather straight-forwardly: we see a function, $N(t)$, such that the first derivative of this function, with respect to time, returns the function itself, multiplied by a constant, λ. This is a classical hallmark of the *exponential function*. We guess that

$$N(t) = Ce^{Dt}. \qquad (4.87)$$

We need to solve for C and D. To solve for C, consider the boundary condition $t = 0$, where we know we have the total original population of particles:

$$N(0) = Ce^{D \cdot 0} = C. \qquad (4.88)$$

We can identify C as the size of the population at $t = 0$, which we denote N_0. To identify D, take the derivative of the solution:

$$\frac{d}{dt}N(t) = \frac{d}{dt}N_0 d^{Dt} = N_0 D e^{Dt} = DN(t) = -\lambda N(t). \tag{4.89}$$

We identify $D = -\lambda$. Thus our solution is of the form:

$$N(t) = N_0 e^{-\lambda t}. \tag{4.90}$$

λ has units of s^{-1}. As such, it can be identified with a *time constant*, $\lambda = 1/\tau$. With a little more work we can see that the decay time of the population, τ, is the time $(t = \tau)$ when 63.2% of the original population has decayed. This can be related to the *half-life*!half-life of the population, the time $t_{1/2}$ at which 50% of the population has decayed, established by the equation:

$$\lambda = \frac{\ln(2)}{t_{1/2}} \longrightarrow \tau = \frac{t_{1/2}}{\ln(2)}. \tag{4.91}$$

The initial number of radioisotopes in a population can be determined by

$$N_0 = \frac{N_A}{m_{isotope}} \times M_{isotope}, \tag{4.92}$$

where N_A is Avogadro's number, $m_{isotope}$ is the atomic mass of the radionuclide, and $M_{isotope}$ is the total mass of the population of the radionuclide. We also introduce the concept of *abundance* of an isotope, which refers to the total amount of that isotope present *relative* to the stable isotopes in a given sample of material.

It is often the case that experiments need to contend with an easily implanted radioactive contaminant (radon, for example) that can decay to other unstable nuclear isotopes. While the radon itself might not cause problems, the isotopes it produces can do so. There will be a competition between the decay of the radon (which we label as A) into a daughter isotope, which we label as B ($A \rightarrow B$), and the subsequent decay of B into another species, $B \rightarrow C$. The equation that describes this situation is known as the *Bateman Equation* and is given as

$$N_B(t) = N_{B,0}e^{-\lambda_B t} + N_{A,0}\frac{\lambda_A}{\lambda_B - \lambda_A}\left[e^{-\lambda_A t} - e^{-\lambda_B t}\right]. \tag{4.93}$$

If $\lambda_A \ll \lambda_B$ (B has a much shorter lifetime than A), it is possible to eventually produce a population level for B that is relatively steady as A will slowly feed particles into the population of B even as B decays away. This situation is known as *secular equilibrium*, and represents the limit that $dN_B/dt = 0$ such that $N_B = (\lambda_A/\lambda_B)N_A$. The population of B is very steady on short timescales, and over very long timescales that typical size of B declines slowly.

The above case of secular equilibrium is a problem for experiments if the decay of B produces final-state particles whose interactions (or other decays) can hide the dark matter signal. Elimination of the original parent species, A, is a critical solution to resolving such conditions.

Event Signatures

When radiation emitted by an unstable isotope impinges on a detector medium, what can be the signatures of these interactions? A diagram representing these major signatures, as well as experiments that employ each by itself, or signatures in combination, is shown in Fig. 4.13. We would expect two major origins for a detector event: interactions with the atomic electrons or with the nucleus. For example, photons (i.e., gamma rays) and electrons (i.e., beta radiation) will tend to scatter off the atomic electrons. Alpha particles will ionize atomic electrons, but they can also interact with the atomic nucleus. Neutrons will interact exclusively with the atomic nucleus. This latter fact is what makes them particularly pernicious: WIMPs and neutrons will both result in nuclear recoils.

The interaction with atomic electrons generates what is known as an *electron recoil*. Gamma rays are the most prevalent source of such recoils, as they are capable of penetrating deeply into material. Beta radiation (fast electrons) is less penetrating and tends to cause many recoils near the surface of detector materials; however, beta radiation can enter the *bulk* of the material to shallower depths than gamma rays and nonetheless induce recoils away from the surface.

Neutrons and WIMPs will both interact with the nucleus and generate the kinds of nuclear recoils discussed earlier. Neutron interactions appear like a 1 GeV WIMP striking a nucleus with a well-defined interaction cross-section; the remainder of their effects defined by the speed with which they strike the nucleus. You can apply Eqns. 4.49

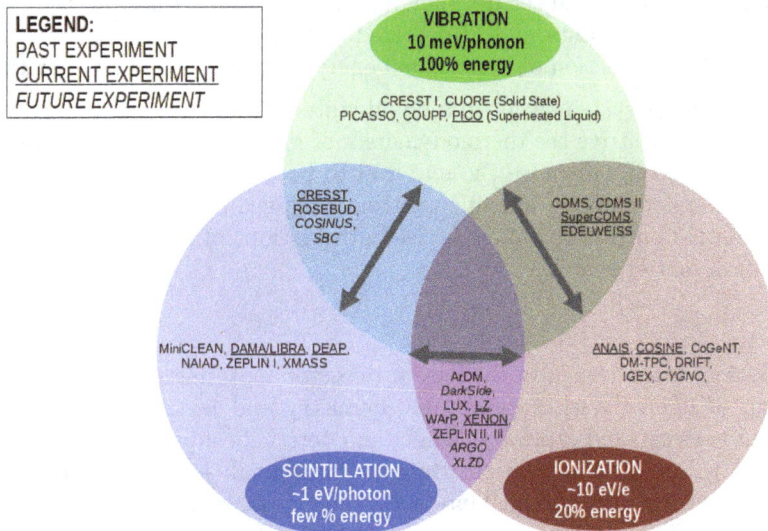

Figure 4.13: Signatures for detecting electron and nuclear recoils in materials and experiments that employ them individually or in combination. This graphic is illustrative, but by no means exhaustive. It surveys the application of signatures in past, current, and proposed/planned/future experiments.

and 4.50 using the neutron mass and its nuclear interaction cross-section, which is very well understood. It is worth noting that neutrons are not WIMPs; their primary means of interacting with a nucleus is by the strong interaction rather than the weak interaction. The probability of an interaction via the strong interaction is much higher than a weak (or weak-like) interaction if a second atom is encountered.

This difference in interactions between neutrons and what is expected for WIMPs provides a key mechanism to distinguish the two. Neutrons tend to interact with more than one atom (*multiple scattering*) when they pass through a detector medium. A WIMP, on the other hand, would be expected to (at most) interact with a single atom and is unlikely to experience another interaction (*single scattering*).

Detector materials are generally selected to emphasize a range of signatures whose relative rates vary by what is interacting in the material. A diagram representing these major signatures, as well as experiments that employ each by itself, or signatures in combination, is shown in Fig. 4.13. These signatures can be summarized broadly as follows:

- Heat

    - Phonons: for systems with a regular order (e.g. solid-state materials) having restoring forces to maintain the structure, phonons (quanta of vibrational energy) are a key signature. For these to be useful, the material must be kept extremely cold to reduce or remove normal thermal noise from the system. In a crystal lattice kept at such low temperatures, a single atomic recoil, whether electronic or nuclear, can set up phonons in the system. Phonons can carry about 10 meV per quanta and, if detected at high efficiency, can be used to infer a significant fraction of the original energy deposited in the material.

    - Superheating: for systems kept in a liquid state and under sufficient pressure, it is possible to contrive the thermodynamics of the system such that a nuclear or electronic recoil will cause the material to boil, generating small bubbles. These can be detected by imaging or sound. This is precisely the operating characteristic of a bubble chamber. Modern versions of these devices play a role in dark matter searches.

- Light

    - Scintillation: radiation absorbed by a detector material will result in the emission of light. With appropriate chemistry and instrumentation, it is possible to observe and capture this light and determine a portion of the energy present in the original interaction. However, in general, scintillation represents only a few percent of the original interaction energy so, even with high detection efficiency, one is sampling only a small portion of the original interaction.

- Ionization

    - The removal of electrons from parent atoms creates a population of free electrons that can then traverse the material. The effects of these electrons will vary depending on the material in which they are liberated. For example, in solid materials (e.g. crystals) the electrons leave behind vacancies in an atomic structure (*holes*) and either the electrons or the holes can be drifted using external electric fields and then read out using modern electronics. In liquids or solids, electrons can induce the creation of light or cause *avalanches* of other electrons and ions. These effects can be induced and amplified through application of external electric fields. Depending on the situation, different materials and technologies can infer varying degrees of the initial interaction energy. Ideally, a detector would attempt to capture the electrons and any energy emissions (photons) they cause before being stopped or captured. This is challenging and even established technologies might infer $\mathcal{O}(10)\%$ of the original interaction energy.

WIMPs and neutrons scatter off of nuclei in nuclear recoils (NR). The response of the nucleus to the primary collision can, in turn, induce effects on the surrounding electrons. These would be electron recoils (ER). Most backgrounds scatter off of electrons and cause ERs. Detectors can have different responses to NR and ER, allowing for potential discrimination of these different kinds of interactions. A problem in interpretation arises when an experiment assumes a signature was due to NR, but the reality was an ER, and vice versa. Experiments are generally sensitive to the movement of electrons and emission of light. These effects tend to be probes of how electrons recoil in material. However, if the underlying interaction was on the nucleus, then there may be (and often is) an inefficient transfer between what happened to the nucleus and what is observed, in response, from the electrons.

This leads to the concept of the *quenching factor* (QF). The QF describes the correspondence between the amount of measurable energy in a detector and these two classes of underlying events (NR and ER). An ER signal will be measured in units of keVee (*keV electron equivalent*), owing to the fact that the underlying phenomenon is the ejection of an electron. The NR signal will be measured in keVnr (*keV nuclear recoil*), corresponding to the movement of a heavy atomic nucleus (even light nuclei are thousands of times heavier than electrons). Each detector material will, in principle, have a distinct QF that allows for mapping measurable energy to the energy of the underlying phenomenon. For NR events,

$$E_{visible}(\text{keVee}) = QF \times E_{recoil}(\text{keVnr}). \tag{4.94}$$

Employing well-defined radiation sources, such as a calibrated gamma or neutron emitter whose emission energies and/or rates are known, can be used to determine the QF that relates these two energy scales. For example, gamma emitters will preferentially induce electron recoils, while neutron sources will induce nuclear recoils. The energy spectra of these two sets of recoils, taken in the same material, and taking into account the energies of the incident gamma and neutron radiation, can be combined to determine the relationship between the ER and NR scales.

Experiments can be designed to measure both ER and NR signatures. Their combination is then employed to identify regions of the ER and NR signatures rich in potential WIMP signatures (e.g. low ER, high NR) or backgrounds (e.g. high ER, low NR). One example of this is a nuclear recoil in an ultra-cold germanium crystal. Such a solid-state device allows for both the production of electron-hole pairs through ionization, and of phonons through recoils of the lattice atoms resulting from striking the nucleus of such an atom. In germanium, we find that $E_{visible} \approx (1/3)E_{recoil}$, which means that the quenching factor for germanium is $\sim 30\%$. For more details and examples of experiments that use these techniques see Section 4.7.

Another good example arises from a high-pressure liquid-state system such as a bubble chamber. In this case, we have a volume of superheated fluid at high pressure that is deliberately placed in a meta-stable thermodynamic state. A particle that interacts with the medium and deposits energy above a threshold within a critical radius will result in the formation of an expanding bubble in the medium (c.f. Section 4.7). An interaction below threshold, or one that is more diffuse in the medium, spreading beyond the critical radius, will form bubbles that immediately collapse. These chambers can be tuned so that nuclear recoils induce expanding bubbles, but electron recoils do not. For more details and examples of experiments that use these techniques see Sec. 4.7.

The rich set of signals that result from the variety of direct detection experiments lends itself to a variety of analysis approaches. These range from simple "box cuts," a series of one-dimensional selection criteria on a set of variables, $\vec{x}$, that each have sensitivity to ER, NR, or combinations of these effects to multidimensional approaches like profile likelihood analysis and machine learning.

4.5   Detector Planning and Design

In this section, we will explain the many choices that have been made in the planning and design of dark matter direct detection experiments in response to the nature of the WIMP signal and the challenges posed by the many possible background processes.

Underground Laboratories

As noted earlier, siting facilities deep underground is the key means by which cosmic ray backgrounds are mitigated. There are presently seventeen sites worldwide that are utilized as, or designed to be, underground detection laboratories (Fig. 4.14a). The cosmic ray muon rate (which leads to spallation neutron rates) is shown in Fig. 4.14b as a function of depth, with laboratory names/sites indicated on their respective points.

The hadronic component of the cosmic ray flux is negligible after only $\mathcal{O}(10 \text{ mwe})$ of overburden. Pions, kaons, and other light hadrons produced in nuclear interactions by cosmic rays, are quickly attenuated by virtue of their electromagnetic and nuclear interactions. However, cosmic ray muons can penetrate far deeper and can produce high-energy neutrons (*fast neutrons*). These, in turn, can enter the detector volume and result in keV nuclear recoils in detector material atoms. Neutron production in the vicinity of the detector is facilitated by a few processes: (a) μ^- capture (an inverse beta decay reaction that can convert a proton to a neutron, which is then ejected from the nucleus); (b) photo-nuclear reactions initiated by electromagnetic showers in the surrounding laboratory

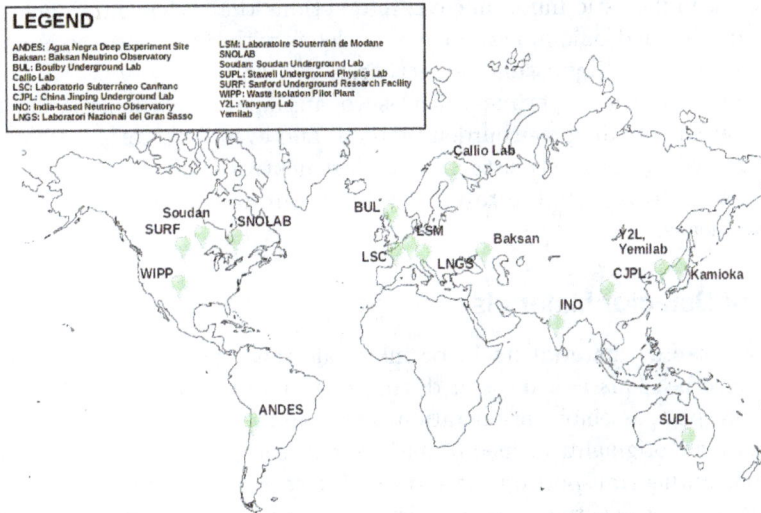

(a) Underground Labs

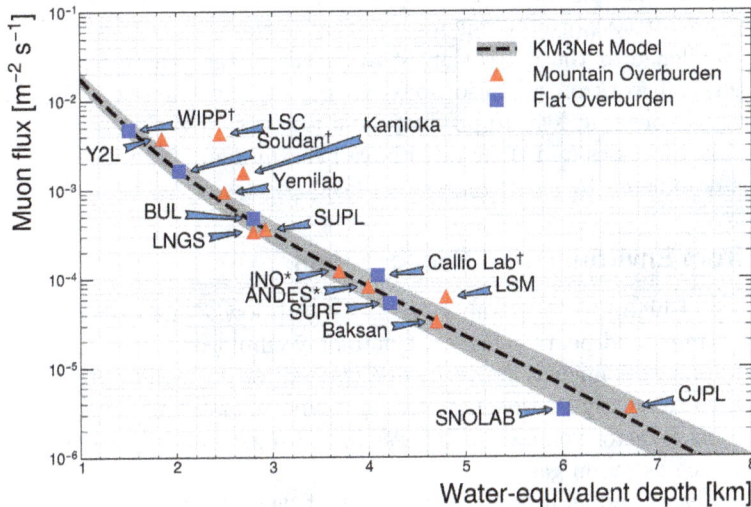

(b) Muon flux vs. Depth

Figure 4.14: (Top) A map with markers showing underground facilities. (Bottom) The intensity of muons vs. depth (in kilometers–water–equivalent) using the map legend. The symbols * and † indicate a planned lab and a lab that formerly hosted experiments, respectively. Image by the authors, with data sourced from Refs. [Bau+22; Bez+18; Dib15; Esc+05; Guo+21; Mon12; Pri+21; Trz+19; WFP24; ZM14]. The muon flux model is obtained from an empirical formula derived by the KM3NET Collaboration from its own data [Age+20]. Map graphic by authors using artwork from Refs [mig12; woo06].

materials; (c) deep-inelastic muon-nucleus interactions that "cleave off" neutrons from the nucleus; and (d) hadronic interactions of nucleons with pions and kaons produced in hadronic showers induced by cosmic ray activities.

The challenge to the experimentalist is to anticipate fast-neutron backgrounds, recognizing that surrounding overburden material and even detector shielding materials (e.g. lead) can present production targets for fast neutrons. The experimentalist must then identify ways to predict the rate of such backgrounds or reject them by passive and/or active means.

Activation of Detector Materials

Activation of detector material itself, or materials that end up close to the detector is another concern. This can happen during one or more phases of detector design, fabrication, transport, assembly, or operation. For example, raw materials used in detector construction, even if originally purified to remove radioisotopes, can be activated by cosmic ray interactions during transport on the surface of the earth. In addition, the routes taken, and their location on the surface, can matter. For example, cosmic ray spectra vary with geomagnetic latitude and flux varies with height above Earth. Choosing a mountainous shipping route, for example, may lead to much higher activation of materials than if the same material were transported along a route that is close to sea level.

Another challenge in the activation of materials is that there is, despite nearly a century of detailed work on radioisotopes, an incomplete understanding of the cross-sections of interactions that lead to production of isotopes. It is simply a fact that not all such production processes and their rates have been measured. In general, radioisotope production is dominated by (n, x) reactions (95%) and (p, x) reactions (5%).

Shielding from Environmental Backgrounds

A combination of high atomic number (*high-Z*) with low-Z materials are employed to diminish the gamma and neutron fluxes. Commonly employed materials are lead (prized for its density of target nuclei), polyethylene (prized for its high neutron capture cross-section owing to the many hydrogen atoms in its molecular structure), and copper[4]. In addition, purging detector volumes with gases like nitrogen can help to drive out airborne radioisotopes such as radon gas.

Active and passive shielding are a key part of dark matter detector design. Some experiments employ large water shields to passively reduce or moderate both environmental radioactivity (e.g. gamma rays from surrounding unstable nuclei) and muon-induced neutrons. In the latter case, water's hydrogen (with its lone, unpaired proton at the nucleus) provides an easy way to absorb neutrons by capturing them as they pass through

---

[4]Even though, per unit volume, it is expensive, copper is often a material of choice for dark matter experiments for at least two reasons. First, it has a high density and provides excellent shielding from external radiation. Second, it has excellent thermal and electrical conductivity. For low-temperature experiments, this can play a key role in the removal of excess heat from the system. The electrical conductivity may also be useful for shielding against external electromagnetic interference. Copper is also less toxic than other shielding materials (lead) and still easy to cut and shape into a desired configuration.

the shield. Current efforts have determined that a water shield with a thickness of 1-3 m can reduce underground fluxes of gamma radiation, or radiogenic fluxes, by a factor of $\sim 10^6$. An active muon shield, employing materials doped with scintillators (e.g., boron or gadolinium) that emit light when traversed by charged particles, can be used to provide a signal coincident with the passage of the background particle. For example, a nuclear recoil (induced by a neutron rather than a WIMP) could occur when a muon passes through the detector and spallates a neutron into the bulk detector material. An active muon shield would "tag" the passage of that muon with high efficiency, allowing experimenters to veto from consideration any nuclear recoil activity within a corresponding time window.

Active dark matter detector volumes, nestled within a layer of active or passive shielding, can also provide their own shielding capabilities. This is known as *self-shielding*. For example, a large volume of liquid xenon employed for ionization and scintillation observations will also generally absorb exterior backgrounds close to the surface of the xenon volume. Experimentalists who design such detectors with good three-dimensional position capabilities can identify these *surface events* and veto them. This is typically done by using external calibration sources to bombard the material, observing where most of the interactions then occur and define a veto region in some outer volume of the active material. Of course, this self-shielding effect comes at a cost: the active mass considered in the dark matter search is reduced by excluding a volume considered to be enriched in potential background. To compensate for the loss of target mass, experimentalists may have to run the experiment much longer (see Eqn. 4.50), as hinted at in the definition of *exposure* (Eqn. 4.46). A lesser amount of available target mass needs an increase in operational time to achieve the same exposure level.

Internal Radioactivity in Detector Material

No amount of external shielding can protect a dark matter experiment from itself. For example, any uranium, thorium, potassium, cesium, cobalt, argon, or krypton (to name just a few) atoms embedded in the active detector material, or in nearby shielding material, can result in radioisotope decays that generate backgrounds. Most of these atoms have naturally occurring unstable isotopes in abundance. Some of them have unstable radioisotopes as a small fraction of their more abundant stable isotope(s).

Elimination of radioisotopes from the detector construction process requires a diverse portfolio of techniques. In some cases, purification methods can be used *in situ*, as is the case for experiments using liquid xenon as a target. Large distillation columns are used to refine the liquid xenon, bringing the levels of krypton and radon into acceptable ranges. When *in situ* purification is not feasible, materials have to be characterized and selected prior to their use in the construction. Care needs to be taken to avoid activation after construction. Detailed handling records (including altitude, duration of time spent outside of shielding, exposure to atmosphere, and so forth) will aid in later recognizing materials that may have become overly contaminated. This challenge is made particularly large because modern dark matter experiments are so sensitive that radioisotope backgrounds have to be reduced below the parts-per-billion (ppb) level.

(a) The XIA Alpha screening facility at Southern Methodist University (SMU) in Dallas, Texas. This instrument is now at SNOLAB.

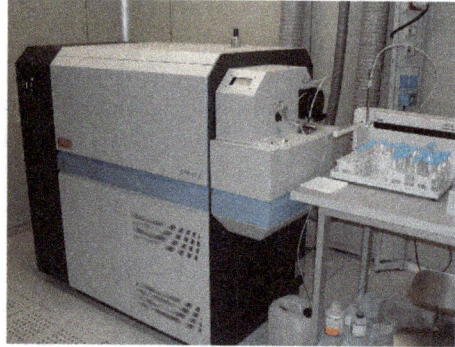

(b) ICMPS at the Gran Sasso Laboratory in Italy, beneath a mountain, approximately 75 miles northeast of Rome [S N+17]

Figure 4.15: Two of the many facilities at universities and laboratories around the world used for pre-screening materials for use in dark matter experiments. Images originally from [Bel04].

Active screening of materials (e.g. using drift chambers or other detector instrumentation that can count radioactive decays) is a part of this process. For example, highly sensitive alpha counters (Fig. 4.15a) can be used to estimate how much activity is implanted at or near the surface of small amounts of detector material. This observation can then be extrapolated to the activity of all detector volumes constructed from that material. Another highly sensitive technique is inductively coupled plasma mass spectrometry (ICP-MS) (Fig. 4.15b) which can detect very low levels of radioisotopes in samples. A combination of these approaches is the most common approach to insure compatibility among independent means of surveying material contamination.

Screening facilities utilize combinations of two major practices to establish a workflow for material assays. One approach is to augment commercially available systems. For example, a commercially purchased high-purity germanium (HPGe) detector can be placed in a custom-designed shield consisting of lead, copper, neutron moderation and capture materials, an active cosmic ray veto, and/or an underground location. Alternatively, a custom system can be developed by a research group or a multi-institution collaboration.

The need to keep detectors very cold means exposing them to *chiller* systems. Such systems are commercially available, but are not always suitable for use in dark matter experiments. For example, off-the-shelf industrial solutions have probably not been constructed from ultra-pure, low-radioactivity materials. Placing these in close proximity to the instrumentation, thus optimizing their cooling abilities, can put the experiment at risk. This necessitates the construction of custom cryostat designs. Attention will be paid to the design and placement of electronics to minimize background sources (such as uranium, thorium, or potassium that is naturally found in metals or materials used in electronics construction).

Isotope/Chain	Standard Size (ppb) | (mBq/kg)		Large Size & Long Count (ppb)
238U	~0.1	~1.0	0.009
232Th	~0.3	~1.5	0.02
40K	~700	~21	87
238U	0.001	0.12	
232Th	0.001	0.004	
40K	1	0.031	

Technique	Sensitivity
Radon Emanation	0.1-10 Bq/kg (Ra)
Immersion Whole Body Counters	10^{-13}-10^{-14} g/g (U/Th)
ICPMS (Inductively Coupled Plasma Mass Spectrometry)	ppt to ppt (U/Th/K)
SIMS/GDMS (Secondary Ion & Glow Discharge Mass Spectroscopy)	1 ppb (SIMS) 10-100ppt (GDMS)
AMS (Accelerator Mass Spectroscopy)	< 1 ppt
Neutron Activation Analysis	100 pg (U), 10 ng (K)

(a) Isotope Sensitivities   (b) Technique Sensitivities

Figure 4.16: (Left) A comparison of material screening capabilities for various isotopes showing the difference between augmented commercial approaches (top three rows) and custom approaches (bottom three rows). Standard-sized samples are compared between both approaches, while large-sized samples with long counting times to integrate radiation exposure to the instrumentation is shown only for an augmented commercial approach. (Right) A comparison of material screening approaches and their current standard sensitivities.

A comparison of commercial and custom approaches is shown in Fig. 4.16a. Many screening options exist (Fig. 4.16b), and, no doubt, many more will be developed, as detector technology sensitivity necessarily advances to meet the challenge of detecting dark matter interactions.

Modeling Radioisotope Backgrounds

Screening materials afford a chance to estimate radioisotope contamination of small samples of the detector. Materials-handling records and tracking affords a chance to estimate exposure to natural radioisotopes such as radon, or activation by cosmic rays. However, it is impossible to monitor the entire instrument *in situ* while it is operating. In this event, a computational model of the possible backgrounds and their impact on the experiment must be developed. These models employ fundamental physics, including the details of particle interactions in material (cross-sections), data from material screening and material handling information, data from the experiment itself, and experience with earlier instrumentation derived from past phases of experimentation.

Three software frameworks exist to calculate the spectra of neutrons produced by $(\alpha - n)$ interactions.

- SOURCES (using the EMPIRE2.19 libraries for cross-section inputs)

- USD WebTool (TENDL 2012 libraries which are validated by TALYS for cross-section inputs)

- NeuCBOT (utilizing TALYS for cross-section inputs)

TENDL is a validated library and EMPIRE is recommended by the International Atomic Energy Agency, but neither can properly calculate all resonant behavior that is experimentally observed. The output energy spectra from these simulations can be used in a full simulation of the experiment to predict the number of background events from neutrons. Validation of these approaches [Coo+18] is essential before applying them to background estimates in current or future experiments. For example, SOURCES-4A and USD agree on the neutron yield (neutrons/s/cm^3) from copper metal within about 20% of each other (including separate predictions of the neutrons from the ^{238}U and ^{232}Th decay chains). However, for a material like stainless steel, these tools only agreed within about 50%.

Another example is to estimate the neutron fluxes from radioisotopes in photomultiplier tube (PMT) glass [WM17]. PMTs are a common technology for detecting and amplifying light emissions in detectors (either as a mechanism for primary detection of dark matter signatures or in an active muon veto system). The NeuCBOT (SOURCES-4C) framework would predict 15 (13) n/year from PMT glass. Within the ranges of constraints on these calculations and uncertainty on the underlying processes, these are fair comparisons.

Predicting neutron fluxes from materials continues to be a circumstance in which significant improvements are needed to keep up with the advances in detector sensitivity. While existing predictions may be sufficient for current or near-generation experiments, that will not continue to be true indefinitely.

Simulation frameworks such as GEANT4 [Ago+03; All+06; All+16] can be used to develop detector and material geometry and response models. Those models can then be subjected to the expected fluxes of radioisotope backgrounds. An example of such a model for the SuperCDMS dark matter search experiment is shown in Fig. 4.17. SuperCDMS is a good example of the type of layout we have thus far described in this text. An exterior series of shields that include lead, polyethylene, and ancient lead depleted of radioisotopes, encloses the cryostat which houses the active dark matter detector volume and maintains it at micro-Kelvin temperatures. The active detector volumes are stacks ("towers") of hockey-puck-sized single crystals of germanium. Each crystal is photolithographically etched with a pattern of sensors for ionization and phonon readout.

This complexity can be laid out in GEANT4, including gaps (absences of any material in a space), inert material regions (electrical wiring or plumbing that must pass through the shield to serve the inner components), and other realities of the detector. Each material can be modeled, along with its expected, or measured, level of radioactivity. A simulation of what the detector would see over a period of time can then be generated. From this, the energy recoil spectrum, derived purely from background processes, can be produced. This can include radiogenic backgrounds, cosmic ray-induced backgrounds, neutrino backgrounds, and others not yet discussed (Fig. 4.18). Signal events (dark matter nuclear recoils) will be defined by experiment—and analysis-specific thresholds. In the example shown, a dark matter nuclear recoil is counted only for $E_R > 0.27$ keVnr. A 10 GeV/c^2 WIMP candidate can deposit up to $\mathcal{O}(10)$ keVnr; for the highest such recoil energy regions, these models predict just fractions of a background event, per kilogram of material, per keVnr, per year.

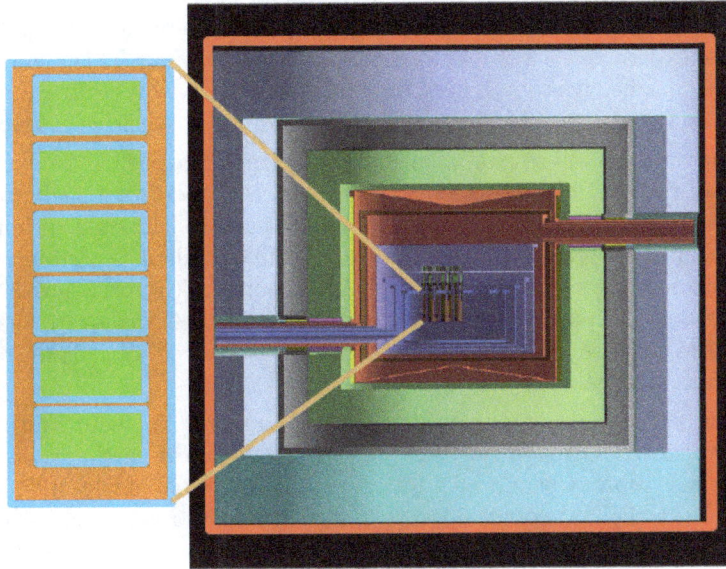

Figure 4.17: A GEANT4 model of the SuperCDMS dark matter detector. Image provided by the SuperCDMS Collaboration.

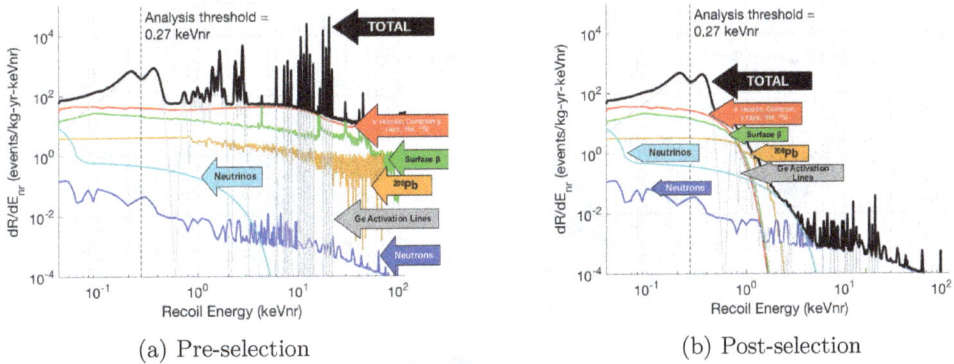

(a) Pre-selection

(b) Post-selection

Figure 4.18: Figures from Ref. [Agn+17a]. Background spectra, before (left) and after (right) analysis cuts in germanium iZIP detectors, shown as a function of nuclear recoil energy (keVnr) to allow direct comparison of the various backgrounds to the natural dark matter interaction energy scale. The highest line represents the total background rate. This is made from components labeled in the figures: electron recoils from Compton gamma-rays, ^{3}H and ^{32}Si; Ge activation lines, convoluted with a 10 eV r.m.s. resolution; surface betas; surface ^{206}Pb recoils; neutrons; and neutrino interactions.

Designing Radiopure Materials

It is increasingly the case that experiments cannot use commercial-grade materials in the construction of experiments. For example, let's consider copper. This metal is essential in cryostat design, since it has excellent thermal conductivity. Its electrical conductivity makes it equally valuable in the construction of shielding for sensitive electronics.

Copper, however, poses several challenges when used in dark matter experiments. First, it is mined from the earth and refined. This means copper is naturally contaminated by radioisotopes, primarily uranium and thorium. Radioactive elements such as radon are known to implant on copper when it is left in open air. While it is certainly possible to identify manufacturers who produce highly refined copper with purity levels of the most stable copper isotope, copper-79, far better than 99% (c.f. Ref. [Bal+21]), such materials may, nonetheless, require additional refinement to further reduce natural radioisotope contamination to acceptable levels.

There is an additional complication regarding copper refinement and purification: nearly all such industrial capability is sited on the surface of the earth, leaving it exposed to cosmic ray radiation. Cosmic rays interacting with copper nuclei induce a nuclear reaction that leaves behind cobalt-60, a potential source of gamma and beta radiation backgrounds. Even ultra-pure copper, exposed to cosmic rays, can be induced by this process to produce radiation far in excess of the tolerances of dark matter experiments. It is estimated that this effect adds 0.4 μBq/kg/day of excess radiation to copper samples that would then be utilized in the construction of sensitive equipment.

A controlled electroforming process can prevent naturally occurring contaminants from playing a significant role during the formation of the metal by using a bath of copper ions (Fig. 4.19) in acid. This manufacturing process results in copper with ≤ 0.1 μBq/kg activity levels from either the uranium or thorium chain.

(a) Copper nuggets prior to electroformation

(b) Copper cylinders from electroformation

Figure 4.19: (Left) Copper nuggets prior to their use in copper electroforming. (Right) Copper cylinders resulting from the electroformation process. Images by PNNL researchers Eric Hoppe, Brian LaFerriere, Jason Merriman and Nicole Overman and made available courtesy of Pacific Northwest National Laboratory.

There are not strong incentives for metals manufacturers to push copper purity to higher levels or to move their production facilities underground to prevent the creation of cobalt-60 (and other shorter-lived isotopes not discussed here). At the time this book is being written, there are not strong market-based incentives for companies to invest the additional capital required to achieve the needs of the dark matter community. The philosophy of physicists in this kind of situation is straightforward: *if you cannot buy it, make it yourself.* For example, a facility at the Pacific Northwest National Laboratory (PNNL) has been constructed to electroform copper in a controlled manner [Ove+12], and another site was constructed at the Sanford Underground Research Facility (SURF) in South Dakota for underground copper electroforming [Hei22]. Other such facilities exist or are being planned in other locations in the world, but there are presently only a few.

Copper is but one material found in the construction of dark matter experiments in which it tends to play only a "supporting role." What about the material for the dark matter target? An example of a problem therein lies in the purification of noble gases and liquids for use in detector volumes.

Xenon is utilized in both dark matter and ultra-sensitive neutrino experiments. High purity is necessary in order to suppress the radioisotope backgrounds that would otherwise render these instruments unusable. The XENON Collaboration designed, built, and operated their own xenon purification facility at Purdue University [and16]. This system can both purify or distill the noble gas. Whereas commercial xenon is contaminated by krypton at the level of 1 ppm–10 ppb, the XENON Collaboration's current-generation 1-ton detector depends on levels at or below 0.2 ppt to meet its design goals. This is approximately 100,000–500,000 times higher purity than commercial-grade xenon. This is only achievable with the Collaboration's custom distillation and refinement approach, consisting of a 5.5 m distillation column that can generate 6.5 kg of refined xenon per hour.

Neutrino Interactions in Dark Matter Detectors

The neutrino is a low-mass, weakly interacting particle. In principle, it is a perfect test of whether a dark matter detector is capable of observing a weak-scale (or weaker) nuclear interaction such as we hope occurs between dark matter and standard model matter. The challenge with neutrinos is that they are relativistic—they travel at nearly the speed of light—because of their extremely low masses. The sum of all three neutrino mass eigenstates is constrained by the CMB to be less than 0.12 eV [Agh+20a][5]. While the individual mass eigenstates' masses are not known, they will clearly behave like low-mass or nearly massless WIMP-like candidates.

The most copious nearby source of neutrinos is the sun. The fusion reactions in its core, particularly the *pp* chain of reactions, generate high fluxes of low-energy neutrinos

---

[5]It merits comment that, if there are more than three neutrino species in the universe, the constraint on the sum of their masses is altered. There is no direct or indirect compelling evidence for more than three low-mass neutrino species. However, it is important to note that heavy neutrinos would elude this constraint (especially depending on their interactions). If stable, such particles could make excellent dark matter candidates.

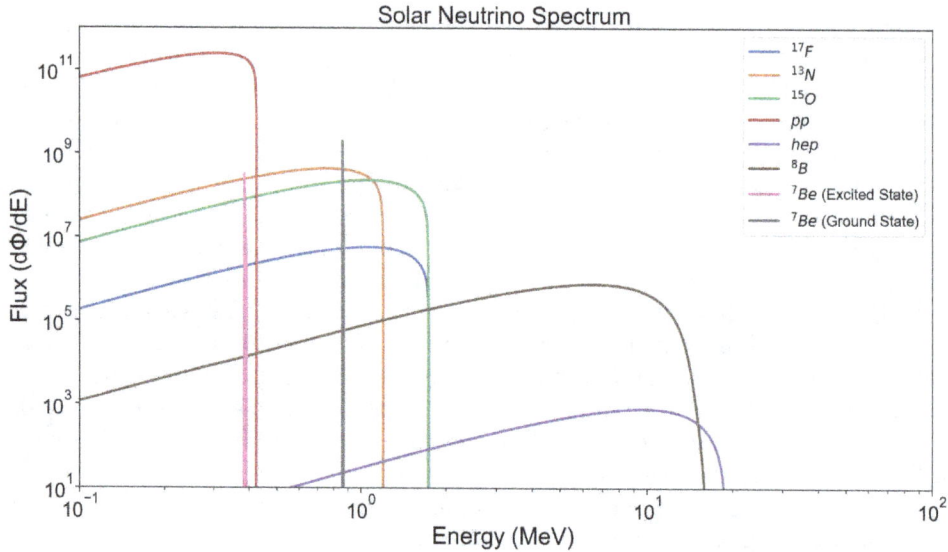

Figure 4.20: A few components of the solar neutrino flux spectrum. Data are from Ref. [BS05].

(Fig. 4.20). These primarily contribute to the ER background via $\nu - e$ scattering at a level of 10 to 25 events per tonne of detector material per year at low energies. The lower-rate ^{8}B fusion neutrinos have higher energies, and, as a result, their NR cannot be distinguished from WIMP signals. Those are expected to lead to 1,000 events per tonne of detector material per year of operation. This will especially affect experiments using heavy targets such as germanium.

There are also atmospheric neutrinos produced by cosmic ray interactions (e.g., muon decay in the atmosphere). Cosmic ray collisions with air molecules produce pions that subsequently lead to the production of muon and electron neutrinos and antineutrinos. Current and upcoming direct detection dark matter experiments are primarily susceptible to the component of the atmospheric neutrino spectrum whose energy is less than 100 MeV.

There are cosmic sources of neutrinos from beyond the solar system. There exists a diffuse background of neutrinos from supernova explosions. There also exist higher energy neutrinos than even those produced in solar cores or stellar deaths, but their flux is even lower. These are expected to contribute at the level of 0.01–0.05 events per tonne per year [BFS14].

As dark matter direct detection technologies continue to advance in sensitivity, the neutrinos will provide a "fog" of electron and nuclear recoils that will have to be modeled with increasing precision. The best approach, of course, would be a dedicated instrumentation or *in situ* estimation of this background. These approaches would allow for the neutrino contributions to be subtracted from the data to isolate any WIMP nuclear recoil signatures. Such backgrounds are expected to be observed in the generation of experiments that began operation in about 2023 and that will continue into the 2030s.

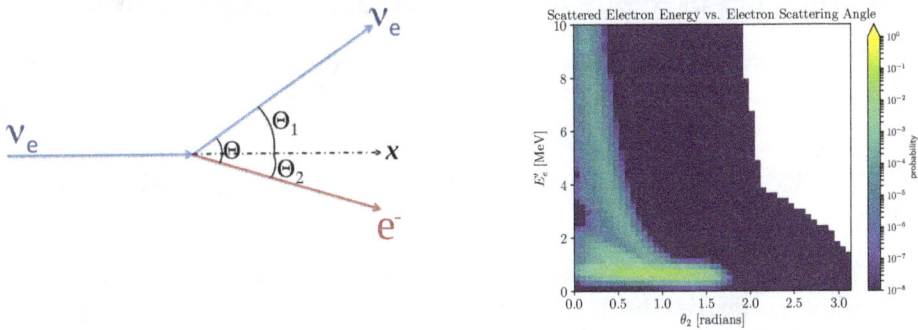

Figure 4.21: (Left) A model of the leading-order scattering of an atomic electron by an incident neutrino (e.g. a solar neutrino). Image from Ref. [HS16]. (Right) The degree of correlation between the ejected electron energy and the scattering angle with respect to the incident neutrino direction (data from private communication).

Coherent neutrino scattering events are clearly on the horizon and the community is fully engaged in working to meet the forthcoming challenge. This effort will nevertheless benefit from new ideas and technologies.

Directional detection capability, which we explore more in Section 7.4, may prove to be a crucial tool for identifying and rejecting the solar neutrino background in future experiments. Neutrinos from the sun would, ideally, result in directional interactions; e.g., a recoil (from an ejected electron or a struck nucleus) that points back toward the sun. The reality of such scattering processes is that a neutrino that ejects an atomic electron will not do so only in a way that guarantees that the the final-state electron points along the direction of the original neutrino.

The degree of directional correlation between a neutrino's trajectory and a scattered atomic electron has been explored at leading order in the weak interaction at the level of mathematical constraint [t H71] and Monte Carlo simulation of electron-scattering distributions [HS16]. The correlation between the scattering angle of the final-state electron and the incident angle of the original neutrino, as a function of the energy of the neutrino and folding in the solar neutrino spectrum, is illustrated in Fig. 4.21. (This plot is based on the cited work. It is not published and is provided with permission via a private communication.) There is a reasonable expectation that, at the 68.3% (95.4%) confidence level, the scattered electron will be within 1 radian (1.4 radians) of the incident neutrino direction. In Gerard t'Hooft's original work on the subject, he calculated that the scattering angle should trend as $\theta \approx \sqrt{m_e/E'_e}$, where E'_e is the energy of the scattered electron (all in natural units). This trend is evident in Fig. 4.21. This relationship may provide sufficient information for future directional detection experiments to veto or categorize such events as being more *neutrino-like* than dark matter-like.

4.6 A Summary of Needs for Direct Detection Experiments

Direct detection experiments will clearly need to establish the following characteristics to make progress:

- The ability to see low-energy dark-matter-induced recoils (< 10 keV). This will require

    - Radiogenic purity
    - Low energy thresholds

- Ability to distinguish nuclear recoils

    - Differentiate between electronic and nuclear interactions
    - Differentiate between alpha radiation interactions and nuclear recoils

- Radiogenic and cosmogenic backgrounds mitigation

    - Passive and/or active shielding from these backgrounds
    - Position reconstruction and fiducialization
    - Characterization of these backgrounds

- Long exposures with stability on similarly long timescales

    - This will be especially true to observe annual and diurnal modulation

The search space for dark matter is, in principle, vast. A variety of theoretical models provide compelling candidates for dark matter. The space of their possible properties (e.g., cross-section vs. mass) is shown in Fig. 4.22. A non-exhaustive list of models includes the Minimal Supersymmetric Standard Model (MSSM) [Cah+12], Constrained MSSM (CMSSM) [CW13], Next-to-Minimal Supersymmetric Standard Model (NMSSM) [Cao+14], and Asymmetric Dark Matter [LYZ12]. Existing experimental constraints can be overlaid on this space to demonstrate what has, and what has not, been excluded by the current portfolio of projects. We caution the reader to not over-interpret the model exclusions visible in illustrations such as Fig. 4.22; theoretical frameworks such as supersymmetry have many free parameters and simplified supersymmetry models ignore the full extent of such theories. The model points on such plots indicate where simplified calculations have, so far, provided potentially promising search spaces. However, these models also sample *very little of the available parameter space* in such frameworks.

The best way to meet the challenge of a vast parameter space is a large suite of experimental approaches. Examples of experiments that have failed to detect dark matter candidates and whose data are then used to exclude parts of a specific model space are shown in Fig. 4.23. We will conclude this chapter with a look at several ongoing or planned dark matter direct detection experiments. This tour will cover most of the technologies that are illustrated in a graph like the one shown in Fig. 4.23.

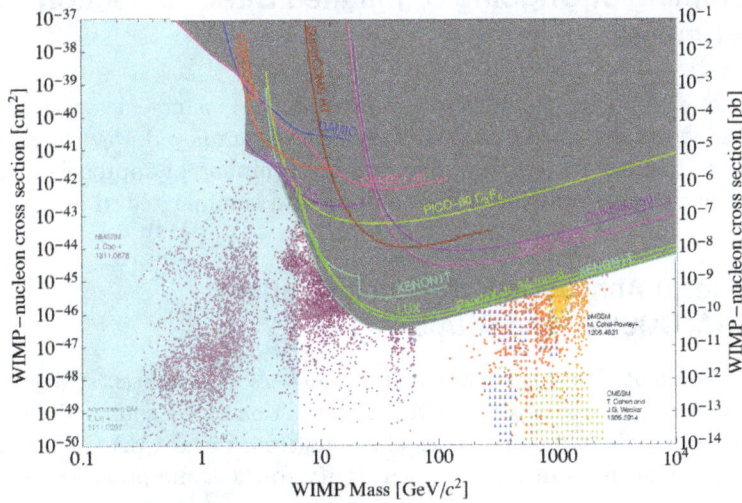

Figure 4.22: An example of the theory-driven search space for dark matter across a portfolio of plausible theoretical models. Superimposed exclusions of that space are based on current experimental efforts. Figure generated using [Tar24].

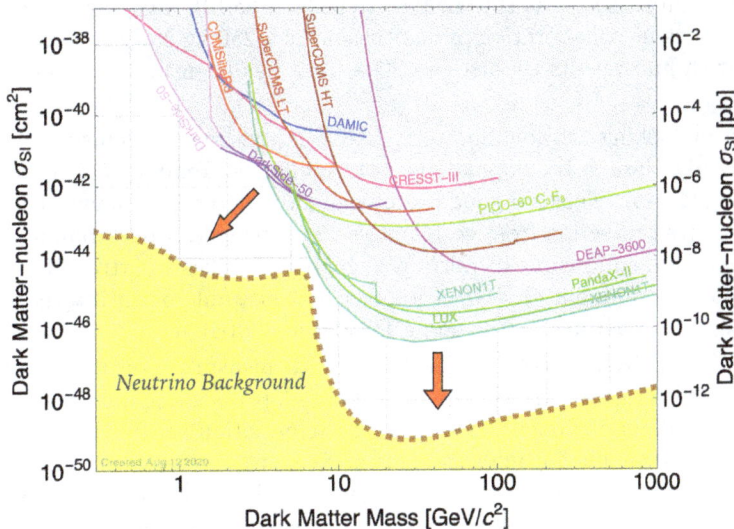

Figure 4.23: An example of an experimental exclusion plot comparing spin-independent cross-section versus candidate mass. The neutrino "fog" background expectation is shown as a shaded contour at the bottom of the plot. The arrows indicate the directions in which forthcoming experiments will push into the parameter space seeking better sensitivity at low mass and increased sensitivity at high mass. Current versions of this plot are available at the end of this chapter. Image generated using [Tar24].

4.7　A Sampling of Ongoing or Planned Direct Detection Experiments

This account of experiments is by no means exhaustive. It is not meant to be an historic record of such efforts nor a road map of every concept employed now or planned for the future. Rather, it is meant to provide the reader a sampling of approaches whose design will echo familiar themes outlined in this chapter and summarized in Sec. 4.6.

Time-Dependent Annual Dark Matter Modulation: DAMA/LIBRA, COSINE, and ANAIS

A very fair question at this stage in the text is: *have we already seen strong evidence of a dark matter signature in any experiment?*. An illustration of the complexity of the subject of direct detection is the unfolding series of results from the DAMA/LIBRA experiment and two independent instruments with sensitivity to the same processes, COSINE-100 and ANAIS.

First DAMA, and, later, DAMA/LIBRA, have reported positive results in the search for an annually modulated dark-matter-like signal going back to 1996. The DAMA phase of this experiment was a 100 kg NaI (sodium iodide) crystal array operated in Laboratori Nazionali del Gran Sasso (LNGS) from 1996 to 2002. The LIBRA phase of the experiment, operated by the same collaboration of institutions, is a 250 kg NaI crystal array operating since 2003, with first results reported in 2008, that has been updated periodically since then.

We begin by looking at the laboratory. LNGS is constructed beneath the Gran Sasso mountain in Italy, close to the populations of L'Aquila and Teramo, at a depth of just over 3000 mwe (Fig. 4.14b). This is a good example of an underground mountain laboratory at which there is less protection from cosmic rays from the sides than from above. However, its location along a subterranean portion of highway makes it extremely accessible for both people and equipment. LNGS is an excellent example of an international facility hosting a range of experiments, including DAMA and LIBRA.

The DAMA/LIBRA instrument is designed to measure scintillation from particle interactions in its detectors, specifically in thallium-doped sodium iodide crystals, denoted NaI(Tl). The thallium dopant provides a strong scintillation effect. However, scintillation is the only signal available for analysis. This prevents experimentalists from being able to discriminate among a wide range of background hypotheses and the possible dark matter signature. The primary effort of their work has been to search for such a time-dependent (annual modulation) effect on top of a large background (e.g. a modulation with the right frequency and phase, see Section 4.3). This means accounting, through simulation, for time dependence in the possible backgrounds. Possible potential backgrounds include a time dependence in the annual cosmic ray rate, and in radioisotope backgrounds due to freeze/thaw cycles around the mountain laboratory that trap/release radioisotopes throughout the year. The experiment has no direct discrimination between nuclear and electron recoils.

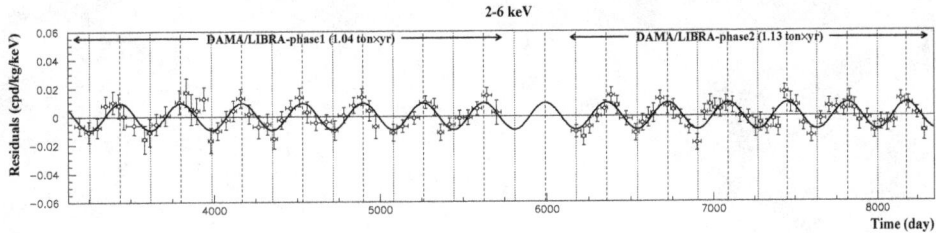

(a) DAMA/LIBRA Annual Modulation

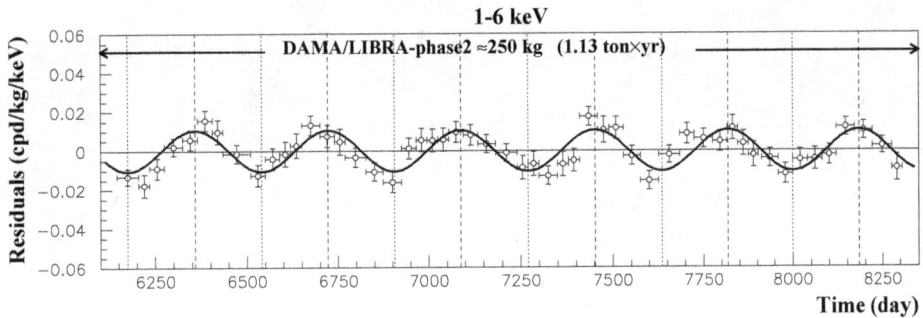

(b) LIBRA-only Annual Modulation

Figure 4.24: Results from the DAMA/LIBRA experiment. (Top) The full fourteen-cycle DAMA/LIBRA annual modulation result after subtracting flat background contributions. This analysis uses a higher energy threshold in the range of $2 - 6$ keV of recoil energy. (Bottom) The LIBRA-only results allowing for a lower-energy threshold of $1 - 6$ keV of recoil energy. Figures extracted from [Ber+18].

They have reported a signal observed over fourteen cycles with a significance above the null (no modulation) hypothesis of 12.9 σ in the $2 - 6$ keV energy range (Fig. 4.24a). The collaboration was also able to lower their energy threshold with the new LIBRA phase of the experiment and have reported from that more limited sample alone a 9.5 σ signal for single-scatter events in the $1 - 6$ keV energy range observed over six cycles (Fig. 4.24b).

Again, the instrument has no *in situ* background discrimination capabilities. The collaboration has interpreted their observations in various ways, resulting in regions of the plane shown in Fig. 4.22 where we would expect to observe, with other independent instruments, those signals. However, independent experiments using very different technologies have not reported observing candidates with the expected masses and cross-sections. This has prompted a robust and extended debate over whether the DAMA/LIBRA background models or analysis methods are comprehensive, or whether the dark matter interpretation is tied directly to the NaI technology of DAMA/LIBRA and is not transferable to other experiments.

Figure 4.25: The COSINE-100 detector. Photograph courtesy of the COSINE-100 Collaboration.

Over the past two decades, with significant activity since 2015, there have been efforts to construct experiments to directly cross-check the claims of the DAMA/LIBRA collaboration. Earlier efforts may have been dismissed using arguments about atomic, nuclear, or technological differences, or through dark matter models that allow for a signal exclusive to the DAMA/LIBRA experiment. More recently, independent collaborations have concentrated on developing high-purity NaI(Tl) crystals (to at least the level claimed by the DAMA/LIBRA collaboration in their own instrumentation) for use in similar scintillation-based experiments. An independent but similar suite of experiments will close out the arguments about competing projects being too unlike DAMA/LIBRA in technology choices.

Although we will focus with recent results on just two experiments whose methodologies are intentionally similar to DAMA/LIBRA, it should be noted that there are other collaborations working on the cross-check, including the Sodium-iodide with Active Background RejEction (SABRE) project which plans to deploy detectors in both the northern and southern hemispheres (*See* [Ant+19; Cal+21]).

The first cross-check is the COSINE-100 experiment [Adh+19] (Fig. 4.25). The instrument was originally located in the Yangyang Underground Laboratory (Y2L) in South Korea, a mountain laboratory. At its most shielded point, the experimental site lies under about 700 m of rock (1800 mwe) [Zhu+05]. It consists of eight copper-encapsulated NaI(Tl) crystals. The total active detector mass is 106 kg. The detection of scintillation light is accomplished by two three-inch photomultiplier tubes per crystal, with an event trigger set at the level of a 0.2 photoelectron threshold. Calibration of the instrument is accomplished through insertion of sources close to the detectors via access tubes. The total background expected in this instrument is at the level of $2-4$ times the DAMA/LIBRA average, or about 2.7 counts/day/kg/keV on average in the same $2-6$ keV energy region. One favorable comparison to DAMA/LIBRA is that the *in situ* uranium, thorium, and potassium contamination in these NaI(Tl) crystals is below the levels reported by DAMA, while the ^{210}Pb contamination is at nearly the same level. This crystal detector design permits for a very high light yield during interactions in the material.

COSINE-100 has performed a similar annual modulation search. The signal is so obvious in the DAMA/LIBRA data that, even with a modest exposure (just a couple of years), one should expect to see this striking effect. COSINE-100 have published results from 2.8 years (173 kg · years) of exposure [Adh+22a]. They perform a global maximum likelihood fit using a cosmogenic background and simple harmonic components fit simultaneously across all of the crystals, treating each crystal as an independent detector subsystem. In regions of the data poor in anticipated dark matter signals ("sideband regions"), they observe an event yield in energy that decreases exponentially. This is consistent with known cosmogenic components.

A key quantity extracted from the data from an experiment such as COSINE-100 is the number of events (of a certain type; e.g., dark matter candidates) per day (cpd), per kilogram of detector material, per unit of energy deposited (or observed). In the case of a time-dependent observation, this quantity will be referred to as the *amplitude* (meaning the maximum excursion of the value away from zero).

The COSINE-100 collaboration reported a best-fit result of their data that prefers an amplitude of 0.0067 ± 0.0042 cpd/kg/keV when the phase and period of the oscillation are fixed in the model to those preferred by the DAMA/LIBRA observation. This result is consistent with both the null hypothesis (no modulation) and the DAMA/LIBRA best-fit value.

This is early data from COSINE-100. The collaboration projects that, within five years of the start of data-taking, they should be able to cover the DAMA/LIBRA signal region with 3σ significance from the null hypothesis. They have projected that future versions of their data analysis will lower the recoil energy threshold to 1 keV and improve on the selection of recoil events to reduce the total exposure time required to achieve the 3σ significance threshold for a DAMA/LIBRA-like signature.

The second experiment we wish to discuss is the Annual Modulation with NaI Scintillators (ANAIS) instrument [Ama+19a; Ama+21]. This is also a NaI(Tl) crystal-based experiment, consisting of 112.5 kg of active material. The experiment is housed in the Canfranc Underground Laboratory (LSC), within a mountain railway tunnel in Spain. It has been operating since 2017. Canfranc is a mountain-based laboratory, providing a depth of 2450 mwe for this experiment. In March, 2021 the ANAIS collaboration made public the results of an analysis of three years of data-taking (313.95 kg · y). They found a best fit in the [1–6] keV ([2–6] keV) energy region, a modulation amplitude of -0.0034 ± 0.0042 cpd/kg/keV (0.0003 ± 0.0037 cpd/kg/keV), and determined that this is consistent with the absence of modulation signals in their data. Their results are compatible with COSINE-100 and incompatible with the DAMA/LIBRA result at 3.3 σ (2.6 σ) for an expected sensitivity to the DAMA/LIBRA-like modulation signature of 2.5 σ

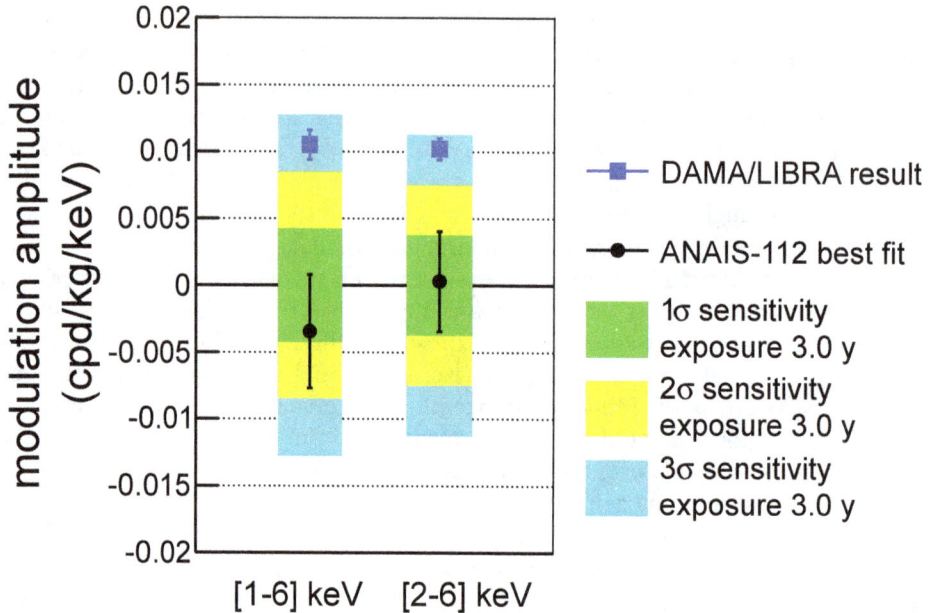

Figure 4.26: The ANAIS best-fit modulation amplitude result compared with the DAMA/LIBRA best-fit result for both recoil energy ranges considered by DAMA/LIBRA. Figure from Ref. [Ama+21].

(2.7 σ) (Fig. 4.26). While not completely definitive, taken together with the COSINE-100 results, there is a growing body of evidence, using NaI(Tl) scintillation-based crystal experiments, that the DAMA/LIBRA result cannot be replicated as a dark matter signal.

The challenges to the DAMA/LIBRA observation in its "home field" of NaI(Tl) technology are only expected to mount in the coming years. An independent experiment called COSINUS [Kah+18] aims to use NaI(Tl)-based detectors to collect multiple signals of a dark matter interaction and perform *in situ* measurements of their quenching factor (Eqn. 4.94). COSINUS will be more akin to the kinds of experiments described in the next section that have at least two physical signals available from the same technology, allowing for multi-dimensional classification of different kinds of interactions (backgrounds vs. dark matter). In addition, there appear to be efforts in the NaI community-at-large to open up their data sets to facilitate more transparency in the handling of the time dependency of the data. This is aimed to improve scrutiny of the choices that are made in looking for time-varying signals. Such choices have been demonstrated in the past to create false positives (the appearance of a time-varying signal where none is actually present). Such moves at openness aim to increase the understanding of techniques and their influence on interpretation of the data.

Liquid Noble Detectors

In general, liquid noble detector technologies use a dual-phase (liquid and gas) approach to broaden the range of signatures at their disposal. This adds background discrimination capabilities, especially for high-mass dark matter candidates, that, presently, are fairly unrivalled in the field. Some experiments that use this approach are XENON (Fig. 4.27), LUX-ZEPLIN (LZ)[6], Darkside, and PandaX. They operate the dual-phase systems as a time projection chamber (TPC), wherein electric fields are used to drift ions that

---

[6]ZEPLIN stands for ZonEd Proportional scintillation in LIquid Noble gases and was a series of detectors operated, starting in the 1990s, at the Boulby Underground Laboratory near Whitby in the United Kingdom. The LUX (Large Underground Xenon) project was constructed and operated in the 2000s at the Sanford Underground Laboratory in the Homestake Mine in Lead, South Dakota, in the United States. LUX-ZEPLIN, or LZ for short, is a follow-on instrument built by the merged collaborations and has been operating in the Sanford Laboratory since the early 2020s.

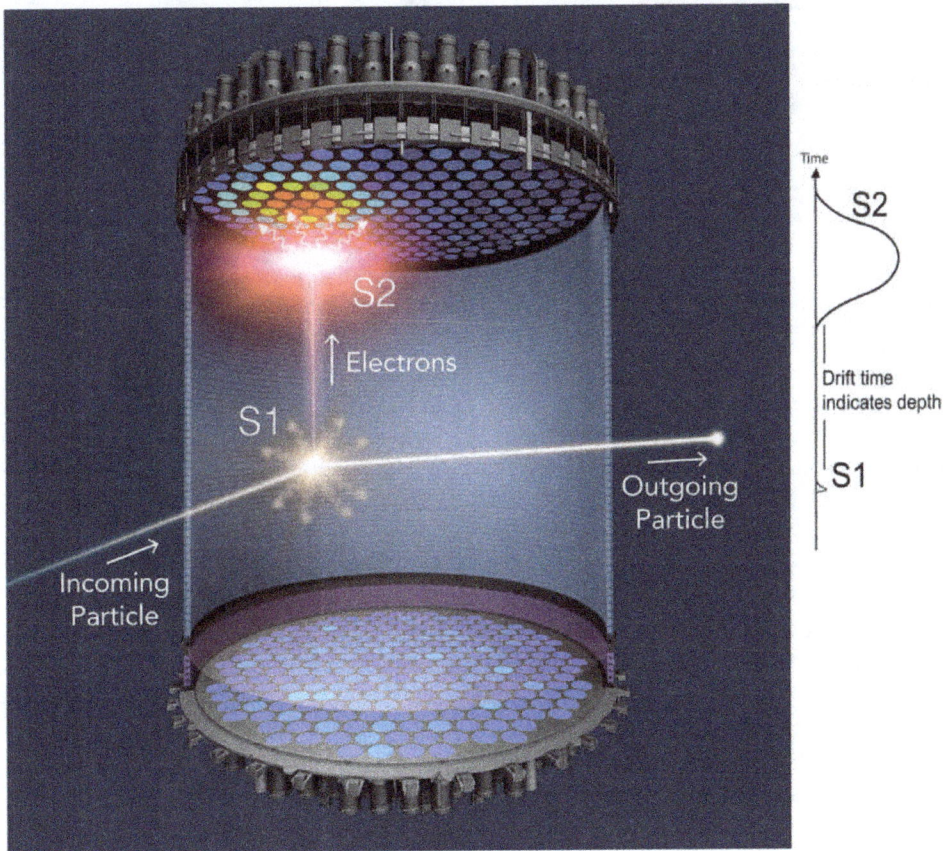

Figure 4.27: The operating principle of a liquid xenon TPC using the XENON-1T detector as an example. Figure from Ref. [Kud19].

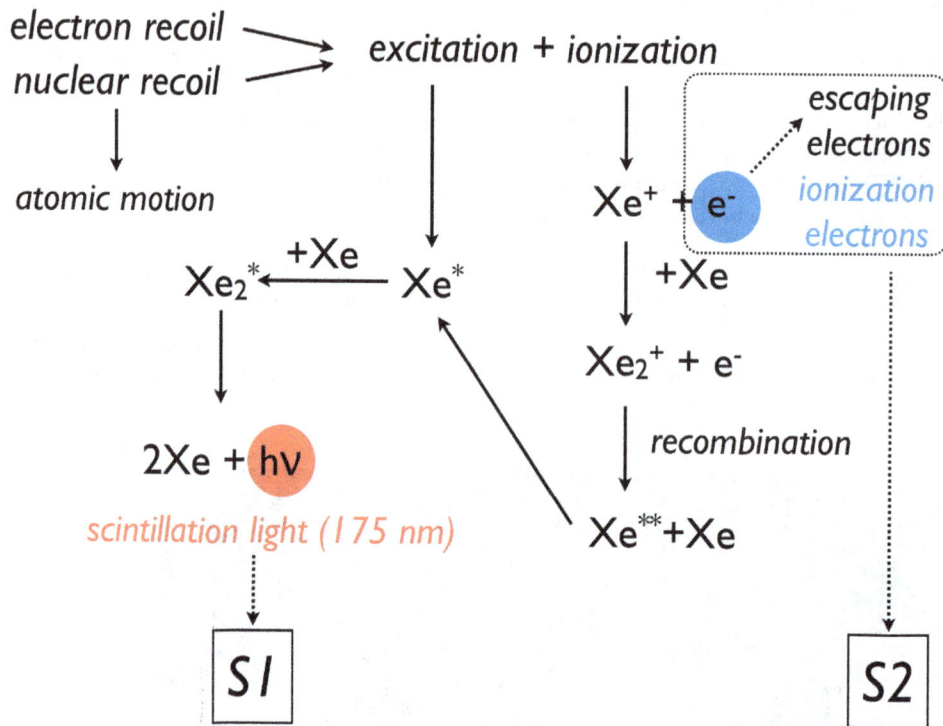

Figure 4.28: Schematic of scintillation production in a two-phase xenon detector [Man+10].

result from nuclear and electronic recoils. A grid of electrodes provides two-dimensional information about the position of the interaction when ions reach the electrodes. The third dimension is provided from the known and reliably stable drift speed of ions in the noble liquid or gas, combined with information about the time of the occurrence of the original primary and secondary interactions.

The operating principles of these instruments are straight-forward and based on significant experience with TPCs. The ideas are illustrated in Fig. 4.27 [Kud19]. Interactions in the liquid phase (the largest single volume in the instrument) produce both atomic excitation and ionization (Fig. 4.28). The relaxation from the excited states leads to photon production in the liquid phase, observed as scintillation light. This effect is immediate and establishes the t_0 of the interaction. This first signal is known as $S1$.

Ionization electrons are drifted with an applied electric field and guided from the liquid phase into the gas phase. In the gas phase, electrons are further accelerated through a stronger electric field producing proportional scintillation. The pulse of energy from this effect is known as $S2$. PMTs on the bottom and top of the chamber record scintillation light. The pulse distribution in the PMTs provides the $x - y$ coordinates of the original interaction, while the drift time of the ions through the phases of the experiment yields the

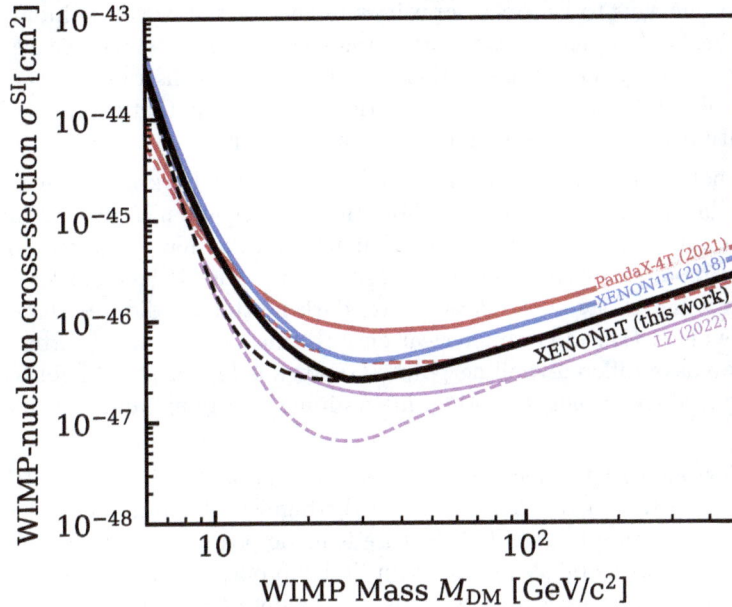

Figure 4.29: WIMP candidate exclusion limits as reported by the XENONnT Collaboration in Ref. [Apr+23].

z coordinate. The ratio of the energy deposited in $S1$ and $S2$ is a powerful discriminant against background processes, separating, especially, electron and nuclear recoils as a population.

For example, a gamma ray that ionizes an atom inside the liquid noble volume will result in a small pulse of initial energy (S1), and a comparatively larger pulse of secondary energy (S2) as the avalanche of ionization occurs when the original ionization electrons drift into the gas phase. A WIMP-like particle striking a nucleus in the liquid phase, however, will cause a large pulse of S1 energy followed by an equally large pulse of energy in S2. For the WIMP, $S2/S1$ will be closer to 1.0 than for the gamma ray.

Noble liquid experiments have come to dominate the search sensitivity for WIMP candidates with large masses $m_\chi \gtrsim 10$ GeV$/c^2$. This is evident from Fig. 4.23 where experiments like XENON1T, LZ, and PandaX define the parameter space exclusions in that mass region.

An excellent example of the sensitivity and reach of these noble-element-based detectors is given by the first results from the XENONnT experiment [Apr+23] (Fig. 4.29). The successor to XENON1T (which housed about 1 tonne of active liquid xenon), XENONnT aims to scale to the multi-tonne active volume level and defines the new current generation of such experiments. At the time of the writing of this text, XENONnT have only released their first preliminary WIMP search results. The results demonstrate overall excellent control and understanding of their backgrounds—from neutrons generated by radioisotope decay and cosmic rays to isotopic impurities in their active xenon volume. No strong evidence for heavy dark matter candidates was seen during this first science-quality run (Science Run 0). They interpret this non-observation as a constraint on the interaction cross-section of WIMP candidates as a function of their hypothetical mass. The results, compared with other similar experiments, are shown in Fig. 4.29.

Xenon is not the only noble elemental target utilized in dark matter searches. There is an entirely independent and parallel effort that utilizes argon [Agn+18b] as the target material, again in liquid and gas phases. Xenon and argon have different strengths and weaknesses. Argon is more naturally abundant than xenon. It has a lower atomic mass, which makes it better suited for lower-mass dark matter candidates than xenon (the benefit of this distinction will only be clear once the mass of any dark matter candidate is known). They have different boiling points (xenon's is higher than argon's), providing different technical challenges in storing, maintaining, purifying, and utilizing the target materials.

Argon offers another technical advantage, pulse-shape discrimination [BH06], where the shape of the S1 and/or S2 signals can distinguish electron-like and neutron-like interactions in the argon target. This feature is made possible by the fact that the time constant associated with argon's prompt scintillation property is longer (6 ns) than xenon's (2.2 ns) and is therefore easier, with efficient light detection, to observe (the late-time effects in argon are also stretched over a longer timescale than in xenon). Even in a single-phase experiment (one with only liquid argon), this shape can be measured and used to reject electron recoil backgrounds. Experiments that utilize argon as the target material have evolved over the past ten to twenty years, culminating in the current single-phase DEAP-3600 [Ama+19b] and soon-to-start (2026) two-phase Darkside-20k [Aal+18].

Within the two xenon and argon technology communities, there are formal plans to merge into large, singular collaborations aimed at constructing a single, large, next-generation dark matter experiment. The consortium pursuing this in the xenon community is referred to as *XLZD*; in the argon community, the consortium is called the Global Argon Dark Matter Collaboration (GADMC), and the future joint experimental program is referred to as *Argo*.

Liquid targets have a distinct advantage that has been well-proven especially in the last decade: *scalability*. The solid detectors discussed in the previous and the next section have well-defined challenges as to their ability to grow in size. Typically, for solid-state technologies, one needs to either increase the size of the material volume (while maintaining its structural integrity) or multiply the number of small volumes collected into a large, single target. Liquid targets have the benefit of being able, simply, to fill the volume they are given.

The challenges in liquid targets have primarily been in acquisition (obtaining sufficient volumes of noble elements) and purification. In addition, there have been challenges in measuring the Q-factor of these targets, though this is less of a challenge than it was at the inception of this technology's growth. At the present time, liquid target dark matter detectors are unrivalled in sensitivity for dark matter candidates with masses of $\mathcal{O}(10-100)$ GeV.

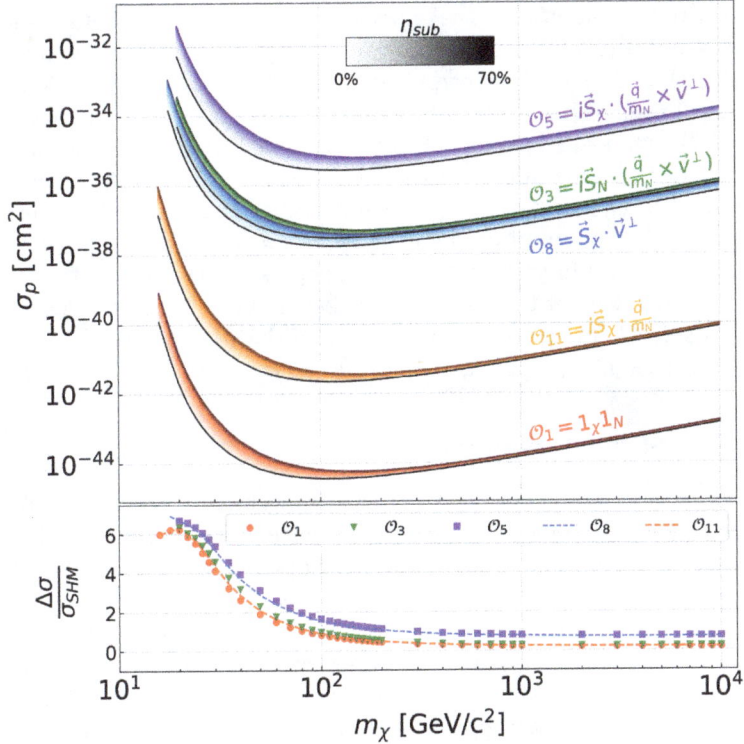

Figure 4.30: Constraints from the DEAP-3600 experiment on operators in the EFT described in Section 4.3. The standard halo model (SHM) is compared to a galactic halo model derived from the Gaia [Pru+16] observations of local stellar distributions, which imply a local stream of dark matter that is anisotropic, in contrast with the assumptions of the SHM.

We close this discussion with an example of the impact that liquid noble detectors have had on our approaches to constraining dark matter interaction and on the impact of galactic model assumption on those interpretations. Effective field theory approaches to describing dark matter/SM matter interactions were briefly discussed in Section 4.3. The DEAP-3600 experiment has demonstrated how projects can use their data to constrain these EFT frameworks [Adh+20; Adh+22b]. We note that other experiments have also made interpretations in the context of EFTs [Agn+20; Ang+19; Apr+17; Apr+19b; Sch+15b].

The DEAP-3600 experiment has previously searched for nuclear recoils from dark matter candidates and observed data consistent with a background-only hypothesis [Aja+19b]. This non-observation is then interpreted as constraints on cross-section contributions of several of the operators in the EFT framework, specifically $\mathcal{O}_1$, $\mathcal{O}_3$, $\mathcal{O}_5$, $\mathcal{O}_8$, and $\mathcal{O}_{11}$ (Fig. 4.30).

This particular work emphasizes the importance of recent observations of the Milky Way using the Gaia satellite [Pru+16], which provides high-precision stellar mapping. Data from Gaia imply that there is a local anisotropic dark matter stream in the region of the Milky Way containing our solar system [Bel+18; Hel+18; Mye+18a; Mye+18b], in contrast to the assumptions of the standard halo model. They demonstrate how constraints on the EFT parameters can depend, especially in some m_χ regions, strongly on the assumptions about the local halo model. The work includes the effect of changing the local dark matter density, relative to the nominal assumption, by a degree $f(\vec{v} = (1 - \eta_{sub})f_{SHM}(\vec{v}) + \eta_{sub}f_{sub}(\vec{v})$ where η_{sub} is allowed to range from 0–0.70 and represents the degree to which the dark matter distribution is part of the anisotropic substructure. The interpretations shown in Fig. 4.30 use a galactic model discussed in Ref. [NLB19], but the DEAP collaboration considers alternatives in their publication.

Cryogenic Solid-State Detectors

Cryogenic solid-state detectors are crystal detectors generally designed to read out both ionization and phonon signals from interactions within the target material. Compared to ionization-based approaches, phonon-based approaches can provide better energy resolution.

Acoustic phonons, on which we will focus here, are in-phase collective oscillations of the ions in a unit cell of the crystal lattice and can propagate across the entire lattice. They travel at the speed of sound in the material, c_s, and have energy scales given by $E_{\text{acoustic}} = \hbar c_s k$, where k is lattice or crystal momentum related to the direction of the displacements of atoms in the lattice. The energy scale of phonons is typically $\mathcal{O}(\text{meV})$, since the sound speed in solids is in the range of several km/s. A process that deposits keV-level energy will produce $\mathcal{O}(10^6)$ phonons, each of which carries a small piece of the total energy. The loss of a small fraction of phonons will not have a large impact on the accurate measurement of the interaction energy. We emphasize that no detection method is fully efficient, so loss has to be built into projections for a detection technology.

Let us contrast this with other energy-loss mechanisms. In semiconducting materials (e.g., silicon or germanium), an energy of $\sim 3 - 5$ eV is required to create an electron-hole (e-h) pair. For a deposited recoil energy at the keV scale, this equates to approximately 300 e-h pairs. A detector that relies on scintillation, in comparison, will yield 20–40 eV per scintillation photon. Those photons must then go on to interact with atoms to produce electrons, which are the particles that result in the detectable signal. Those electrons liberated by photons (called *photoelectrons*) then provide detection of the interaction. However, this detection process is not fully efficient. For example, not all scintillation photons will have sufficient energy to release an electron, and not all electrons will be detected. When combined with the quantum efficiency of a photodetector, this equates to just a few photoelectrons per keV. The same recoil energy, deposited in two different media (one that scintillates/ionizes, and one that vibrates), will lead to very different qualities of signal, and pose different degrees of challenge to a detection strategy.

There are two families of sensors for reading out phonon signals: thermal and athermal sensors. Thermal sensors measure an increase in temperature. A thermal sensor must

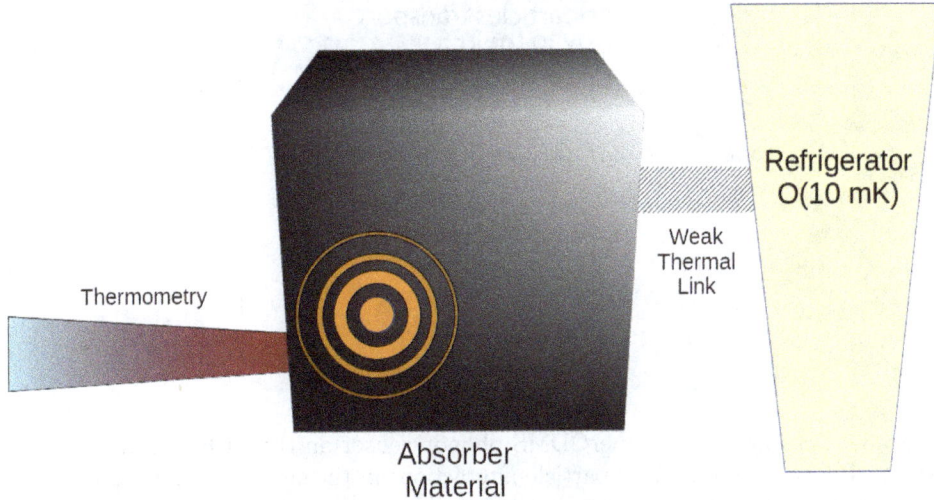

Figure 4.31: Schematic of a generic cryogenic detector. A thermometer measures the temperature of an absorber that has a weak thermal link to a refrigerator. The measured temperature increase is proportional to the deposited energy divided by the heat capacity of the absorber. Figure concept courtesy of Professor Matt Pyle (University of California, Berkeley).

wait until the phonons reach thermal equilibrium (*thermalization*) within the bulk of the detector and then within the sensor itself. Thermalization requires milliseconds. Athermal sensors are sensitive to fast, non-equilibrium phonons that contain information on the location and type of recoil that occurred. Schematically, the detector is comprised of an absorber with a weak thermal link to a refrigerator (Fig. 4.31).

The maximum temperature change in the cryogenic detector is given by

$$\Delta T = \frac{E}{C(T)} \tag{4.95}$$

where E is the deposited energy, and $C(T)$ is the temperature-dependent heat capacity of the absorber. Pure dielectric crystals and superconductors at temperatures lower than their critical temperatures have heat capacities given by

$$C(T) \sim \frac{n}{M} \left(\frac{T}{\Theta_D} \right)^3 \tag{4.96}$$

where m is the absorber mass, M is the molecular weight of the absorber, and Θ_D is the Debye temperature. From this we see that the lower the crystal's temperature, the larger the ΔT per unit of absorbed energy. For a 100 g detector at 10 mK, a 1 keV energy deposition increases the temperature by $\sim 1~\mu K$. Such a temperature increase is readily measurable by such devices.

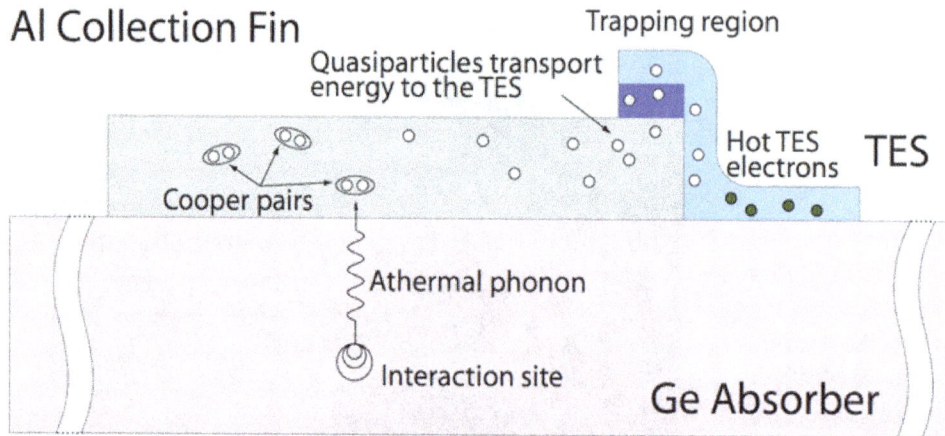

Figure 4.32: Schematic of a SuperCDMS phonon sensor on the surface of a germanium substrate. Phonons produced by particle interactions in the substrate break Cooper pairs in the aluminum fin producing quasiparticles. The quasiparticles diffuse into the TES, increasing its temperature. Image provided by the SuperCDMS Collaboration, originally from Ref. [Saa02].

The two most widely used technologies to measure these signals are semiconductor thermistors such as neutron-doped germanium sensors (Neutron Transmutation Doped, or NTD, technology) and transition edge sensors (TESs). In both NTDs and TESs, an energy deposition produces a change in the electrical resistance that can be measured. Both kinds of devices are built on the feature that their resistance is a strong function of temperature.

NTDs are small germanium semiconductor crystals that have been exposed to a neutron flux to create a large, controlled density of impurities. The resistance in these devices is a strong function of temperature. NTDs measure small temperature variations relative to a T_0. The resistance is continuously measured by flowing current through the crystals and measuring the resulting voltage. The sensors are affixed (via an adhesive) so that they make direct physical contact with the absorber.

A TES is a thin superconducting film that is operated near its critical temperature. TESs are directly fabricated onto crystal substrates using photolithography, thus avoiding a separate process for affixing them to the absorber. A heater with an electrothermal feedback system maintains the temperature of the TES at its superconducting threshold. When used as a calorimeter, a TES is set to a temperature that causes its resistance to be somewhere between its superconducting (zero resistance) and normal conducting states. A slight upward change in temperature, as through the deposit of energy, will rapidly increase the resistance.

An example of an application of a TES as a calorimeter is when it is paired with absorbing and interfacing materials (Fig. 4.32). When a particle interacts in the absorber, phonons can be generated. An example absorber might be a semiconductor crystal. The phonons propagate to the surface of the absorber where they can be collected by another material that is interfaced to the TES or NTD. For example, the collectors might be made from aluminum, which is known to become superconducting at 1.2 K. The TES is operated at a much lower mK-level temperature, so the aluminum will remain overall superconducting even as energy is transferred from the absorber, through the aluminum, to the TES.

When phonons enter the aluminum fins some Cooper pairs are broken, creating quasiparticles[7]. Those quasiparticles diffuse into the TES, releasing their binding energy and increasing the temperature of the TES. This rise in temperature increases the resistance of the voltage-biased TES. Temperature changes are then detected by a change in the feedback current and collected by a Superconducting Quantum Interference Device (SQUID). This process is illustrated in Fig. 4.32.

It is valuable to illustrate a few experiments in this space to help understand the kinds of technologies and strategies common to the cryogenic solid state effort.

Ionization and Phonons: SuperCDMS and EDELWEISS

Multi-generation experimental programs utilizing the detection of both ionization and phonons are represented by the SuperCDMS and EDELWEISS experiments. The SuperCDMS (Super Cryogenic Dark Matter Search) collaboration has a long history of specifically using TES technology for detector readout. The collaboration's technology focus has been on the combined detection of ionization and heat (vibration) in order to reject non-dark-matter events in their detector. The EDELWEISS (Expérience pour DEtecter Les WIMPs En Site Souterrain) collaboration has a long history of using bolometer-style readout technology.

The SuperCDMS collaboration is presently constructing an upgraded Generation 2 (G2) experiment within SNOLAB in Sudbury, Canada [Agn+17b]. The cryostat will be able to accommodate up to seven towers and operate at a temperature of 15 mK. The initial payload will contain twenty-four detectors amounting to a mass of 30 kg. The detectors are configured into four towers, with each tower containing six detectors. Each detector will be a 100 mm diameter, 33.3 mm-thick germanium or silicon crystal with mass 1.39 kg (Ge) or 0.61 kg (Si). Two of the detector towers will contain detectors operated in high voltage (HV) mode. Each HV tower will contain four germanium and two silicon detectors. The detectors in the remaining towers will be operated in interdigitated, z-sensitive ionization, and phonon-mediated (iZIP) mode. One of those towers will contain six germanium detectors; the other tower will contain four germanium and two silicon

---

[7]We feel it is useful here to clarify what is meant, in this case, by a *quasiparticle*. Unlike a particle in empty space (e.g., an electron) or even a conduction electron from a metal ion in a crystal, quasiparticles have additional complexity. In the specific case of Cooper pairs, which are formed from weakly bound electron pairs spaced at macroscopic distances in a semiconducting material, breaking the pair forms *Bogoliubov quasiparticles*. Each quasiparticle is formed by an electron and corresponding hole in the semiconductor, functioning somewhat like an electric dipole with a binding energy. The e-h pair can propagate like a particle, but is a collective phenomenon.

detectors. The background from gamma particles is expected to be less than 0.1 dru, where a dru (differential rate unit) equates to 1 event/kg/day/keV).

iZIP technology permits the use of both crystal vibrations (quantized into phonons) and ionization as the basis for building discrimination algorithms. Each iZIP detector will have six phonon channels and two charge channels on each detector face. The ionization sensors are interleaved on each face between the phonon sensors, as illustrated in Fig. 4.33. When a particle interacts in these detectors, simultaneous measurements of prompt phonon and ionization signals provides a means to combine two sources of information to discriminate between nuclear-recoiling and electron-recoiling interactions. The total energy involved in the recoil is the sum of the energies deposited in ionization and phonons. The *ionization yield* or *yield* is defined as the ratio of ionization energy to recoil energy.

This technology provides an *in situ* means to reject backgrounds caused by beta radiation occurring near the edges of the crystal (known as a "surface event"). In past designs, such decays led to incomplete energy collection and mimicked the signature of a dark matter particle interacting with a nucleus in the bulk of the crystal, away from the edges. In the present design, beta radiation near the edge of the crystal will result in all charge from ionization being drifted to and collected on only one side of the crystal. In contrast, a dark matter particle interacting in the bulk of the crystal will cause charges and holes to drift to both sides of the crystal. Therefore, this detector allows for surface beta radiation rejection when an event involved one-sided charge collection (Fig. 4.34).

This capability is a result of years of refinement of detector design. All surface and electron recoil events above a few keV can be removed using appropriate selection criteria. These detectors have sensitivity, in a "background-free" mode, to dark matter candidates with masses above ~ 5 GeV and are sensitive to dark matter candidates with masses above ~ 1 GeV under specific detector-operating conditions.

In contrast, the HV detectors emphasize phonon collection over ionization collection and are designed to be operated under a much higher voltage (up to 100 V). This allows them to take advantage of the Neganov-Trofimov-Luke (NTL) effect. When electrons from ionization are drifted across a potential (V), the NTL effect results in a large number of phonons being generated. The total phonon energy (E_t) is a combination of the primary recoil energy (E_r) and the NLT phonon energy. The NTL phonon energy is given by

$$E_{NTL} = N_{eh} e V_b \tag{4.97}$$

where V_b is the bias voltage across the detector, the number of electron-hole pairs generated is $N_{eh} = E_r/\epsilon_\gamma$, and ϵ_γ is the average energy needed to generate an electron-hole pair. In germanium, ϵ_γ is 3 eV. Thus the total phonon energy can be written as

$$E_t = E_r + N_{eh} e V_b. \tag{4.98}$$

When V_b is very large, the NTL phonons dominate the signal and permit lower energy thresholds to be achieved than would be possible from the iZIP design.

HV detectors provide ultra-high resolution by indirect charge measurements. The use of the *pulse shape* of the signal (the time structure of that signal when it arrives at the surface and is read out by the TESs) can provide discrimination between surface and

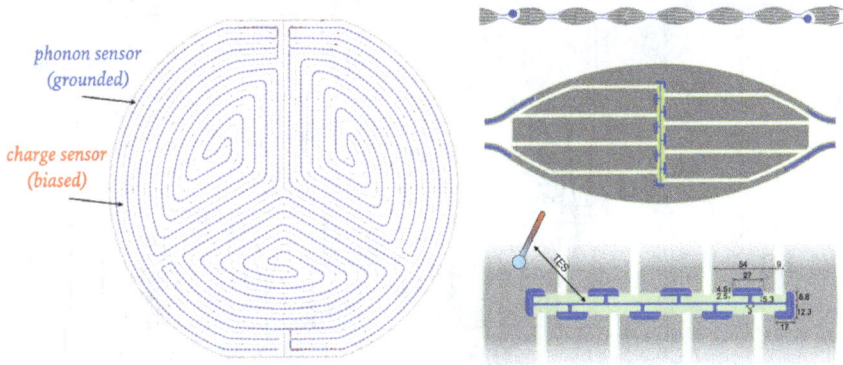

Figure 4.33: Schematic of a SuperCDMS iZIP detector. Left: Lines of ionization sensors (fainter shading) are interleaved between grounded phonon sensors (darker shading). Right: Each phonon sensor consists of an aluminum superconducting fin (fainter shading) and a TES (darker shading.) Figure by the SuperCDMS Collaboration.

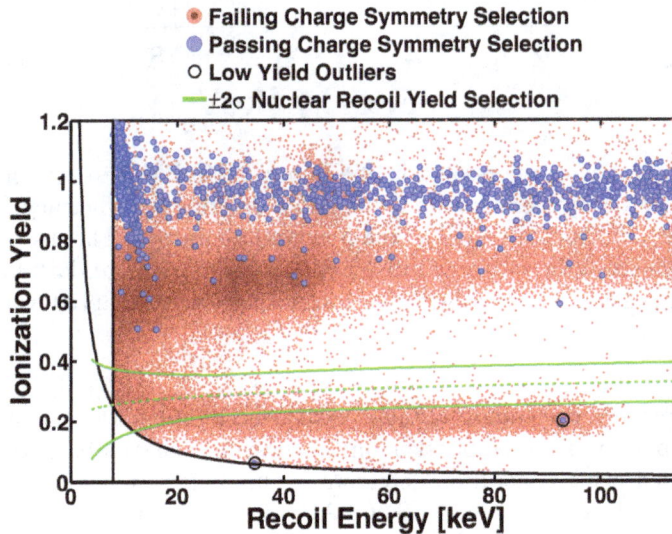

Figure 4.34: Ionization yield versus phonon recoil energy with the $\pm 2\sigma$ ionization yield range of neutrons indicated (area between the fainter, solid, horizontal lines). Results were obtained using 900 live hours of calibration data having a ^{210}Pb source facing one side of a detector. The asymptotic dark, solid curve is the ionization threshold (2 keVee); the vertical solid line is the recoil energy threshold (8 keVnr). Symmetric charge events (darker dots) in the interior of the crystal and the events that fail the symmetric charge cut (small, lighter dots) include surface events from beta decay (electrons), gammas, and lead nuclei, and are clearly separated from the nuclear recoil band. Figure from Ref. [Agn+13].

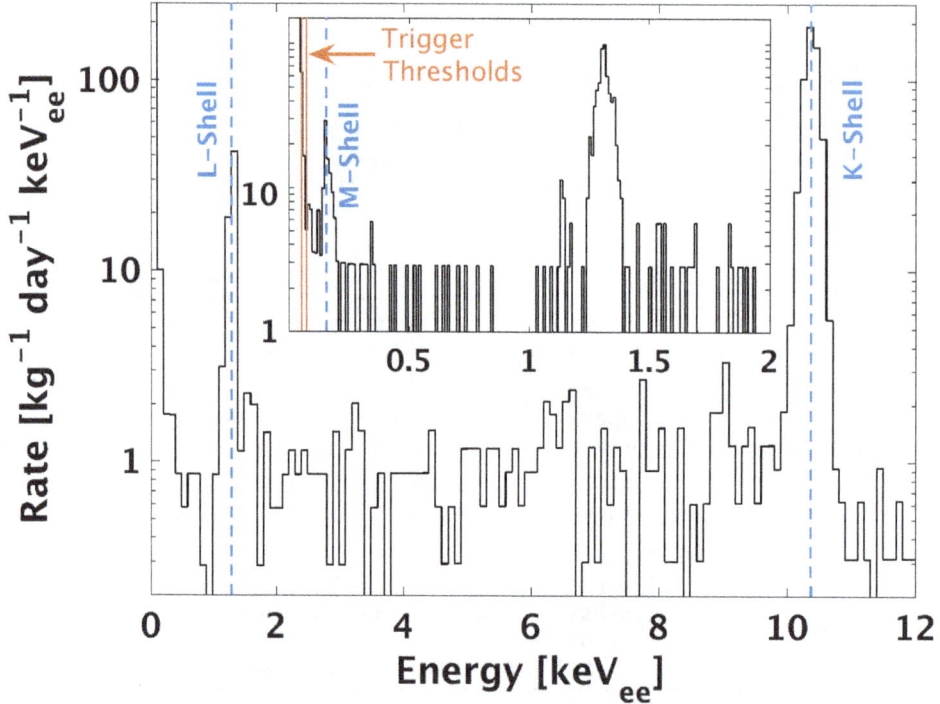

Figure 4.35: Illustration of the resolution of a SuperCDMS HV detector using the WIMP-search spectrum after application of all cuts and correcting for efficiency (except trigger efficiency) in the detectors operated at the Soudan Laboratory. The inset shows a zoom (with smaller bin size) of the energy range actually used in a final dark matter analysis. ^{71}Ge activation peaks are marked with vertical dashed lines. Adapted from [Agn+16].

bulk events. However, the computation of yield is no longer performed as independent ionization information is not emphasized in this design. The capability of HV detectors was demonstrated using generation detectors in the Soudan Underground Laboratory. The HV detectors to be operated in SNOLAB will be able to search for dark matter candidates with masses down to ~ 0.3 GeV, since they achieved thresholds of 75 eVee and 56 eVee as illustrated in Fig. 4.35.

As noted earlier, nuclear recoils produce electron-hole pairs less efficiently than electron recoils. As such, the energy and interaction type can be defined [Alb+22] through the yield (Y) where $Y \equiv 1$ for electron recoils (where ionization comprises essentially all of the energy in the recoil event).

$$N_{eh} = Y(E_r)\frac{E_r}{\epsilon_\gamma}. \tag{4.99}$$

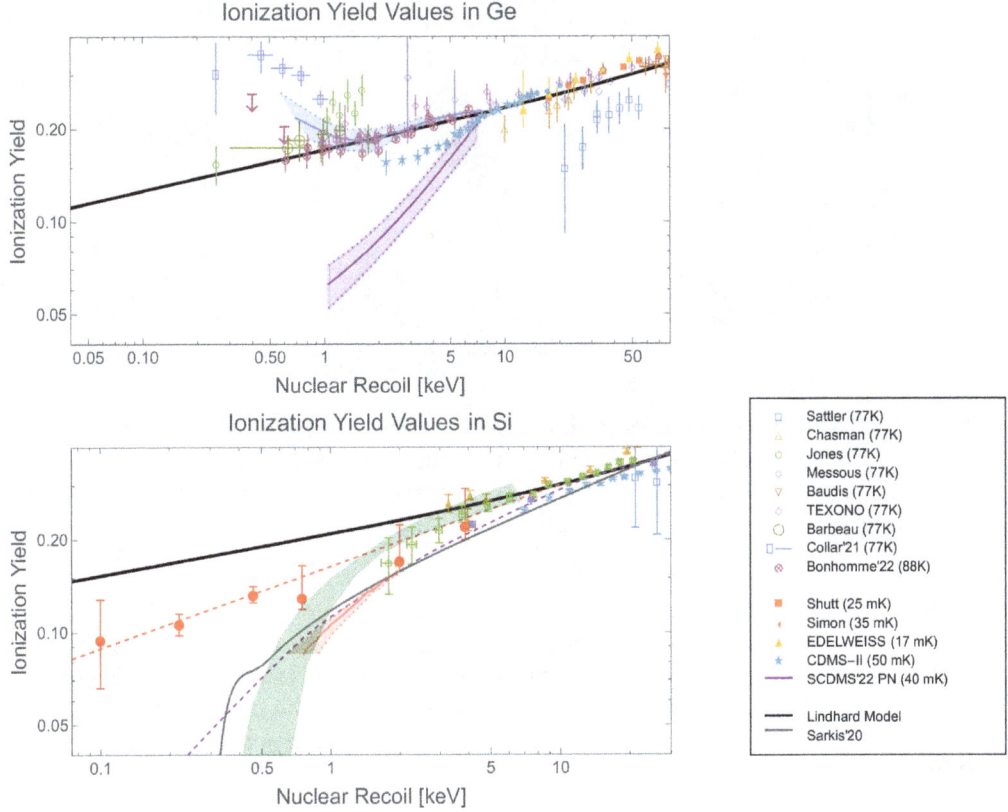

Figure 4.36: Comparison of ionization yield vs. recoil energy for models and experimental data, including at low nuclear recoil energies. The top left figure shows results for germanium and the bottom left figure for silicon. A common legend is shown at the right. Data are shown with points while models are shown using lines. Figures provided by Professor Tarek Saab (University of Florida).

Examples of measurements of yield vs. recoil energy are shown for germanium and silicon (Fig. 4.36). The total recoil energy can be written as

$$E_t = E_r \left(1 + Y(E_r) \frac{eV_b}{\epsilon_\gamma} \right). \tag{4.100}$$

The energy scale is also dependent on the interaction type. The detectors are calibrated using electron recoils and the resulting energy scale is labeled as keV$_{ee}$. It is possible to then convert the electron-equivalent energy scale to the nuclear-equivalent energy scale,

$$E_{nr} = E_{ee} \left(\frac{1 + eV_b/\epsilon_{gamma}}{1 + Y(E_{nr})eV_b/\epsilon_{gamma}} \right) \tag{4.101}$$

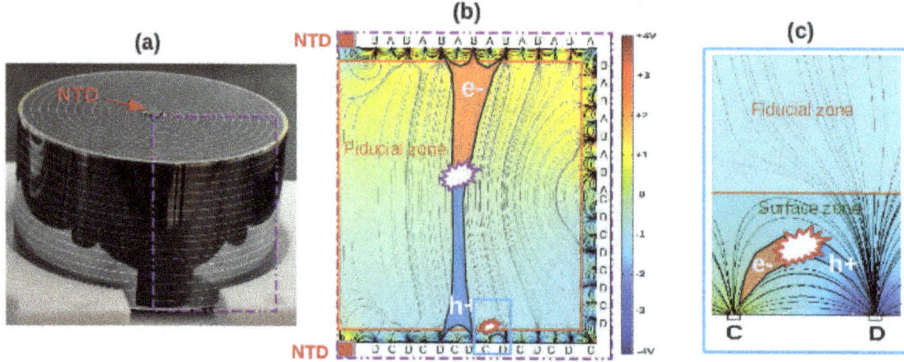

Figure 4.37: (a): An FID detector showing the interleaved ionization sensors and NTD sensor. (b): Cross-section of an FID detector illustrating the fiducial zone for an event generated in the bulk of the detector. (c): Illustration of charge collection for an event near the surface of the detector. Adapted from [Arn+18c].

where $Y(E_{nr})$ must be determined by either direct measurement or with a model. In places where data must be interpolated or, more challenging, extrapolated, then a model must be employed. For example, in both cases the data at lower energies is sparse or non-existence. This necessitates an assumption about how the material behaves as one lowers the threshold. The most common model used for ionization yield is the Lindhard model [J L63a; J L63b; J L68]:

$$Y(E_{nr}) = \frac{k \cdot g(\epsilon)}{1 + k \cdot g(\epsilon)} \qquad (4.102)$$

where $g(\epsilon) = 3\epsilon^{0.15} + 0.7\epsilon^{0.6} + \epsilon$, $\epsilon = 11.5 E_{nr}(\text{keV})Z^{7/3}$, Z is the atomic number of the material, and $k = 0.157$ in germanium. Although this value of k agrees with experimental measurements above 1 keV$_{nr}$, there are few measurements at lower energies. As such, for the lowest energy events, an uncertainty must be taken into account. This is typically done by varying k uniformly through the majority of experimental results [BCT07; BM12; JK75].

Also, the SuperCDMS Collaboration is pursuing the development of HVeV detectors. These detectors have demonstrated single electron-hole pair resolution and have sensitivity to a variety of sub-GeV dark matter models using just gram-day exposures [Agn+18c; Alk+20; Ama+20].

We close this section with a summary of the EDELWEISS-III experiment (Fig. 4.37). Recently decommissioned, the experiment was located in Laboratoire Souterrain de Modane (LSM) in Modane, France. EDELWEISS III consisted of twenty-four cylindrical, high-purity germanium crystals, each with mass ∼860 g operated at 18 mK [Arn+18c]. Their fully interdigitated (FID) detectors collected charge via concentric aluminum ionizatiation sensors that were interleaved on all absorber surfaces and two NTD Ge sensors that were glued onto the top and bottom face of the detectors. The configuration of the ionization sensors allowed for the fiducialization of bulk events in the inner region of the detectors.

There were three issues that somewhat limited the EDELWEISS III performance. The most concerning for dark matter searches was that the neutron background was observed to be five times what was expected. This limits sensitivity for searches aiming for masses ≥ 10 GeV/c^2, while affecting the lower mass region far less. Similar to SuperCDMS, the EDELWEISS collaboration worked to develop a new detector concept in order to pursue very light, electron-interacting dark matter [Arn+18a]. The NTD phonon resolution and an unknown background source of "heat-only" events (events that deposit energy in the detector without the creation of electron-hole pairs) were identified as limiting factors in their searches [Gas+21]. These issues made it difficult to disentangle noise-triggered events, possible events from leakage currents, and other such backgrounds from potential dark matter signal candidates.

CRESST

We now look at a cryogenic detector that combines ionization and scintillation as the key detector signals: the Cryogenic Rare Event Search with Superconducting Thermometers (CRESST). The CRESST experiment is operated in LNGS, most recently in its third phase (CRESST-III). The experiment uses an array of ten cryogenic scintillating CaWO$_4$ crystals that are custom-grown. Each crystal has dimensions of 20 mm × 20 mm × 10 mm and is instrumented with tungsten TESs to read out phonon signals. Each detector module consists of a scintillating crystal with mass of ∼25 g equipped with a TES next to a silicon-on-sapphire light absorber. The absorber is also equipped with a TES. The modules have fully scintillating housing and are instrumented special holders that provide the ability to veto radiogenic backgrounds from surrounding surfaces and suppress thermal signals from particle interactions in surrounding materials [Man+20]. Similar to SuperCDMS and EDELWEISS, the CRESST detectors allow for the discrimination between ER and NR. A schematic of the detectors is shown in Fig. 4.38.

As an example, we consider results [Pet+20] from the CRESST detector that were obtained with a single module. The combination of light sensors and the element oxygen in the detector target allows sensitivity to dark matter particle masses as low as 160 MeV/c^2. An AmBe neutron calibration source is used to define the NR band for each element in the crystal. The live-time for the period of operation, after all data quality and signal selection criteria were applied, was 3.64 kg days. The energy and yield of the events passing all selection criteria are shown in Fig. 4.39.

All 441 events in the acceptance region were considered to be dark matter candidates with no background subtraction when setting the upper-limit on the WIMP-nucleon scattering cross-section using the optimum interval method [Yel02]. This result excluded new parameter space down to 0.16 GeV/c^2 and improved exclusion limits by a factor of six over the previous best limit.

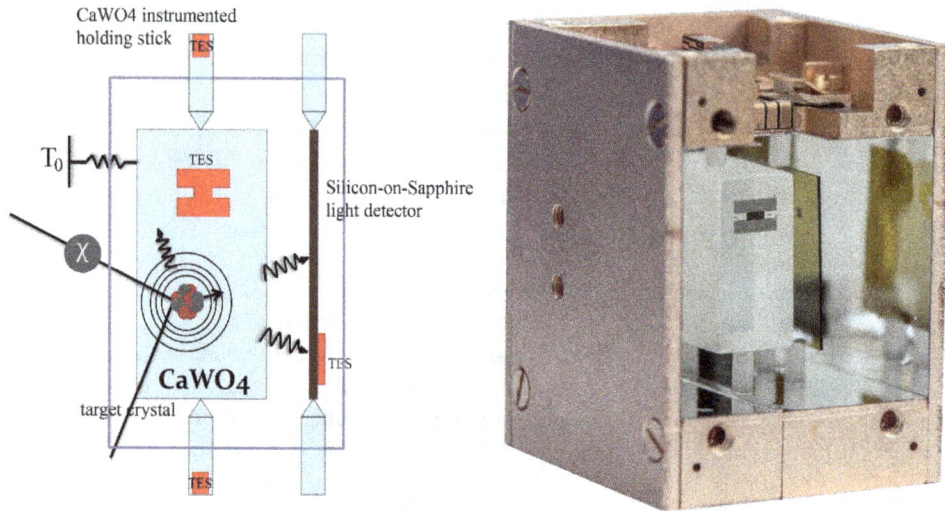

Figure 4.38: Left: Schematic of the CRESST-III detectors highlighting aspects of the improved detector design which includes fully scintillating housing and CaWO$_4$ instrumented-holding sticks. Right: Picture of a CRESST-III detector module. Adapted from [Man+20].

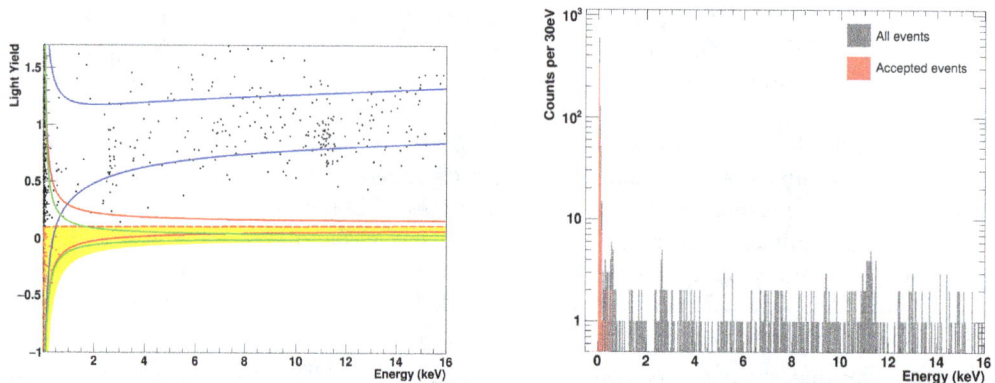

Figure 4.39: Left: Light yield vs. phonon energy for events passing selection criteria in the dark matter data set. The top pair of curves enclose the 90% C.L. electron recoil band, the bottom two pairs of lines enclose the 90% C.L. nuclear recoil bands for oxygen and tungstenshaded band encloses the dark matter candidate acceptance region. All events in this region are taken as dark matter candidates. Right: Energy spectrum of all events passing selection criteria in the dark matter data set. Adapted from [Pet+20]

Bubble Chambers and Superheated Liquids

Bubble chambers are sealed vessels filled with a liquid under pressure. The pressure and temperature are set carefully so that the volume of liquid is in a super-heated state, meaning it is above its boiling temperature, but boiling is not allowed to initiate in the volume. Achieving this requires careful thermodynamic and mechanical control. In addition, design practices are applied to prevent bubble nucleation at the liquid-vessel interface. When a particle interacts with a nucleus in the liquid and deposits sufficient energy, the medium transitions at the point of the interaction from liquid to gas, resulting in a bubble. The location and formation details of the bubble provide information about the particle that initiated it.

A bubble is a finite volume of the liquid that has transitioned to the gas state while the surrounding region remains liquid; there is a boundary between the two phases. To understand the principles of bubble nucleation [Sei58; SSD19], consider first a bubble in thermal and chemical equilibrium. To form the bubble, one must deposit a sufficient amount of energy within a given volume such that this volume of liquid can be converted to the gas phase. In the super-heated state, one need only input the energy required to move the system from its local, meta-stable liquid state to the nearby lower-energy gas state. The energy threshold for this can be denoted E_{thr} and the radius of a spherical region into which this minimum energy must be injected is referred to as the *critical radius*, r_{crit}.

In that case, the temperature of the liquid is equal to the temperature of the bubble, $T_l = T_b$. Assuming that there is no surface tension, if the pressure of the gas in the bubble, P_b, is greater than the inward pressure from the liquid, P_l, the bubble would be expected to expand. However, there generally should be additional forces between the molecules or atoms at the interface that hold the structure of the bubble, generally referred to as *surface tension*. This tension exerts a second pressure, P_s. To be in this non-equilibrium situation, where the interior volume of the bubble expands with time, the bubble will grow when

$$P_b > P_l + P_s \tag{4.103}$$

and the radius of the bubble is bigger than a critical radius. The pressure due to surface tension can be written as

$$P_s = \frac{2\sigma}{r}. \tag{4.104}$$

where σ is the surface tension and is material- and temperature-dependent. Any volume that is thus converted to gas, at or above the critical radius, will nucleate a bubble. Thus, we write the radius condition for bubble growth as

$$r > r_{crit} = \frac{2\sigma}{P_b - P_l}. \tag{4.105}$$

Bubbles that do not meet these conditions collapse and return to liquid state. The energy threshold for bubble nucleation is given by

$$E_{thr} = 4\pi r_{crit}^2 \left(\sigma - T \left[\frac{d\sigma}{dT} \right]_\mu \right) + \frac{4\pi}{3} r_{crit}^3 \rho_b (h_b - h_l) - \frac{4\pi}{3} r_{crit}^3 (P_b - P_l) \tag{4.106}$$

where ρ_b is the bubble density and h is the specific heat of the liquid (h_l) or bubble (h_b). The first term describes the free energy of the surface and heat transfer from the

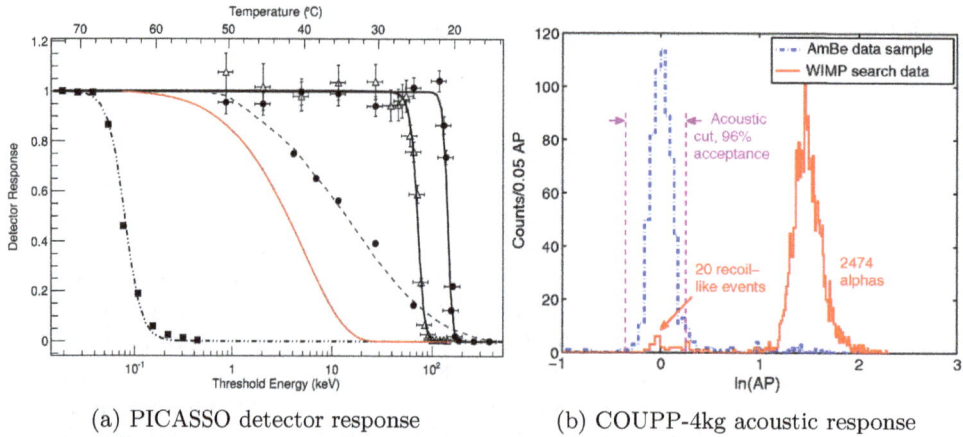

(a) PICASSO detector response   (b) COUPP-4kg acoustic response

Figure 4.40: (Left) PICASSO Detector response to different particles in superheated C_4F_{10}. From left to right: 1.75 MeV γ-rays and minimum ionizing particles (dot-dashed); model of 50 GeV/c^2 WIMP (red); polyenergetic neutrons from an AcBe source (dotted); α particles at the Bragg peak from ^{241}Am decays (open triangles); and ^{210}Pb recoil nuclei from ^{226}Ra spikes (dots). Figure is from Ref. [Arc+12]. (Right) Detector response to alpha particles and nuclear recoils as characterized by an acoustic parameter (AP) in the COUPP-4kg dark matter experiment operated in SNOLAB. Adapted from [Beh+12].

surroundings; the second term describes the energy required to convert the volume of gas to liquid; the third term describes the energy required to generate the surface. A bubble forms in equilibrium at this threshold, grows larger above this threshold, and collapses below this threshold.

Bubble chambers as dark matter detectors are attractive because of their unique ability to discriminate between different particle interactions [Arc+12]. Figure 4.40a shows the response of super-heated C_4F_{10} to various particles as determined by the PICASSO (Project in CAnada to Search for Super-symmetric Objects). The major backgrounds we have explored arise from particles that interact with the electrons instead of the nucleus, such as gamma and beta rays. Electron-recoiling events deposit less energy in the target medium than those that generate nuclear recoils. Since the threshold for bubble formation is entirely tuneable in these chambers using temperature and pressure, it is possible to set the threshold higher than the energy deposited by many common background interactions, such as beta or gamma radiation ionizing the medium. This allows such detectors to be tuned to respond *only* to nuclear recoil interactions.

Since bubble chambers are generally insensitive to electrons and photons, this leaves alpha particles and neutrons as potential backgrounds. To mitigate nuclear recoils from neutrons, the experiments can be located underground using active and/or passive shielding. Alpha particles are emitted by a wide range of radioisotope decays, notably in the radon decay chain. Radon could implant in the target liquid or on the surfaces surrounding that liquid. Careful purification and construction techniques can limit this risk. Further, bubble chambers have found that nuclear recoils from alpha events can be isolated using acoustic techniques. For example, alpha particles deposit their energy over tens of microns whereas nuclear recoils deposit their energy over tens of nanometers. The sound waves that result from the bubble formation differences quite literally lead to neutrons sounding different from alpha particles (alpha particles are about four times louder than neutrons for the same energy). Acoustic sensors, combined with cameras that detect the bubble formation, can be used to reject alpha backgrounds. An example of measuring this effect in the COUPP (Chicagoland Observatory for Underground Particle Physics) experiment, which used piezoelectric sensors, is illustrated in Fig. 4.40b.

The PICASSO and COUPP experimental collaborations joined forces to create a new collaboration, and instrument, called PICO. They name their detectors either by the volume of the target material (in which case the name includes the letter "L" for "liters") or by target mass (in which case the name excluded the "L"). The collaboration is presently completing the design and beginning the construction of a 250 kg bubble chamber, which, for historical reasons, is named PICO-500. As we write this text, that same collaboration is operating the PICO-40L experiment which deployed many of the same designs and technologies that are incorporated into PICO-500. The collaboration has already produced results from two runs with PICO-2L [Amo+15; Amo+16] and PICO-60 [Amo+17]. The detectors have revealed challenges related to a range of issues such as the interfaces between materials in the chamber design, thermal management of the chamber, and the cleaning of surfaces in contact with the target liquid.

Charge-Coupled Device-Based Detectors

We close this survey with a look at another approach that has led to an emerging generation of experiments: the use of charge-coupled devices (CCDs) as dark matter detectors. CCDs are the backbone of modern commercial camera technology (Fig. 4.41), having a long history of use in the scientific community. For example, they are central

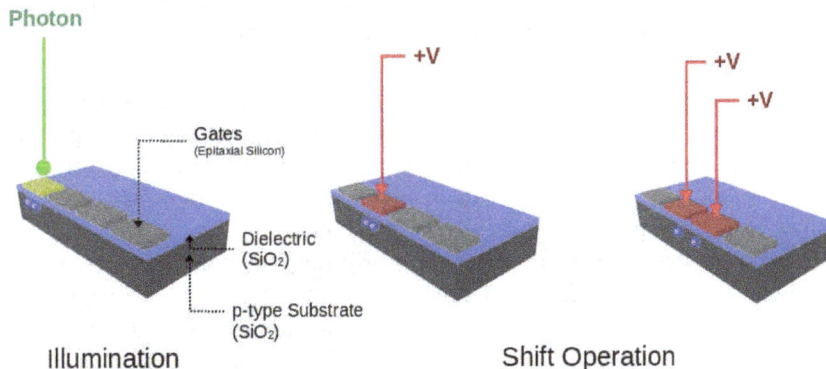

Figure 4.41: An illustration of the basic operation of a CCD in a camera.

to modern telescope camera systems, including those deployed as survey instruments such as the Dark Energy Survey's DECam [Abb+05] and the forthcoming Vera Rubin Telescope [Ive+19] imaging system. We will begin by briefly reviewing CCD technology and then discuss a few examples of the evolution of CCD-based dark matter detectors.

The operation of a CCD in a camera is shown in Fig. 4.41. An array of pixels is constructed in a plane by creating small capacitors, each one a sandwich of a gate material (e.g., silicon deposited by epitaxy), a dielectric material beneath (e.g., silica, SiO_2), and then a semiconducting substrate composed of p-type silicon. The substrate is electrically grounded at its base. The p-type silicon is created by chemically doping normal silicon to create a small excess of vacancies in the valence band (absences of electrons where they could have been present, or "holes"). The application of a voltage (known as a *bias voltage*) on the gates generates an electric field vertically down from the gate to the grounded region at the bottom of the substrate. This repels holes away from the gates by drawing electrons from deeper below, up toward the dielectric layer. The deep bulk becomes enriched with holes (vacancies) while the original holes near the dielectric boundary are filled with electrons. This creates a zone with no free-moving charges—a *depletion region*—just below the dielectric-substrate interface. Under this condition, any new electron-hole pairs created near the substrate-silica surface (by an external influence beyond the described system) will result in the liberated electrons remaining near the surface while the holes are pulled deeper into the substrate by the gate's electric field.

In such a state, the CCD is ready to collect an image. Light is permitted to hit the surface of the CCD layer, which serves as the focal plane of the camera. By design, visible-light photons (with wavelengths in the $\mathcal{O}(100)$ nm range) striking the system will penetrate into the substrate. The silica and gate layers are comparable in size to those wavelengths, permitting photons to reach and interact in the bulk. This will create new electron-hole pairs in the bulk via the photoelectric effect. The electrons remain trapped near or under the closest gate, held in place by the bias voltage. The camera aperture is closed after the exposure so that no more light can induce electron-hole pairs.

Where light illuminated the plane there will now be a trapped charge, and where no light struck the device there will be little or no trapped charge (there are noise effects that we will discuss below). The gate voltages are then shifted so that each row of gates shifts their charge sequentially until it reaches the end of the row, where the signal is amplified and read out.

In this way, an entire row of charge can be read out with each pixel's charge preserved in the sequence. Electronics then convert this signal, pixel-by-pixel, into the representation of an image. The amount of charge stored in each pixel being proportional to the intensity of the light shone upon it. The surface can be arranged with pixels sensitive to different wavelengths of light, thus permitting sensitivity to color.

(We note that the concept presented here is overly simplified and there are alternative read-out schemes in modern devices that greatly speed taking the next image.)

A key background process that confounds the collection of a reliable image is a set of phenomena collectively known as *dark current*. Consider a CCD that is completely shielded from light. In a perfect device, reading out the CCD in this situation should obtain zero charge in each column. This would result in a perfectly black image. However, in a real-life CCD, what often occurs is a *non-zero charge readout*, where pixels have collected electrons under their gates even though no light reached the CCD. This results in a "noise image" prompted by an apparent electron-generating process. Such dark current can readily be induced by the random thermal motion of atoms (heat) in the CCD. Two key effects result from such random motion.

The first occurs in the depletion region in the substrate, induced by the electric field from the gate. Electrons drift closer to the gate and holes go deeper into the bulk; the region between is depleted in charge carriers. This recombination of electrons and holes will occur due to random motion, thus generating a current.

The second effect results from *diffusion* of charge carriers. The volume of substrate below the depletion region is not entirely emptied of its mobile charges. The electric field from the gate, which acted to create the depletion region by moving holes toward the grounded side of the substrate, results in a non-uniform charge distribution with more holes accumulated deep in the substrate and fewer holes closer to the depletion region. This non-linear charge population causes charges to diffuse from regions where they are more highly concentrated (closer to ground) to where they are less concentrated (closer to the depletion region). The effect is similar to how molecules in a gas will tend to move from regions of higher density to those of lower density. This movement results in a net current.

The band gap, combined with the temperature of the material, acts to control which process dominates. As these are quantum mechanical phenomena, the heat energy scale need not be exactly equal to the band gap to induce such effects. Diffusion of charge across an energy barrier can happen spontaneously via quantum tunneling. The kinetic energy of the charges determines the probability that an interaction with the barrier will result in charges appearing on the other side, and $T \propto \langle K \rangle$. This process can be adequately described, at the pixel level, by a combination of the Arrhenius law and the Meyer-Neldel Rules [Wid+02],

$$I_{dark} = I_0 \exp\left[\Delta E \left(\frac{1}{E_{MN}} - \frac{1}{k_B T}\right)\right]. \tag{4.107}$$

For silicon, a fit of the above model to the observed dark noise from a 0.2 Mpix device, operated in the temperature range $T = [222, 291]$ K, yields the parameters $I_0 = 1685$ e/s, $E_{MN} = 25.3$ meV [Wid+02]. The energy barrier ΔE is observed to vary with temperature in the device, with a typical value of 1.14 eV above 260 K and $\Delta E \approx 0.57$ eV below that temperature (half the silicon band gap). In this silicon device, T=260 K corresponds to the point at which the two major contributors to dark current—depletion current and diffusion current—swap dominance (diffusion current dominates at low temperature). From this, one can predict, at room temperature, a dark current of about 1600 e/s, per pixel.

This dark current rate is why scientific-grade CCD instrumentation is cooled below room temperature. Reducing thermal noise allows for a more "true black" image to be produced when no light actually shines on the focal plane. CCDs used in dark matter experiments are cooled to about 130 K, yielding an expected dark current rate of about 10^{-9} e/s. This yields $\approx 10^{-4}$ e/pix/day and permits much greater sensitivity to rare processes on a one-day or multi-day timescale. Improving CCD design and readout, as we see below, leads to even greater sensitivity.

CCDs are used as dark matter detectors by taking light entirely out of the process. CCDs can be enclosed within cold and extremely light-tight environments. The CCDs can be placed into an operational mode capable of collecting electrons beneath the gates when electron-hole pairs are produced.

In the absence of light, the same processes that induce signals in cryogenic solid-state, noble liquid, or super-heated, liquid detectors will similarly have the potential to create electron-hole pairs in the CCD. These can be caused by alpha, beta, gamma, or neutron radiation.

Dark matter nuclear recoils can also produce signals under these conditions. The usual strategies would then apply: the need for radiopure-material construction as well as screening (c.f. Ref. [Agu+15b]) and layers of shielding to protect the CCDs.

A dark matter signature in a CCD will look similar to its expectation in other technologies: a single-site (a single pixel) nuclear or electron interaction (where interactions with the electrons are favored for very light dark matter candidates below the nuclear or nucleon mass scales). Neutrons would similarly produce such single-site (nuclear) interactions, while betas, gammas, and alphas will ionize over longer lengths of the CCD (many pixels).

CCD experiments thus look for single-pixel or very-few-pixel interactions with as few photoelectrons as can be reasonably detected by such an instrument. The lower the mass of the dark matter, the more likely a single-electron interaction will occur, necessitating single-electron-per pixel sensitivity.

The decade of 2010–2019 saw a rising generation of CCD experiments. This began with the DArk Matter In CCDs (DAMIC) experiment [Agu+19]. The design then evolved to DAMIC-M at the LSM [Nor+22] and the Sub-Electron-Noise Skipper-CCD Experimental Instrument (SENSEI) [Abr+19; Ada+23; Bar+20b; Tif+17] at Fermi National Accelerator Laboratory (SENSEI@MINOS) as well as at SNOLAB (SENSEI@SNOLAB). The dark matter CCD program will soon culminate in a large-scale apparatus, the Observatory of Skipper CCDs Unveiling Recoiling Atoms (OSCURA), intended to be sited at SNOLAB [Agu+22].

While each generation of the CCD detector design varies in complexity and size, all of them operate on the basic principles outlined above. The collaborations designing, building, and operating these experiments have (as with all other dark matter programs) learned lessons from each successive instrument (and have often utilized new information gleaned from a next-generation detector). SENSEI, for example, was the first operating dark matter experiment to utilize skipper CCD technology, a more recent CCD design that suppresses noise and enables the experiment to read out individual photoelectrons from the pixels. This technology has demonstrated single photoelectron sensitivity.

OSCURA will represent a significant upscaling of the designs used in DAMIC, DAMIC-M, and SENSEI.

We can illustrate the impact of this technology using the SENSEI experiment. When all other external sources of background have been controlled (e.g., radiation from materials

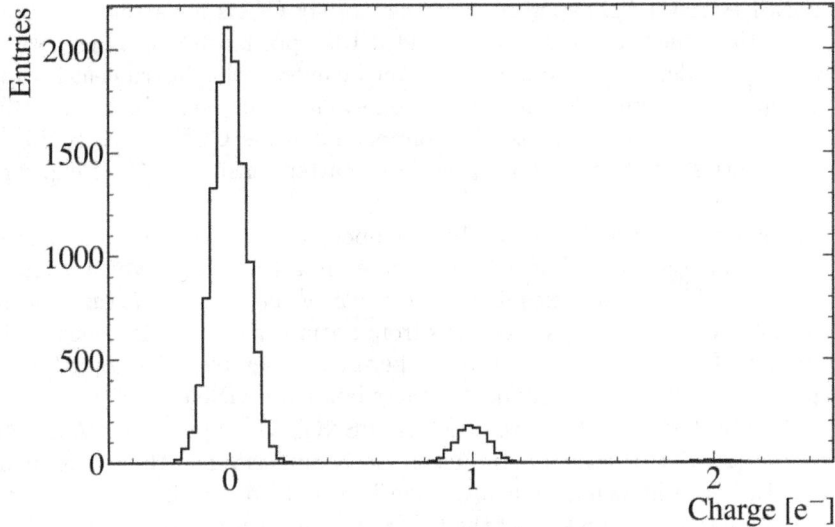

Figure 4.42: A demonstration of single-electron sensitivity in the SENSEI Skipper CCD design. Under low-light conditions, the CCD collected 4000 samples per pixel. The binning of the histogram is 0.03 e^- and demonstrates excellent resolution on the integer charge expectations. Figure from Ref. [Ada+23].

surrounding the CCDs), the key problem that limits sensitivity of these experiments is dark current.

For example, in traditional scientific-grade CCDs prior to the skipper CCD design, dark current in individual pixels was capable of generating typical readout noises of two electrons. This meant that it was impossible to distinguish an interaction in a single CCD pixel from noise, unless the interaction was capable of producing more than two photoelectrons per interaction. Such requirements would ultimately, due to the recoil energy consideration, limit the low-mass sensitivity of an experiment. DAMIC, for example, reported in its first results a sensitivity limit that corresponded to a dark current of 10^{-3} e^-/pixel/day [Agu+17], translating into 100 million pixels expected to contain at least one photoelectron. This dark current level is consistent with the scale of the prediction from the Arrhenius and Meyer-Neldel approach. For the original SENSEI design, with 280 million pixels, this was an implausibly high single-electron background [Tif+17].

The Skipper CCD design suppresses this noise and permits single-electron signal readout [Cha+90; Jan+90]. A demonstration of this capability is shown in Fig. 4.42.

SENSEI reports a dark current of $(1.459 \pm 0.020) \times 10^{-4}$ e^-/pix/day, or (419 ± 6) e^-/g · day given their target mass (see below). In their existing published results, however, they do not yet utilize the single-electron channel, instead focusing on the analysis of events in the $[2-10]$ e^- region of their capability. This has to do with the level of noise still present in this generation of instrument for the lowest multiplicity electron yield.

The SENSEI@SNOLAB experiment consists of six CCDs for a total of about 13 g of mass. Together, these devices form a 6144 × 1024 pixel = 6.3 Mpix camera. While well below the pixel density of modern commercial cameras, this instrument is capable of observing a single electron-hole pair production event in the region below a gate. Such performance is unnecessarily precise for commercial-grade CCDs (e.g., mobile phone cameras) but entirely critical to low-mass dark matter sensitivity. The experiment is shown in Fig. 4.43.

The CCDs are cooled and shielded by a copper box and, surrounding part of that box, a thicker copper shield. That is then enclosed in a lead shield which is enclosed in a polyethylene shield. Water boxes line the exterior of the entire structure and provide the outer-most protection against neutrons from natural fission by isotopes in the rock surrounding SNOLAB. Fully enclosed, the experiment resembles a large and featureless cube (Fig. 4.43). Only the exteriors of the water boxes are visible.

The first published data from the SENSEI@SNOLAB experiment [Ada+23] used 534.89 g-days (293.53 of which were used for commissioning the experiment; the remaining 241.35 were hidden from analysis until the data processing methods were fully commissioned), with each exposure of the CCDs lasting twenty hours (Fig. 4.44a). After applying quality requirements on exposures and removing those parts of the images having "hot pixels" (a common problem in CCDs, where pixels misbehave at the manufacturing level and must be masked out) or bad columns of pixels (due to a failure of the readout at the edge of the column), SENSEI@SNOLAB was left with 100.72 g-days of high-quality data (55.13 for commissioning, of which 45.59 were hidden for final analysis). These data formed the basis of their published dark matter search.

The SENSEI collaboration defined "signal events" as single-pixel hits consistent with the liberation of 2–10 electrons (above 10, which corresponds to a higher energy threshold, they are not competitive with other experiments that are sensitive to that same region). They reported $\mathcal{O}(1-10)$ observed events after all selection criteria. They do not yet employ a background model, so they obtain limits only on the dark matter interaction cross-section by treating the background contribution to their data at a nuisance parameter. As a result, their best description of the data occurs when no signal is assumed and all observed events are treated as background. They then interpret the results in difference scenarios, one of which we briefly explore (illustrated in Fig. 4.44b).

A recent interpretative framework for many dark matter experiments has been the use of the Migdal effect [Ibe+18; Liu+20; Mig41]. This effect occurs when the nucleus of an atom is perturbed (as by the emission of an alpha or beta particle, or by an interaction with a particle that transfers a low amount of energy to the nucleus). This results in electronic excitation (electrons perturbed into new energy states, with transitions possible), nuclear recoil, and lattice vibration in a solid-state device (phonon creation).

The quantum mechanical effect of this perturbation is to alter the nominal solutions to the Schroedinger Wave Equation for a multi-electron atom, and, in particular, to alter the allowed energy levels of the atom. If the entire atom moved as one, this would not be a problem; however, there is a brief time delay between the interaction that perturbs the nucleus and the response of the electron cloud (as a whole) to that perturbation. As a result of this time displacement in reactions, electrons in orbit around the nucleus

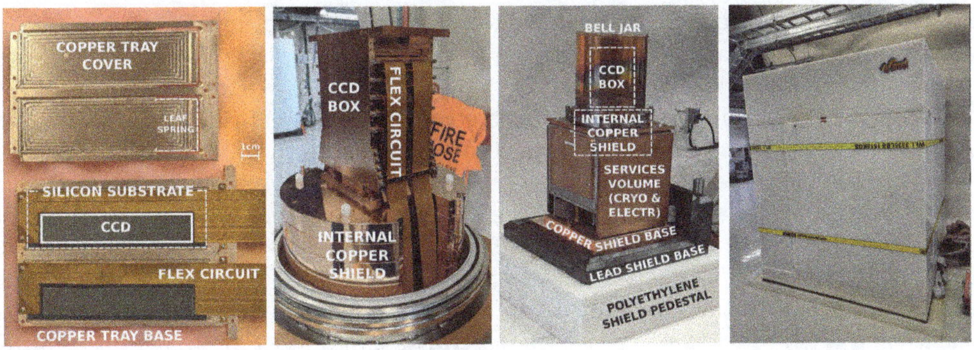

Figure 4.43: The SENSEI detector as originally constructed at SNOLAB. The left view shows two CCD modules in their copper tray. The second-from-the-left view shows the copper housing that cools and shields the CCD payload. The second-from-the-right view shows the sealed vacuum vessel housing the components mentioned earlier, but prior to the final installation of the outer copper, then lead, and then polyethylene and water shields. The right view shows the fully constructed SENSEI detector showing that only the water boxes are visible. The left three images are from from Ref. [Ada+23] while the right image is by the authors.

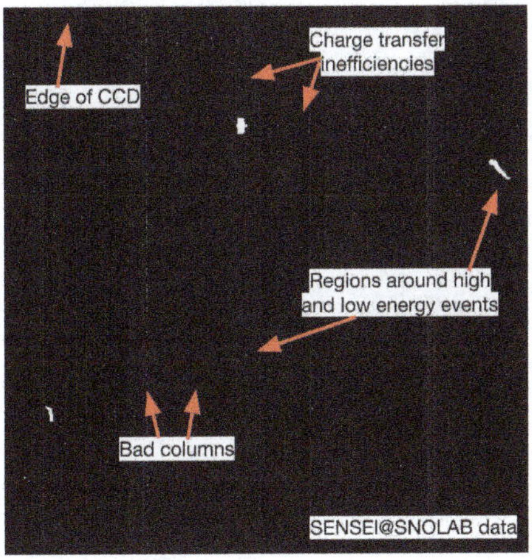

(a) SENSEI 20-hour exposure

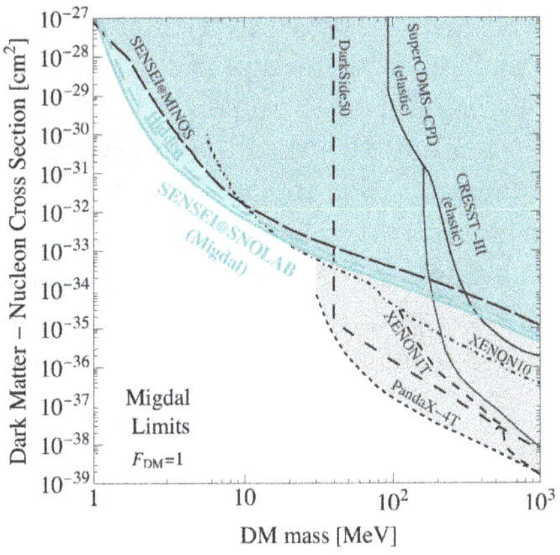

(b) Data interpretation using the Migdal effect

Figure 4.44: (Left) A twenty-hour CCD exposure taken by SENSEI. The most prominent features are bad pixels and some bad columns of pixels. High-energy particle collisions, as from alpha or neutron radiation, are exceedingly rare in a twenty-hour window owing to the good design of the experimental enclosure and use of radiopure materials. (Right) An example of the interpretation of first results from the SENSEI@SNOLAB program, assuming dark matter interactions with nuclei that induce the Migdal effect. Figures from Ref. [Ada+23; Sti23].

can experience transitions to new energy levels (resulting in the emission of one or more photons) and/or ionization (removal from atomic orbit). This represents an *inelastic* collision of a particle with the nucleus, e.g. $\chi + A \rightarrow \chi + A + n\gamma + me^-$, where $m, n \geq 0$. The Migdal effect has never been directly observed, but it is a natural consequence of the application of non-relativistic quantum mechanics with a perturbed atomic potential.

SENSEI is one of the most recent experiments to utilize the Migdal effect as a means to interpret scattering data. However, such interpretations have earlier been used by multiple experiments including the XENON [Apr+19a] and SuperCDMS [Alb+23] Collaborations.

The ejection of an electron from one of the atomic shells can, in principle, result from any of the unperturbed electronic states. However, the valence electrons in atoms bounds in a crystal structure (e.g. silicon or germanium lattices typical in semiconductor growth) are not part of any single atom but are shared between atoms bound in the lattice. This complicates the expected effects of a nuclear recoil inducing the Migdal effect on these outer-most electrons. The electrons in the inner shells are easier to deal with as they can be considered solely bound to the parent nucleus. Interpretations attempt to accommodate these complications, although such efforts are still in their early stages.

We close this section by noting that a wide array of technology has been brought to bear on the question of the nature of dark matter. The use of scientific-grade CCDs for this is one of the most recent developments. Compared to super-heated liquids, cryogenic liquids, and large semiconductor crystals, this technology is among the newest. It has passed well beyond a demonstrator phase and completed at least one significant analysis.

CCDs have proven particularly useful at beginning to further constrain the parameter space of very low-mass particle dark matter candidates, owing to the particular sensitivity to few-electron scattering.

Our use of the Migdal effect at this stage in the text is meant to suggest that there is a potentially wide range of subtle phenomena, not only in solid-state devices, but also in their liquid counterparts, that could be evaluated from the literature (some of it published decades ago in the context of fundamental phenomena in specific subjects) and utilized as novel experimental approaches to underlie a new generation of experiment. We have explored some of those possibilities in Chapter 7.

4.8 Conclusion

The next decade will be very exciting for dark matter direct detection. The next stages of multiple large-target experiments (called Generation 2 or "G2" experiments) with sensitivity even to solar neutrinos will be coming online, collecting new data sets, and improving sensitivity across a wide mass range to various dark matter models. Research and development for even more massive xenon and argon detectors is occurring within the XLZD (XENON, LUX-ZEPLIN, and DARWIN) and GADM (Global Argon Dark Matter) collaborations. As the names imply, these proposals represent the merging of past-generation experimental groups into large single collaborations aimed at constructing single-technology large-scale detectors at the multi-tonne scale.

The solid-state detector program is making fast progress in pushing its technologies to sensitivities to the lowest mass dark matter candidates. Programs like SuperCDMS and CRESST are aiming for next-generation facilities and designs while CCD-based dark matter detectors have evolved rapidly and begun major operations. Super-heated liquid detectors are also making strides. They are attractive because they provide the unique ability to change target liquids (such as out C_3F_8 or CF_3I). In the event of a discovery, this allows the experiment to select another target to favor interactions with certain nuclei, or take advantage of differing spin structures of nuclei (e.g., nuclei with unpaired nucleon spins, enhancing sensitivity to spin-dependent dark matter interactions). These designs aim for large-target volumes with low energy thresholds, allowing them to both lower the mass range in which they can search while allowing for spin-dependent and spin-independent searches.

We conclude this chapter on the direct detection of dark matter by showing the current state of the constraints on dark matter's particle properties. These constraints are limited to a mass range of $m_\chi = [0.3, 1000]$ GeV and to cross-sections with either nucleons or electrons generally below $\sigma_\chi = 10^{-29}$ cm^2. An average Milky Way dark matter halo density of 0.3 GeV/cm^3 is assumed. Due to the absence of a positive and confirmed observation of a signal in direct detection experiments, it is then only possible to make mathematical and statistical statements about the degree to which certain dark matter properties, defined within certain broad scenarios, have been excluded by non-observation.

For example, let us imagine an experiment that should have easily detected a dark matter candidate with a mass of 1 GeV and a nucleon interaction cross section of 10^{-40} cm^2. If such an experiment sees no evidence for such a particle, even with careful (or even ideal) accounting for background processes in the data, then one can say with a certain degree of statistical confidence that a dark matter particle with this set of properties is "ruled out." This is known as an *exclusion* in this scenario (dark matter interacting with a nucleon). However, if for that very same experiment, it was not possible to observe a particle with mass 1 GeV and a nucleon interaction cross-section of 10^{-42} cm^2, even given ideal operating conditions, one would then say that this hypothesis was not excluded by the measurement. It would require a future experiment, with more sensitivity to rarer interactions, to make a statement about the validity (or not) of this rarer process. Generally, experiments scan over hypotheses (in pairs of values of m_χ and σ_χ) and then build *exclusion contours*, graphs of interaction cross-section vs. mass, where regions indicated in solid shading (or above or below a contour line, as indicated by the legend in the plot) are excluded and those left unshaded are unexcluded.

Constraint plots can be generated by a variety of tools, and we demonstrate current exclusions using the Dark Matter Limit Plotter [Tar24]. These constraints range over a few scenarios: the spin-independent (SI) scattering case (Fig. 4.45), the spin-dependent (SD) scattering off the neutron (SDn, Fig. 4.46), and electronic recoils assuming either no dark matter form factor ($F_{DM} = 1$, Fig. 4.47), or one that goes as $F_{DM} \propto 1/q^2$ (Fig. 4.48), where q is the momentum transfer in the collision. Only experiments that have reported results since 2018 are included. The tool was last updated with new data on October 12, 2024. The cross-sections at which the neutrino background would become visible to experiments, mapped into this interpretive framework, are also shown. Since the neutrino background rate depends on the target material (silicon, xenon, etc.) that effect is shown using the target that has gotten closest to the neutrino fog in some mass region of dark matter candidate space.

Some general conclusions can be drawn from the constraint plots. First, there is still considerable work to be done at both high and low mass to close the gap between current experimental sensitivity and the neutrino interaction region. As we have emphasized, encountering the latter bound is not the end of the search for dark matter, but opens a new and challenging era that might be addressed by ideas in Chapter 7. There is an especially great deal of work needing to be done in the spin-dependent scattering interpretation, where more of the cross-section space is still open to constraint or discovery. We anticipate that the coming generation of detectors will make significant progress in this area, though opportunities abound for new ideas.

In the electronic recoil interpretation, it is clear that a great amount of progress has been made in a short time. While nuclear recoil experiments have results going back well before 2018, there are very few such constraints from experiments prior to that era. Nevertheless, the technology has advanced quickly and is making progress in mapping out the space of possibilities.

Of particular note is that the freeze-out production scenario is beginning to be probed by the current generation of experiments and will definitely be a wide target for this type of experiment and their near-term successors. The freeze-in scenario is still about two orders of magnitude (or more) in cross-section sensitivity below the existing results, but it is clear that, as collaborations make rapid progress, that space is expected to close during the next decades.

Although WIMPs, as a general class, remain a viable and interesting candidate, other dark matter scenarios are gaining traction in the theoretical community as technological advances allow for the exploration of these ideas. This has opened a new window of exploration for all detector technologies and spurred the creation of other technologies, which are summarized in Chapter 7. Given the wealth of possibilities, a diverse set of experimental designs and targets are needed to constrain the theory and couplings of any discovered signal.

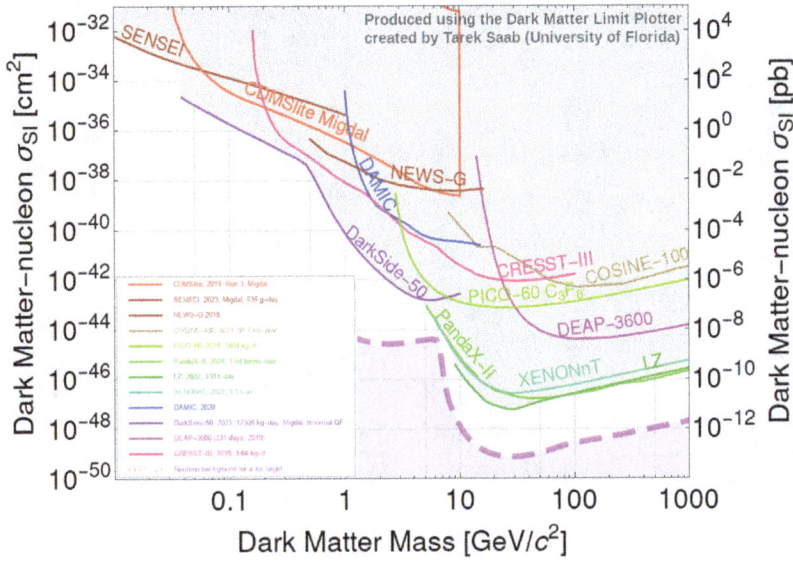

Figure 4.45: Exclusion limits in the spin-independent nuclear scattering scenario. Data used are from Refs. [Aal+22; Abd+19; Ada+23; Adh+21; Agn+23; Agu+20; Aja+19a; Alb+23; Amo+19; Apr+23; Arm+22; Arn+18b; Bo+24; Col18; Liu+22; Rup+14].

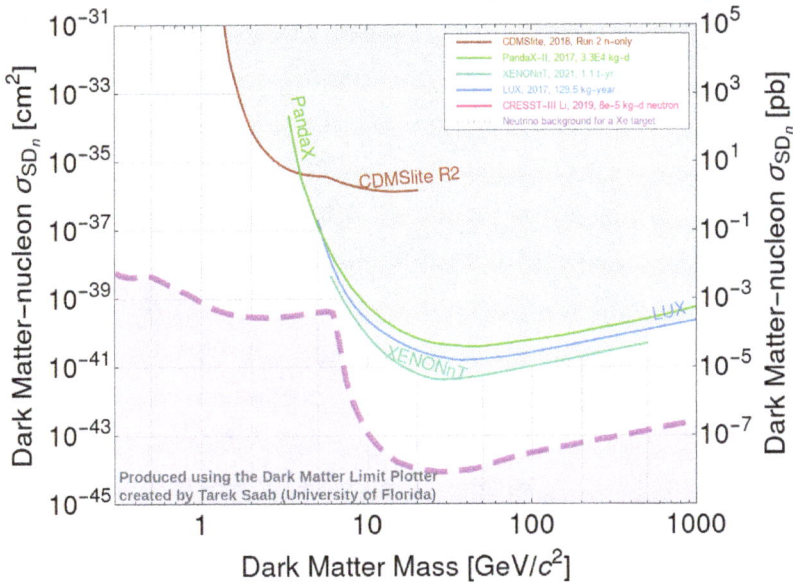

Figure 4.46: Exclusion limits in the spin-dependent neutron scattering scenario. Data used are from Refs. [Agn+18d; Ake+17; Apr+23; Fu+17; Rup+14].

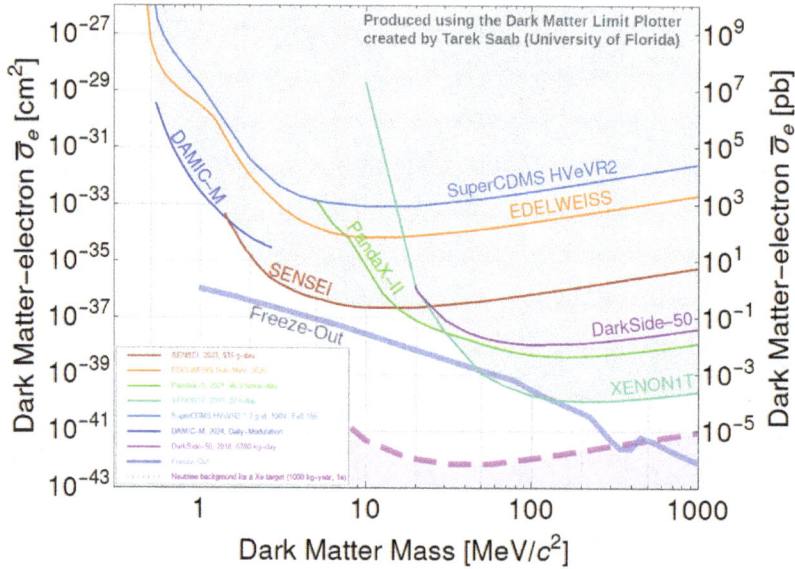

Figure 4.47: Exclusion limits in the electronic recoil scenario, assuming no dark matter form factor ($F_{DM} = 1$). Data used are from Refs. [Ada+23; Agn+18a; Ama+20; Apr+19c; Arn+20; Arn+24; Che+21; EMV12; ESY18; EVY17].

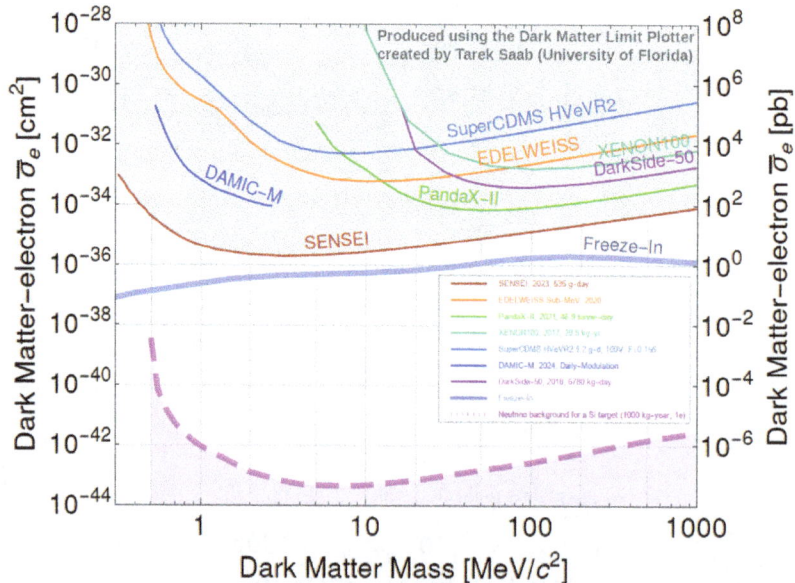

Figure 4.48: Exclusion limits in the electronic recoil scenario, assuming a dark matter form factor $F_{DM} \propto 1/q^2$. Data used are from Refs. [Ada+23; Agn+18a; Ama+20; Apr+19c; Arn+20; Arn+24; Che+21; EMV12; ESY18; EVY17].

CHAPTER 5

Dark Matter Detection
by Annihilation and Production

We have thus far focused on finding evidence of non-gravitational dark matter interactions by using galactic dark matter as a "beam" that can strike a stationary target on the earth. Direct observation, in a laboratory setting, of a non-background nuclear or electron recoil would be strong evidence for dark matter having interactions with its standard model (SM) counterparts.

There are other places in the universe, beyond the neighborhood of the Milky Way, where dark matter will congregate in greater amount in potentially far denser populations. This brings with it the possibility for observing dark matter interacting not just with SM matter, but interacting with itself. Since these interactions can be observed only if SM particles result from these processes, the signatures would be expected to appear as the *anomalous* production of particles such as photons, cosmic rays, neutrinos, and so forth, that arise in places where a population of such particle output cannot be explained by SM means. This is what is meant by *indirect detection* of dark matter interactions.

Indirect detection is a method that brings to bear a synthesis of astrophysical observations to look for excesses in the universe that might explain a dense or intense population of dark matter that is experiencing non-gravitational interactions that produce radiation. In this chapter, we will review approaches and observations that use this indirect means of searching for dark matter properties.

Direct detection by scattering is, of itself, a subject fraught with theoretical challenges, including the degree of, and the mechanism for coupling dark matter to normal matter, the spin structure of the interaction, and the model of the Milky Way's dark matter halo. Indirect detection is challenged by additional considerations such as the specific nature of the coupling between dark matter and the exact particles that can ultimately be detected over long distances in the cosmos. The halo models play a role here, too, as astronomers cannot search everywhere all at once and must target places where dark matter is expected to most accumulate. In addition, as in ultra-pure materials for direct detection, there is a degree of uncertainty for indirect detection due to the incomplete knowledge of all nominal sources of SM particles from astrophysical objects and phenomena other than dark matter.

Finally, we will also explore attempts to produce dark matter directly, using particle colliders. While indirect detection relies on annihilation into SM particles, collider approaches rely on the opposite mechanism: annihilating SM particles in the hope of producing dark matter. This approach has significant challenges as well, not the least of which is that the specific interactions between dark and normal matter are completely unknown, except for their gravitational interaction.

Since dark matter is "out there" in the cosmos, when its constituents interact with a terrestrial detector or cause excessive SM-based signatures in astronomical observations, there is a general confidence that one is probing the extant dark matter population. However, it is more challenging to tease from direct detection experiments the exact properties of the dark matter particles. The separation of signal from background is also made challenging by the uncertainties in both detector material properties and local astrophysical phenomena.

A particle collider might produce a new and rare particle. Such a discovery alone does not guarantee a correspondence to dark matter. This is because the particle is not being observed in the context of a host astrophysical population. Collider experiments offer a high degree of control over experimental conditions, which can make much easier the measurement of the properties of particles; but the fact that it makes particles derived from other particles does not directly tell scientists about the role of the new particle. Nevertheless, in this hunt, it is important that all approaches be pursued. Ultimately, all three approaches—direct, indirect, and collider—will play an additive role in the final understanding of dark matter.

5.1 Mechanisms for the Indirect Detection of Dark Matter Interactions

Mechanisms that can begin with one or more dark matter particles and end in SM particles are shown in Fig. 5.1. The dark matter that structures galaxies and galaxy clusters must have a long lifetime (stable on the timescale of the present universe) but it does not

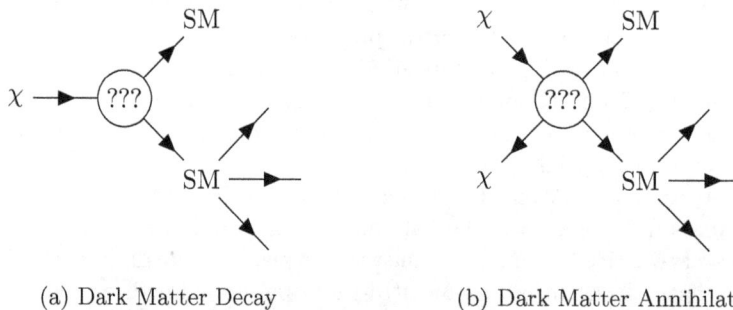

(a) Dark Matter Decay   (b) Dark Matter Annihilation

Figure 5.1: Examples of processes involving dark matter that can end entirely in SM particles.

have to be stable forever. A long, but finite, lifetime is not ruled out by our current cosmological and astrophysical understanding of dark matter processes. If that is the case, dark matter will decay (Fig. 5.1a). Dark matter may also be able to interact with itself through non-gravitational means, resulting in dark matter *annihilation* that results in SM particles (Fig. 5.1b). Annihilation will likely be observable in places where the density of dark matter is great. Decay can, in principle, happen in any population of dark matter, but it would be easiest to observe in places where dark matter's density is naturally high.

There are a few ingredients to look for when conducting indirect searches for dark matter:

- Production rates of the relevant SM particles (which are referred to as "messengers"). These will be related to the annihilation or decay of dark matter.

- Energy scale of the SM messengers. This is set by the mass of the dark matter particle.

- Annihilation (or decay) products will be largely dependent on the model that explains how, precisely, these interactions actually produce SM particles.

Decaying Dark Matter

If dark matter is heavier than some, or many, SM particles, then there must be some principle that *stabilizes* dark matter for long periods of time. This will be a very model-dependent effect. We can look to a specific model, such as a supersymmetric SM, that provides a mechanism on which we can focus attention. This is by no means the only way to approach this subject.

In supersymmetry, one can invoke a new discrete symmetry (analogous to C, P, and T described in Section 1.3, but related to an internal property of matter) referred to as R-parity, that can prevent (if conserved) the decay of supersymmetric partners to their SM counterparts. For example, imagine such a parity operator whose quantum number is defined as:

$$P_R |X\rangle = (-1)^{3B+L+2s} |X\rangle \equiv (-1)^{3(B-L)+2s} |X\rangle \qquad (5.1)$$

where B is the baryon number of a state, L is the lepton number of a state, and s is the spin quantum number of a state. Since superpartners are spin-partners of their SM counterparts (the superpartner of a SM scalar or vector is a spin-1/2 fermion; the superpartner of a SM spin-1/2 fermion is a scalar), then we will find that

- R-parity is +1 for SM particles (they are R-parity even)

- R-parity is −1 for superpartners (they are R-parity odd).

Let's try some examples. Consider a scalar SM state, denoted $|0\rangle$ (the Higgs boson). This has $B = 0$ (not a baryon), $L = 0$ (not a lepton), and $s = 0$. When acted upon by the R-parity operator:

$$P_R |0\rangle = (-1)^{0+0+0} |0\rangle = (+1) |0\rangle. \qquad (5.2)$$

Consider an electron, $|e\rangle$, with $B = 0$, $L = 1$, and $s = 1/2$:

$$P_R |e\rangle = (-1)^{0+1+2\cdot\frac{1}{2}} |e\rangle = (-1)^{1+1} |e\rangle = (+1) |e\rangle. \tag{5.3}$$

Now consider the selectron, the scalar superpartner of the electron, $|\tilde{e}\rangle$, with $B = 0$, $L = 1$, and $s = 0$:

$$P_R |\tilde{e}\rangle = (-1)^{0+1+0} |\tilde{e}\rangle = (-1)^{1} |\tilde{e}\rangle = (-1) |\tilde{e}\rangle. \tag{5.4}$$

From this, we see that any initial state that is composed only of SM particles or only of superparticles, the R-parity of the initial state will be preserved in a final state similarly composed. However, if the initial and final states have different numbers of each kind of particle, then R-parity will not be conserved. The question of whether supersymmetry is a valid natural symmetry, and the further question of whether R-parity is a conserved symmetry in such theories, is wholly unresolved.

In supersymmetric theories, there is a lightest electrically neutral particle, the *neutralino*, denoted χ^0. This is a superposition of several superpartners of SM particles, specifically the H, Z, γ, and g (gluon). Those partners, the higgsino, zino, photino, and gluino, are all spin-1/2 fermions that superpose into a single neutralino state. The neutralino makes an excellent dark matter candidate. If R-parity is conserved in all interactions, then the neutralino will be stable. This is excellent news for dark matter, as we would need dark matter to generally be stable on the scale of the lifetime of the universe.

However, symmetries of a low-energy theory are often broken by high-scale physics. As an example, assume that a dark matter particle with mass $m_\chi = \mathcal{O}(10 \text{ TeV})$ decays via a dimension-5 quantum field theory operator that is normally suppressed below the grand unification (GUT) energy scale ($M_{GUT} \approx 10^{16}$ GeV)[1]. This corresponds to

$$\Gamma_5 \sim \frac{1}{M_{GUT}^2} m_\chi^3 \longrightarrow \tau_\chi = \frac{\hbar}{\Gamma_5} \sim \hbar \cdot M_{GUT}^2/m_\chi^3. \tag{5.5}$$

This yields a lifetime of about one second. Clearly, that is far too short. What if, instead, the suppression was at the Planck energy scale ($\sim 10^{19}$ GeV)? This yields a lifetime for dark matter of only about three years. This is still far too short.

We can instead presume that the decay proceeds though a higher-order dimension-6 operator. We can again assume the suppression occurs at the GUT scale. Here

$$\Gamma_6 \sim \frac{1}{M_{GUT}^4} m_\chi^5 \longrightarrow \tau = \frac{\hbar}{\Gamma_6} \sim \hbar \cdot M_{GUT}^4/m_\chi^5. \tag{5.6}$$

---

[1]The choice of a dimension-5 operator is motivated in, for example, extensions of the SM Lagrangian that incorporate lepton-number violation and accommodate a Majorana fermion description of the neutrino. This would be necessary if, for example, the neutrino is determined by experiment to be its own antiparticle (through the lepton-number-violating process of neutrinoless double beta decay) [BP78; Par22]. Such fermions defy description in the Dirac fermion model, which applies to all known fermions except where it is unresolved in the case of the neutrino. Should such an extension be necessary, there is a simplest dimension-5 operator that provides lepton-number violation and accommodates Majorana fermions.

It is worth noting that it is exciting that the question of the nature of the neutrino and the nature of dark matter could be linked puzzles.

This results in a lifetime of $\sim 10^{26}$ s, which is far in excess of the lifetime of the universe ($\sim 10^{17}$ s). Such an addition to the SM Lagrangian would essentially preserve the relic abundance of dark matter left from the beginning of the universe. In addition, no dark matter would ever decay into visible particles to be observed at a collider experiment, where the observing times are extremely short (far less than one second). If dark matter is a non-standard-model particle, the number of such particles in our galaxy is expected to be large enough that, despite this long lifetime, we would, nonetheless, observe them decaying frequently. If, in addition, the decay is to SM particles, we could observe the SM products of the decay.

How many SM particles might be produced by the decay of unstable, long-lived dark matter particles? Consider a volume element dV at a distance r from the earth. In spherical coordinates, this element would be written $dV = r^2 \sin\theta dr d\theta d\phi$. Let n be the number density of dark matter in the volume dV and assume the lifetime of our dark matter particle, χ, satisfies $\tau_\chi \gg t_{universe}$, the age of the universe. At a given time, t, the population of dark matter particles in a region of space will be

$$N = N_0 e^{-t/\tau_\chi} \rightarrow n = n_0 e^{-t/\tau_\chi}. \tag{5.7}$$

where we have implicitly divided the left-hand equation by the common observing volume to obtain the right-hand equation. The rate of change in number density will be given by:

$$-\frac{dn}{dt} = \frac{n_0}{\tau_\chi} e^{-t/\tau_\chi} = n(t)/\tau_\chi. \tag{5.8}$$

We can write the rate of decay at any time using the number density and the differential volume as:

$$-\frac{dN}{dt} = \frac{n(t)\,dV}{\tau_\chi}. \tag{5.9}$$

Assume the system is steady-state, and each decay produces one observable particle. Let A be the detector area and r the distance from dV. The flux of particles observed will be

$$\frac{dN}{dt} = \frac{A}{4\pi r^2} \times \frac{n(t)\,dV}{\tau_\chi} \tag{5.10}$$

$$= \frac{A}{4\pi r^2} \frac{n(\vec{r},t)r^2 \sin\theta d\theta d\phi dr}{\tau_\chi} \tag{5.11}$$

$$= An(\vec{r},t)\frac{d\Omega}{4\pi}\frac{dr}{\tau_\chi}. \tag{5.12}$$

Let's use this to estimate the signal from the Milky Way halo in a 1 m^2 detector. The number density $n(\vec{r},t)$ is that which is within a 1 kpc radius of the earth at a time, t. The density of dark matter in our region of the Milky Way is presently about 0.3 GeV/cm^3 and $n = \rho/m_\chi$. We assume a spherically symmetric distribution and integrate this over solid angle and radius to obtain

$$\frac{dN}{dt} = \frac{Anr}{\tau_\chi} = \frac{A\rho r}{m_\chi \tau_\chi}. \tag{5.13}$$

Using our test case, where $m_\chi = 1$ TeV$/c^2$ and $\tau_\chi = 10^{26}$ s, we find that $dN/dt \approx 10^{-4}$ s$^{-1} \sim 3000$ per year. The challenge, of course, is to somehow discriminate that from the billions (or more) SM particles produced all the time by solar wind, cosmic rays, etc. Such background SM particles continuously rain down on the earth.

Annihilating Dark Matter

Dark matter is not forbidden from self-annihilation by whatever makes it stable. For example, considering our toy supersymmetry model, each of a pair of dark matter particles has R-parity of -1. As such, they can annihilate to a pair of SM particles, each with R-parity of $+1$, without violating parity conservation laws. To demonstrate this simply, consider an initial state of two dark matter particles that can be written in Dirac notation

$$|\chi_1, \chi_2\rangle = |\chi_1\rangle |\chi_2\rangle. \tag{5.14}$$

Consider then acting on this state with a pair of R-parity operators, one for each of the two particles:

$$P_{R_2} P_{R_1} |\chi_1\rangle |\chi_2\rangle = (-1)(-1) |\chi_1\rangle |\chi_2\rangle = |\chi_1\rangle |\chi_2\rangle. \tag{5.15}$$

If we act similarly on the final-state pair of SM particles,

$$P_{R_2} P_{R_1} |SM_1\rangle |SM_2\rangle = (+1)(+1) |SM_1\rangle |SM_2\rangle = |SM_1\rangle |SM_2\rangle. \tag{5.16}$$

R-parity is conserved in this annihilation process, so dark matter can be both stable and self-interacting.

Depending on the model, dark matter can annihilate to either a pair of photons or to other SM states that produce photons in secondary interactions. In general, photons couple to most of the SM fermions (all but the neutrinos) and to some of the gauge bosons (at tree level to the $W^\pm$). Those photons would emerge from dark matter annihilation as gamma rays for any dark matter candidate with mass above the MeV$/c^2$ level. Those gamma rays propagate, essentially unperturbed, from their origin, to the earth, and could be readily detected by satellite- or ground-based telescopes on Earth.

The number of annihilations per unit volume for two distinguishable annihilating particles with number densities n_1 and n_2 is given by

$$\sigma v_{rel} n_1 n_2 \tag{5.17}$$

where σ is the annihilation cross-section and v_{rel} the relative velocity of the two annihilating particles. You can think then of $n_2 v_{rel}$ as the flux of particles "on-target" and n_1 the number density of targets. Of course, if the particles are identical and indistinguishable, then $\sigma v_{rel} n_1 n_2 \to \sigma v_{rel} n^2/2$. The factor of $1/2$ appears to avoid double-counting the indistinguishable particles in the interactions.

This means that, for a fixed annihilation cross-section, the annihilation rate scales merely as ρ^2/m_χ^2. Of course, the cross-section will be a highly model-dependent statement; there are models in which this varies. For example, if the cross-section goes as $\sigma \propto 1/m_\chi^2$, then the annihilation rate goes, instead, as ρ^2/m_χ^4, and is suppressed more so by the dark matter mass. A clear take-home message from this is, however, that one should generally expect the annihilation signals to *weaken* as one considers heavier dark matter-candidate particles.

We can again consider how many SM particles could be expected from dark matter annihilation reactions. The annihilation rate per unit volume will be given by

$$-\frac{dN}{dt} = \langle \sigma v_{rel} \rangle \frac{n^2}{2}. \qquad (5.18)$$

Using the very same other assumptions, as in the case of dark matter decay, we arrive at the production rate of SM particles:

$$\frac{dN}{dt} = \frac{A \langle \sigma v_{rel} \rangle r}{2} \frac{\rho^2}{m_\chi^2}. \qquad (5.19)$$

Recall that the velocity-averaged cross-section is about 3×10^{-26} cm^{1}s^{-1}. Considering a 1 m^2 detector and, again, $m_\chi = 1$ TeV/c^2, we find that $dN/dt \approx 5 \times 10^{-8}$ s^{-1}, or about one event per year.

Constraining ourselves to look at dark matter annihilations only in the local neighborhood of our solar system (within 1 kpc) is clearly a losing strategy, especially given the need to disentangle these signals (gamma rays) from all other sources of gamma rays in the vicinity of Earth. Even assuming that there are no significant gamma ray sources within 1 kpc of Earth, one is then presumably looking for a diffuse source of high-energy gamma rays, and to observe it, assuming no background, would require counting for many years to integrate sufficient signal to claim an observation.

These are very simplistic approaches. For example, it is fair to ask: is the vicinity of Earth the best place to aim telescopes to look for high-energy gamma rays (or other particles) resulting from dark matter annihilations? Are there better places in the universe where dark matter may collect in much higher concentrations that would enhance the rates we have estimated here?

5.2 Standard Model Particles from Annihilations and Decay

Up to this point we have made some simple estimates. It is time to fill in more details. Dark matter annihilations, or decays, could, in principle, produce any SM particle. Most of the particles produced would be expected to decay on short time scales. For example, if dark matter annihilates to weak bosons, these have lifetimes of order 10^{-25} s and can decay to many lighter SM fermions, which, themselves, can also decay. In principle, the SM final states could provide a rich phenomenological portfolio for experimentalists, ... but also, in being so diverse, present a significant challenge to observing, and then interpreting, such signals.

At the end of any complex decay-chain of SM particles, one expects to find a reliable portfolio of stable particles: electrons, positrons, protons, anti-protons, photons, and neutrinos. It is these particles to which astronomers and astroparticle physicists must look if they hope to observe dark matter decay or annihilation.

Photons and Neutrinos

Photons and neutrinos travel to us in straight lines. (Or, more accurately, photons and, to a slightly lesser extent, neutrinos, follow geodesics in space-time.) We observe the number of photons or neutrinos arriving at the detector from a particular solid angle in the sky, within a particular time interval.

Let us again consider our detector of area A and the signal that arises from a volume dV that is located at coordinates (r, θ, ϕ) relative to Earth, which resides at $r = 0$. The spectrum of photons received at Earth, per volume per time, is then given by equations we have already derived,

$$\frac{dN_\gamma}{dE \, dt \, dV} = \left(\frac{dN_\gamma}{dE}\right)_0 \frac{A}{4\pi r^2} \times \begin{cases} \frac{1}{2}\langle \sigma v_{rel}\rangle n(\vec{r})^2, & \text{(annihilation)} \\ \frac{n(\vec{r})}{\tau_\chi}, & \text{(decay)} \end{cases} \tag{5.20}$$

where $\left(\frac{dN_\gamma}{dE}\right)_0$ is the energy spectrum of photons (or neutrinos) produced by each annihilation or decay. Integrating along the line of sight yields

$$\frac{dN_\gamma}{dE \, dt \, d\Omega} = \left(\frac{dN_\gamma}{dE}\right)_0 \frac{A}{4\pi r^2} \times \begin{cases} \frac{\langle \sigma v_{rel}\rangle}{2m_\chi} \int_0^\infty \rho(\vec{r})^2 dr, & \text{(annihilation)} \\ \frac{1}{m_\chi \tau_\chi} \int_0^\infty \rho(\vec{r}) dr, & \text{(decay)}. \end{cases} \tag{5.21}$$

Then we should integrate over the solid angle subtended by the object to obtain the full signal from the source.

$$\frac{dN_\gamma}{dE \, dt} = \left(\frac{dN_\gamma}{dE}\right)_0 \frac{A}{4\pi r^2} \times \begin{cases} \frac{\langle \sigma v_{rel}\rangle}{2m_\chi} \int \int_0^\infty \rho(\vec{r})^2 dr d\Omega, & \text{(annihilation)} \\ \frac{1}{m_\chi \tau_\chi} \int \int_0^\infty \rho(\vec{r}) dr d\Omega, & \text{(decay)} \end{cases} \tag{5.22}$$

The piece of the annihilation spectrum dependent on particle physics that is determined by the distribution of dark matter mass density $\rho(\vec{r})$ in the relevant environment is called the *J-factor*,

$$J_{ann} \equiv \frac{1}{8\pi} \int \int_0^\infty \rho(\vec{r})^2 dr d\Omega. \tag{5.23}$$

Thus, we can write

$$\frac{1}{A} \frac{dN_\gamma}{dE \, dt} = \frac{\langle \sigma v_{rel}\rangle}{2m_\chi} \left(\frac{dN_\gamma}{dE}\right)_0 J_{ann}. \tag{5.24}$$

J-factors of different sources characterize the relative size of the expected annihilation signal. They depend upon the presumed density profile. We explored some cases of these in Section 2.7. The NFW profile (Eqn. 2.148) is a standard choice. Using this assumption for the Milky Way, we have $J_{ann} \approx 10^{17-20} \text{GeV}^2 \text{cm}^{-5}$ for dwarf satellite galaxies like the Large Magellenic Cloud, the Small Magellenic Cloud, and Sculptor, etc. For a region that is within $1°$ of the galactic center of the Milky Way, we have $J_{ann} \approx 10^{22} \text{GeV}^2 \text{cm}^{-5}$.

Based solely on the *J*-factors, it seems we would expect the center of the galaxy to be more promising for annihilation searches, but we need to take potential backgrounds into account. The centers of galaxies are difficult because there are many astrophysical sources of particles given the denser clustering of baryonic matter (stars and dust) in such regions. In addition, expected signal depends strongly on the density profile model. Dwarf galaxies contain few baryons and would, thus, make cleaner targets for searches.

The dark matter velocity profile also contributes to the detection equations above. For example, dark matter particle velocity is generally slower in dwarf galaxies than in our own galactic center. As such, the expected signal is reduced in models where annihilation is suppressed at low velocities and enhanced in models where the converse is true.

The J-factors we estimated are based on a smooth NFW profile. Halo substructure could increase J_{ann} substantially. Large halos are thought to have grown by accretion of many smaller halos (as evidenced by large-scale structure simulations), and, so, there are reasons to expect halo "clumpiness" to play an important role in determining the location of promising signals. Annihilation can be enhanced in these small dense structures where $\langle \rho^2 \rangle \neq \langle \rho \rangle^2$ (because in these regions there may be substantial deviations from the mean dark matter density).

Let's consider an example: *an isotropic signal originating from dark matter annihilations in the intergalactic medium.* Let's begin by assuming that the density in this intergalactic region is equal to the overall cosmological dark matter density. We will integrate all photons (or neutrinos) originating from all distances (all red shifts). We will note that we are specifically interested in photon density and spectrum in a present-day volume, dV_0, arising from annihilations at *earlier* times. Earlier in time, the original volume that contained the interactions that emitted observable particles ("messenger particles") was smaller than it is now. That volume was also closer, originally, than it is now, also owing to the expansion of space.

Let us recall the fundamental relationships:

$$a = \frac{1}{1+z} \tag{5.25}$$

$$\frac{dV_z}{dV_0} = \left(\frac{a}{a_0} \right)^3 = \frac{1}{(1+z)^3}. \tag{5.26}$$

where dV_z is the physical volume at a red shift of z. Recall that a red shift of $z = 0$ indicates the present day. We will introduce cosmological red shift, z, into our detection equation by employing the chain rule:

$$\frac{dN}{dE\,dV} = \int_\infty^0 dz \frac{dt}{dz} \left(\frac{dN(z)}{dE} \right)_0 \frac{A}{4\pi r^2} \times \frac{\langle \sigma v_{rel} \rangle}{2} n(z)^2 \frac{dV_z}{dV_0} \tag{5.27}$$

where $dN(z)/dE$ is the spectrum of photons produced at red shift z that have energy E today (since cosmic expansion will act to red-shift photons, the energy of the photons, by the time they reach the earth, will be lower than when they started out from their original volume). In fact, we can easily define the energy of that red-shifted messenger from its original distance, at z, to its energy today,

$$E_z = E(1+z), \tag{5.28}$$

so that

$$\frac{dN(z)}{dE} = (1+z)\frac{dN(z)}{dE_z} = (1+z)\left(\frac{dN(z)}{dE'}\right)\bigg|_{E'=E(1+z)}. \tag{5.29}$$

$$\frac{dN}{dE\,dV_0} = \int_\infty^0 dz \left(\frac{(1+z)^3}{H(z)}\rho(z=0)^2\right)\left[\left(\frac{dN(z)}{dE'}\right)\bigg|_{E'=E(z+1)}\right] \times \frac{\langle\sigma v_{rel}\rangle}{2}2m_\chi^2 \tag{5.30}$$

where the mathematics in square brackets contains all the model-dependent particle physics that would predict the specific spectrum of messenger particles and the cross-section for annihilations based on the details of the underlying quantum field theory. Here, $H(z)$ refers to the Hubble rate (Section 3.3) but written in terms of redshift instead of time (c.f. Eqn. 3.9 and the discussion thereafter). It is also left to the reader to think about how this equation would change for the case of dark matter decay, rather than mutual annihilation.

We have left the photon/neutrino spectrum-per-annihilation unspecified up to this point. While we have safely encapsulated the unknown particle physics in a singular place in our equations, this approach is not particularly realistic. At some point, we have to speculate on the underlying physics, or we will be unable, even, to make estimates to guide the design of, or plan for, our instruments. What are our options?

Assuming annihilation to SM particles *can* occur (which may already be a significant assumption), we can reasonably assume that they would be dominated by 2-body final states (as depicted in Fig. 5.1b). The spectra of photons and neutrinos produced by dark matter annihilations to pairs of quarks, gauge bosons, leptons, and so forth, and their subsequent decays, fall into three broad categories:

- Hadronic/photon-rich continuum: dark matter annihilates to τ leptons, gauge bosons, or any combination of quarks. These inevitably lead, through well-established physics in electroweak and strong interactions, to numerous neutral pions produced in decays. The π^0 almost entirely decays to pairs of photons.

- Leptonic/photon-poor: dark matter annihilates to mostly electrons or muons. In this case, photons are produced only as part of 3-body final states by final-state radiation or internal bremsstrahlung.

- Lines: dark matter annihilates directly to $\gamma\gamma$ (or $\nu\bar{\nu}$) or a mono-energetic photon and one other particle.

Backgrounds to Indirect Dark Matter Searches

We have hinted at this earlier, but it is time to confront the major issue in these indirect methods of searching for dark matter by decay or annihilation—astrophysical backgrounds. These will include, but are not limited to, cosmic rays, supernovas, pulsars, black holes, and any other phenomena that can output charged or neutral messenger particles with extremely high energies (and which may be compact enough, and in large enough populations, to then appear as a "diffuse source" of such particles). We are generally less concerned about definable point sources of high energy gamma rays and

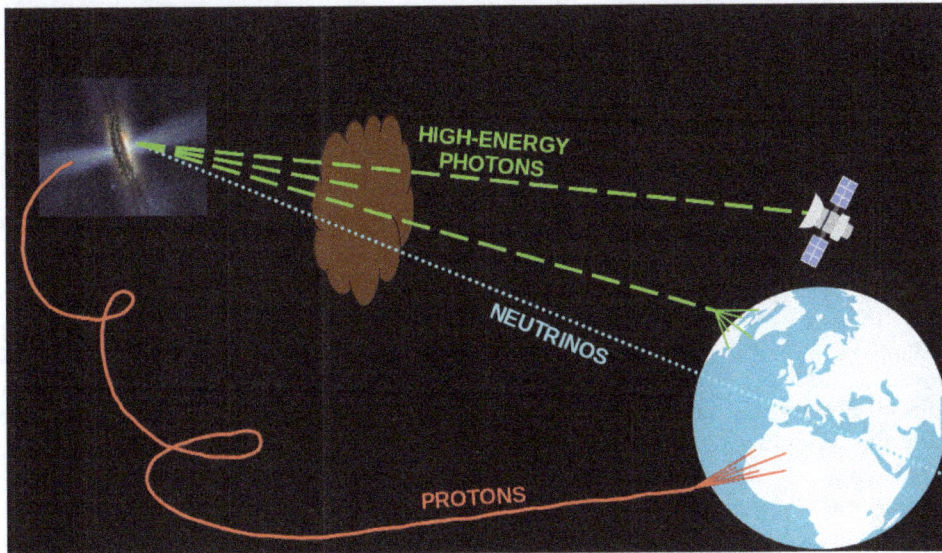

Figure 5.2: An illustration of potential backgrounds to indirect dark matter searches. Active galactic nuclei (supermassive black holes) can accelerate particles to high energies, producing protons, gamma rays, and neutrinos. Some take direct routes to Earth, but others may not. Components of the figure are from Refs. [bas16; sna12; Wik23f].

neutrinos and more concerned about astrophysical populations that, aggregated across the sky, imitate a broad distribution of such particles.

Cosmic rays are the most potentially abundant source that meet these criteria. They primarily originate from beyond our own galaxy, they are diffuse, and they span a vast range of energies. If cosmic rays were entirely localizable to specific point sources, that would make the job of eliminating them from consideration much easier. However, since cosmic rays consist of charged particles, those particles are deflected by stellar, interstellar, galactic, and intergalactic magnetic fields. Their observed trajectories are thus uncorrelated with their original ones and make pinpointing the origin of charged cosmic rays impossible (Fig. 5.2). On the other hand, photons and neutrinos will essentially be undeflected and usable, with improved instrumental data, to help us better understand the specific origins of cosmic rays (assuming that the photons and neutrinos in cosmic rays arise from the same astrophysical sources as the charged particles, which is a plausible hypothesis with supporting evidence).

The charged particle components of cosmic rays are not only deflected, but through interactions with the interstellar and intergalactic media, they lose energy. This means that their spectrum is a function of the distance from Earth to the original source. All of this, together, makes discerning possible dark matter signals from the cosmic ray backgrounds extremely challenging. A primary tool for this, driven by original data but relying on computational models, is an algorithm that provides cosmic ray propagation through the universe from source to target. DRAGON [Vit+18] and GALPROP [PJM22] are good examples of such frameworks and are in common use for experiments performing indirect dark matter searches.

The number densities of cosmic rays can be written as

$$\frac{dn_{CR}}{dE} = \psi(\vec{x}, E, t). \tag{5.31}$$

The evolution of the number density is approximately governed by the diffusion equation,

$$\frac{d\psi}{dt} = D(E)\nabla^2\psi + \frac{\partial}{\partial E}\left(b(E)\,\psi\right) + Q(\vec{x}, E, t), \tag{5.32}$$

where $D(E)$ is the *diffusion coefficient*, $b(E)$ describes the energy loss by the cosmic ray species, and $Q(\vec{x}, E, t)$ characterizes the source feeding cosmic rays into the universe. This equation, however, is already over-simplified. A more accurate treatment will include convection of cosmic rays out of the galactic plane, decay or fragmentation of cosmic ray particles, and diffusive re-acceleration of the cosmic ray particles (rather than just assuming energy loss).

Solutions to this equation require the imposition of boundary conditions. We can treat the galaxy as a cylindrical slab of height h (a few kpc in scale) and radius R (a few tens of kpc in scale). Let us impose a free-escape condition at the slab boundaries (that is, once a particle reaches the edge of our model galaxy, it becomes, for all intents and purposes, completely free of external forces). Let us then parameterize the diffusion coefficient:

$$D(E) = D_0 \times (E/E_0)^\delta, \tag{5.33}$$

where $E_0 = 1$ GeV, $D_0 \approx 10^{28}$ cm^{2}s^{-1}, and the *spectral index* $\delta \in (0.3, 0.7)$ (where current measurements using this model as an hypothesis suggest that $\delta = [0.4, 0.5]$). Values of R, h, and the diffusion parameters can be tuned to match the cosmic-ray data. Having done so, we would then have a reasonable, if empirically driven, model of this portion of the cosmic ray number density.

The time for diffusion of cosmic rays is

$$\tau_{diffusion} \approx \frac{R^2}{D(E)} \tag{5.34}$$

and the time scale for energy loss is

$$\tau_{loss} \approx \frac{E}{b(E)}. \tag{5.35}$$

Let's assume a steady-state regime where

$$\frac{\partial\psi}{\partial t} = 0, \tag{5.36}$$

and make the approximation that

$$\nabla^2\psi \approx \frac{\psi}{R^2}. \tag{5.37}$$

Cosmic rays observed experimentally have an energy spectrum that follows a power law, so we can reasonably assume that

$$\frac{\partial}{\partial E} \approx \frac{1}{E}.$$ (5.38)

With the assumptions in Eqns. 5.33–5.38 in mind, we can write Eqn. 5.32:

$$0 \approx -\frac{R^2 \psi}{R^2 \tau_{diffusion}} - \frac{b(E)\psi}{E} + Q.$$ (5.39)

This leads to an equation for ψ

$$-\frac{\psi}{\tau_{diffusion}} - \frac{\psi}{\tau_{loss}} + Q \approx 0.$$ (5.40)

The solution to this equation has the approximate form

$$\psi \approx Q \min(\tau_{diffusion}, \tau_{loss})$$ (5.41)

since whichever of the two timescales is smaller dictates the dominant term of the approximate diffusion equation. We can then consider two limiting cases.

- $\tau_{diffusion} \ll \tau_{loss}$ (diffusion-dominated regime)

  - In this case,

    $$\psi \propto Q(E)E^{-\delta}.$$

    As noted earlier, $Q(E)$ describes the source of the cosmic rays while δ acts to soften the energy spectrum. Protons observed in cosmic rays have number densities that follow a specific power law, $dn/dE \approx E^{-2.7}$. If we assume that the particles injected into the cosmic ray population have been accelerated by strong shocks, then $dn/dE \propto E^{-2}$, which implies that $\delta = 0.7$.

- $\tau_{loss} \ll \tau_{diffusion}$ (cooling-dominated regime)

  - In this case, the energy losses are quicker than the diffusion time. The main mechanisms for energy loss are synchrotron radiation due to their bending in ambient magnetic fields (which triggers the emission of photons from charged particles) or inverse Compton scattering off of ambient photons (e.g. $p^+ + \gamma \to p^+ + \gamma'$). Photons produced from this mechanism can interact with other matter and produce electron-positron pairs.

  - For inverse Compton scattering, $dE/dt \propto E^{-2}$. The steady-state proton energy spectrum will be proportional to $E^{-(2+\delta)}$. Positrons in the cosmic ray population will arise from two sources. The first are primary positrons that were produced in the original cosmic ray progenitor and travel from the source to the detector on Earth. These will have a spectrum that is also proportional to $E^{-(2+\delta)}$ at low energies, but proportional to E^{-3} at high energies. The second component are secondary positrons that arise from the interactions of cosmic rays and produce positrons as a result. These will have a spectrum proportional to $E^{-(2+\delta)}$ at low energies and proportional to $E^{-(3+\delta)}$ at high energies.

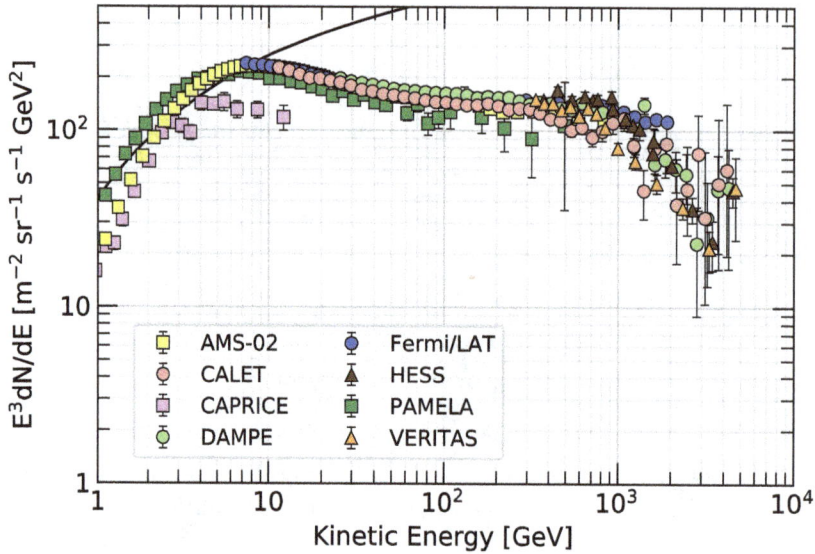

Figure 5.3: Differential spectrum of electrons plus positrons (except PAMELA data, which are electrons only) multiplied by E^3. The line shows the proton spectrum multiplied by 0.01. Figure from Ref. [Wor+22].

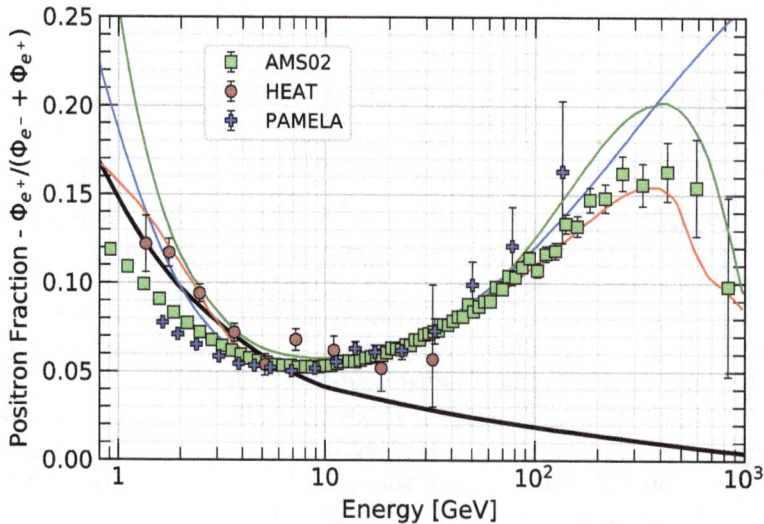

Figure 5.4: The positron fraction (ratio of the flux of e^+ to the total flux of e^+ and e^-). The thick curve is a model of pure secondary production. The three thin lines show three representative attempts to model the positron excess with different phenomena: dark matter decay; propagation physics; production in pulsars. The ratio below 10 GeV is dependent on the polarity of the solar magnetic field. Figure from Ref. [Wor+22].

What do observations of the cosmic ray particle energy spectra tell us? The combined positron and electron spectrum, multiplied by E^3, is shown in Fig. 5.3. We see that, above 10 GeV, this energy-scaled spectrum changes by a factor of only two (appearing near-flat on a logarithmic vertical scale), implying it goes by a power law whose exponent is just shy of a 3 (thus motivating the 2.7 number mentioned above). We can see from the ratio of the positron to the combined energy spectrum (Fig. 5.4), that the positron spectrum remains high at the highest energies (as in the spectrum observed by the PAMELA satellite) while the electron spectrum appears to fall off (as observed in the AMS-02 data). This motivates the statement that positrons at high energies will follow a slightly flatter spectrum than E^{-3}.

In order to detect particles from dark matter decay and/or annihilation, as well as to observe cosmic rays and study the properties of the populations of long-lived SM particles, we will need a broad range of instrumentation capable of sampling these particles across many powers of ten in energy.

5.3 Detection Technologies

In this section, we will survey a variety of astrophysical detection technologies that can be used for indirect searches for dark matter.

Gamma Ray Detectors

Gamma ray detection can be considered in two major categories: satellite-based and ground-based approaches. The challenge of the former is that they are expensive to launch and maintain in orbit near the earth. The challenge of the latter is that gamma rays are strongly attenuated by the earth's atmosphere due to the high interaction cross-section with air molecules; meaning you don't detect the gamma ray, but, rather, the particle shower it induces in the atmosphere. For photons, the span of current and future-proposed x-ray and gamma-ray telescopes is shown in Fig. 5.5.

Ground-based gamma ray detectors have strengths and weaknesses. Examples of such instruments are HESS (The High Energy Stereoscopic System) [Abd+18a], VERITAS (Very Energetic Radiation Imaging Telescope Array System, Fig. 5.6) [Acc+08], and HAWC (The High Altitude Water Čerenkov Observatory) [Abe+23]. While utilizing a range of technologies, these instruments have some common features. Each is built from smaller detector units spread out over a spatial area. This is to allow the air shower of the high-energy gamma rays to be sampled across the active elements of the instrument. In addition, the arrival times at each site of elements of the shower provide key timing information that allows for the flight direction of the original gamma ray to be estimated.

The challenge of these ground-based instruments is that they are best suited for the highest energy gamma rays, but the flux of those rays decreases sharply with energy. For the very highest energy gamma rays, water Čerenkov detectors are necessary (such as HAWC, the Pierre Auger Observatory and others). While these instruments provide source-pointing and energy spectrum information, they are subject to higher backgrounds and

X-Ray and Gamma-Ray Telescopes

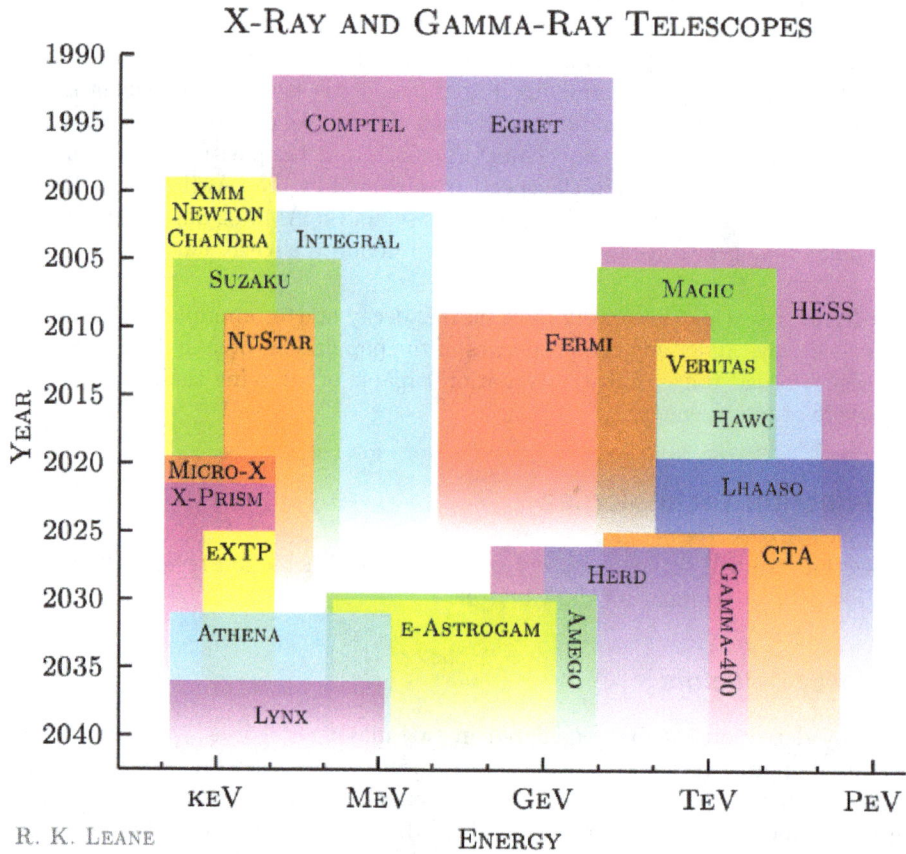

Figure 5.5: Current and proposed future x-ray and gamma-ray telescopes. Graphic by R. K. Leane.

Figure 5.6: View of the Fred Lawrence Whipple Observatory basecamp and the VERITAS array. Image credit: VERITAS.

Figure 5.7: The Fermi LAT instrument (silvery box at the top) is shown integrated with the spacecraft at Spectrum Astro Space Systems in December, 2006. The trio of white and silver structures at the center of the image are some of the sodium-iodide detectors comprising the Gamma Ray Burst instrument. Image Credits: NASA/General Dynamics Advanced Information Systems.

the problem of attenuation of the gamma ray and its shower products by the atmosphere, between the initial shower point and the ground array.

An example of a satellite-based gamma ray detector is the Fermi Gamma Ray Space Telescope, or, simply, the Fermi Telescope. The detection principle for gamma rays is based on particle physics detector technology. Gamma rays enter the active detector region of the instrument (the Large Area Telescope, or LAT [Atw+09]), which is designed to induce those gamma rays to pair-produce in material (Fig. 5.7). The pair production yields an electron-positron pair, which pairs are then detected by their ionization and, finally, their energy shower in material. The ionization produces spatial pixels that can be used to establish the origin and direction of the photon conversion. This yields directional information about the original gamma ray. The energy shower after the tracking system is designed to capture as much of the electron-positron energy as possible. This allows for the determination of the energy of the original gamma ray. Establishing total energy (frequency and wavelength) and direction in the sky are the two key pieces of information for gamma ray astronomy.

In principle, a high-energy electron and positron, or electron and proton, could (coincidentally) enter the instrument at the same time, providing for a population of fake gamma rays. These are reduced by looking at the first layer of the LAT (the entry point for particles) and verifying that there is no activity in that region. Layers of metal foil, a high-Z material selected for inducing photon conversions, come after this opening layer of the LAT. Electrons, positrons, and protons will cause activity in that opening layer, but gamma rays are not expected to do so. An anti-coincidence veto is applied (requiring activity in the tracking chambers and calorimeter, but no activity in the opening layer of the LAT) to reject fake gamma ray conversions.

The LAT is capable of detecting gamma rays with energies from 20 MeV to more than 300 GeV. While impressive, given the pointing and energy reconstruction capabilities of such an instrument, we can see from Fig. 5.5 that this barely scratches the surface of the possible energy ranges of gamma rays.

Neutrino Detectors

Neutrinos can be produced from dark matter decay or annihilation. The benefit of the neutrino is its extremely low interaction rate with matter and its lack of electric charge, meaning it can travel from its origin nearly unaffected in energy or direction. That same advantage is the key weakness for telescopes that would use these messengers as an astrophysical probe: detectors have to be large and dense if they hope to induce neutrinos to interact. They also need an active medium with high efficiency if they hope to observe neutrino interactions and infer neutrino properties (e.g., energy, momentum, and flavor). An example of a neutrino telescope is the IceCube [Aar+17] facility located at the geographic south pole (Fig. 5.8). The instrument is constructed from many strands of PMTs arranged in long lines of optical modules, like strings of pearls. The strings are distributed over a cubic kilometer of volume, providing the name for this instrument.

Neutrinos will interact in the earth's crust beneath the ice or, less likely, in the ice itself. Since neutrino interactions preserve lepton flavor, electron neutrinos will make electrons, muon neutrinos, muons, and so forth. Each of these charged particles will produce Čerenkov radiation as they traverse the ice, and it is this light that IceCube observes. The time structure of the arrival of the light and the intensity of that light provides directional and energy information about the original neutrino.

The advantage of this and other instruments, like Super-Kamiokande [Fuk+03] in Japan or ANTARES [Age+11] in the Mediterranean Sea is that they provide vital directional and spectral information about the cosmic neutrinos. Using this information, one can look for dark matter decay or annihilations from places in the universe where it might clump, such as in the center of the sun (due to gravitational in-fall over five billion years), or even the center of the earth or the Milky Way's galactic center. Of course, the drawbacks are also fairly clear. Non-astrophysical neutrinos (neutrinos from the sun that interact in

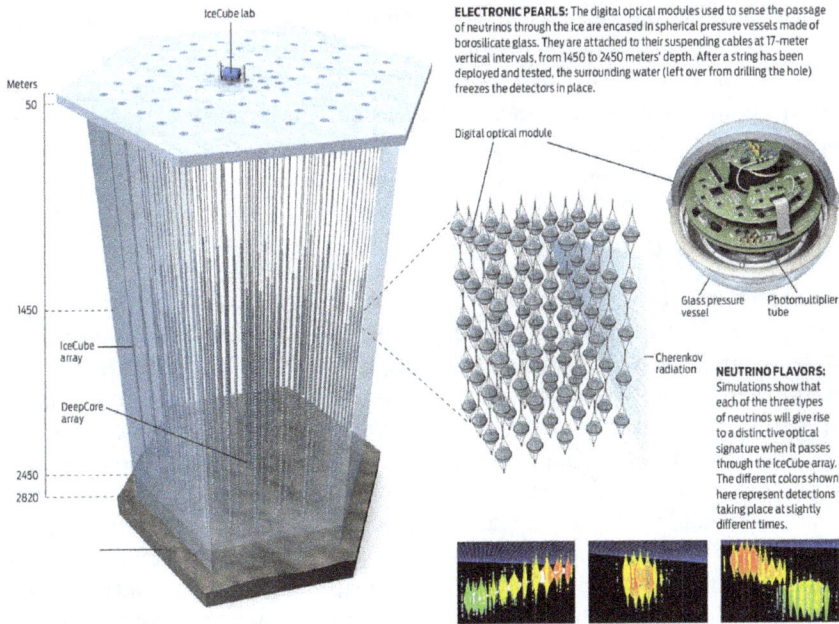

Figure 5.8: An illustration of the IceCube instrument located under the geographic south pole of the earth. Image credit: Emily Cooper.

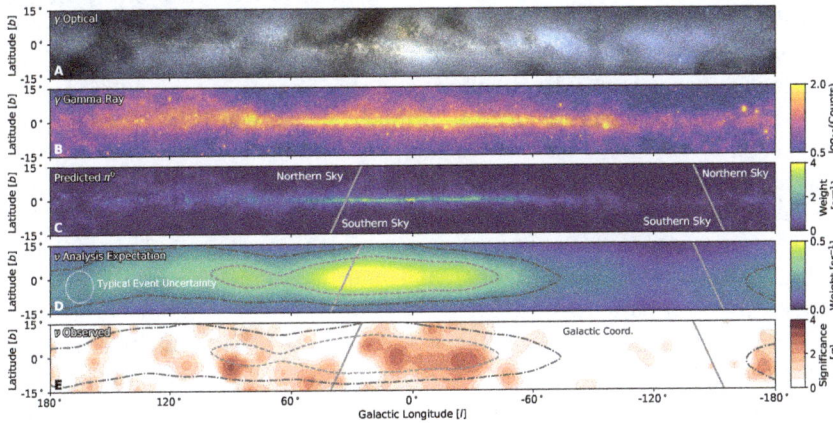

Figure 5.9: The plane of the Milky Way galaxy is shown using a combination of photons. Visible light in (A), shows dust lanes and stars; (B) gamma rays above 1 GeV as observed by the Fermi Telescope; (C) shows predictions of neutrino fluxes based on gamma ray source hypothesis that applies to the Fermi Telescope data. It is the same as shown in (C), but includes IceCube detection efficiencies shown in (D). (E), the observed neutrino signal significances. Figure from Ref. [Abb+23a].

the earth's atmosphere and rain particles on the telescope) provide a background to these observations. In addition, the low statistics induced by the neutrino's weak interactions and low cross-sections with nuclei provide a great challenge to precision measurements.

Despite the challenge of working with the neutrino, neutrino telescopes provide valuable information about where in the sky sources of neutrinos appear to exist. An example of the Milky Way's galactic plane in photons and neutrinos is shown in Fig. 5.9 using a combination of data from many telescopes, including IceCube.

Other Cosmic Ray Detectors

We have focused so far on photons and neutrinos as they provide the best astrophysical messengers given their low attenuation and deflection over the vast distances they must travel from their sources. However, studying other kinds of particles like electrons, positrons, protons, anti-protons, and heavy ions is also essential in the indirect dark matter detection effort. Such particles may also result from the long chain of decays of the final-state SM particles that result from dark matter decay or annihilation. These particles (even if not from dark matter) would also arise from the same astrophysical background sources that produce confounding signals in gamma ray and neutrino observations. Mapping their populations allows us to infer, for example, the expected rate of neutrinos and/or gamma rays from those same astrophysical sources. This helps constrain observations that appear to be an "excess" over what might be expected from purely SM astrophysical sources.

The highest energy, electrically charged cosmic ray particles are accessible to ground-based instruments like the Pierre Auger Observatory [Abr+04] (Fig. 5.10). The use of a large and distributed network of Ĉerenkov detectors allows for a map of the air shower to be developed and to determine the energy spectrum (Fig. 5.11) of the cosmic ray primary (a charged particle striking the atmosphere). The background in this kind of high-energy reconstruction is rather low. The drawbacks of this approach are as noted above: the momentum information of the cosmic ray primary, which can be determined, does not, itself, point back to the source of that cosmic ray.

A lower energy component of the charged cosmic ray spectrum is accessible to space-based instrumentation. A good example of this is the Alpha Magnetic Spectrometer [Agu+02; Agu+21] (AMS) experiment housed on the International Space Station (Fig. 5.12a). AMS is a quintessential multi-component particle detector—one that operates in space (Fig. 5.12b). It consists of transition radiation detection and time-of-flight systems to establish the identity of particles that enter the instrument, as well as charged-particle tracking systems (including a magnetic field for bending those particles to determine their momentum and charge), a Ĉerenkov system to aid in particle identification, and calorimetry for precise energy measurements of cosmic rays.

An example of measurements by AMS is shown in Fig. 5.13 [Agu+15a]. AMS has made precision measurements of protons and anti-protons, electrons and positrons, and heavy ions (including antimatter ions) in an energy range from about 1 GeV up to a few TeV.

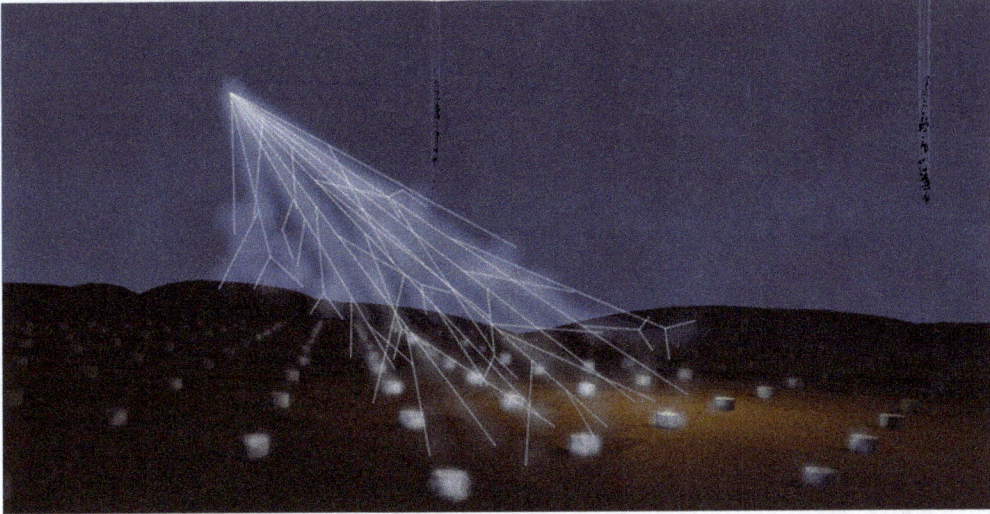

Figure 5.10: An illustration of an air shower event and the array of Ĉerenkov detectors at Auger. Image credit: ASPERA/G.Toma/A.Saftoiu.

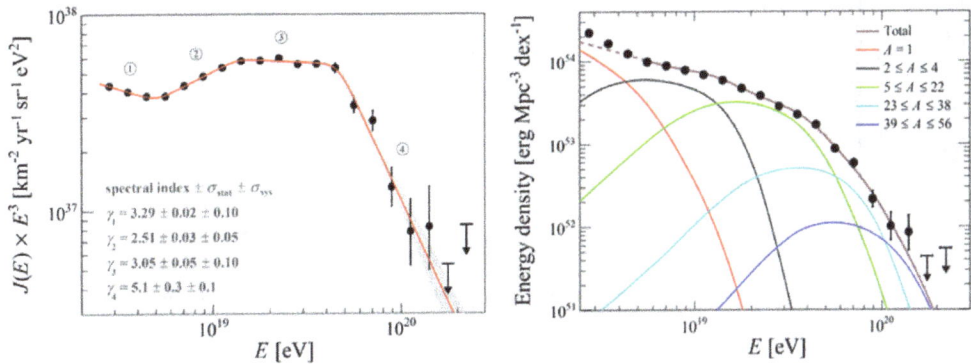

Figure 5.11: (Left) The cosmic ray energy spectrum (multiplied by E^3). (Right) The energy density in cosmic rays. These results are distilled from observations taken from 2004 to 2018 by the Pierre Auger Observatory. Figures from Ref. [Aab+20].

(a) The AMS experiment affixed to its operating location on the International Space Station. The space shuttle that delivered it to the ISS, the last ever to fly, can be seen in the background. Image courtesy of AMS/NASA.

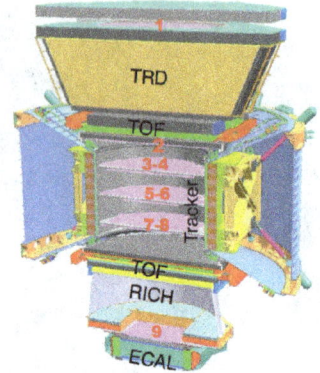

(b) A schematic showing the layout of the instrumentation that comprises the AMS experiment. Figure from Ref. [Spa14].

Figure 5.12: Two views of the AMS instrument (left) in space and (right) in schematic view.

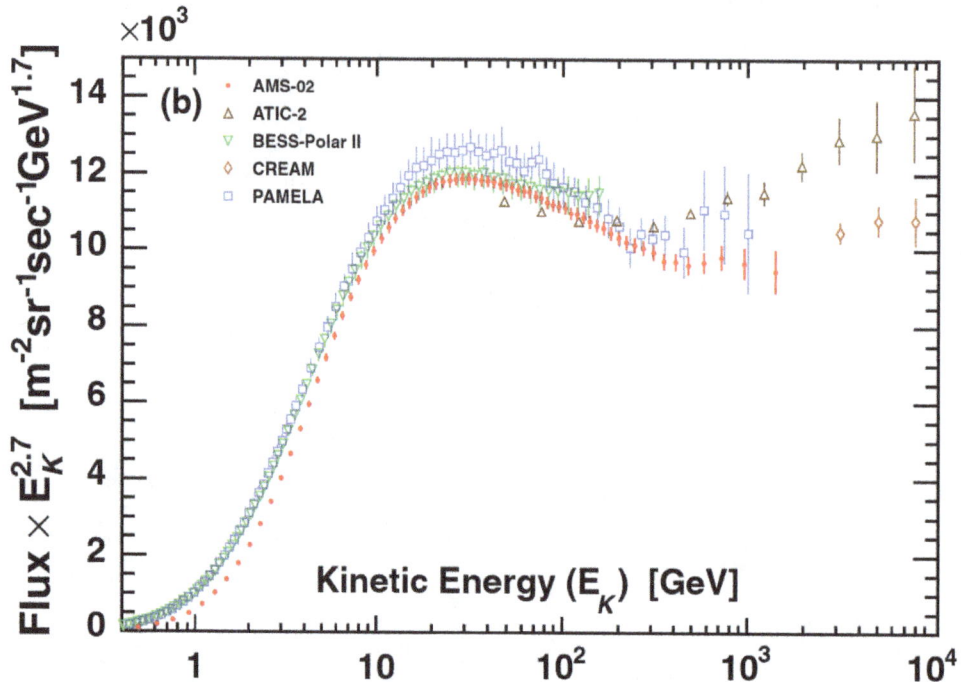

Figure 5.13: The energy spectrum of protons between 1 GeV and 1.8 TeV obtained from the AMS-02 instrument and compared against measurements from independent cosmic ray observatories. Figure from Ref. [Agu+15a].

5.4 The Gamma Ray Sky

High energy γ rays (Fig. 5.14) are produced through the propagation of cosmic rays in the galaxy. The gamma rays arise from various sources (Fig. 5.15) such as bremsstrahlung, neutral pion decay, and inverse Compton scattering. The all-sky map in Fig. 5.14 contains several key features. First is the clear galactic plane, which fills the horizontal long axis at the center of the map with a dense region of gamma ray emissions. Above and below the galactic plane we can see small "hot spots" that are point-sources in those regions. The haze behind the galaxy is the extragalactic background. Difficult to see in this image, above and below the middle of the galactic plane, are "bubbles"—large and extended regions of gamma ray emission that extend out above and below the plane like inflating balloons. These are the "Fermi Bubbles."

The origin of the Fermi Bubbles (Fig. 5.16) is believed to be the remains of the previous active cycle of the supermassive black hole at the center of the Milky Way designated Sagittarius A*. The gamma ray sky is rich and complex, including structures previously unknown before the Fermi Gamma-ray Space Telescope.

Models of the gamma ray sky must include or reproduce its complex range of observations if physicists are to determine the degree of contributions from dark matter annihilation and/or decay. Each component contributing to the overall γ ray emission is modeled as a template. There are primary templates for the galaxy's diffuse emission (e.g., the galactic plane), gamma-ray point sources, isotropic extragalactic sources that are red-shifted as the galaxy moves through the local group and the Virgo supercluster, the

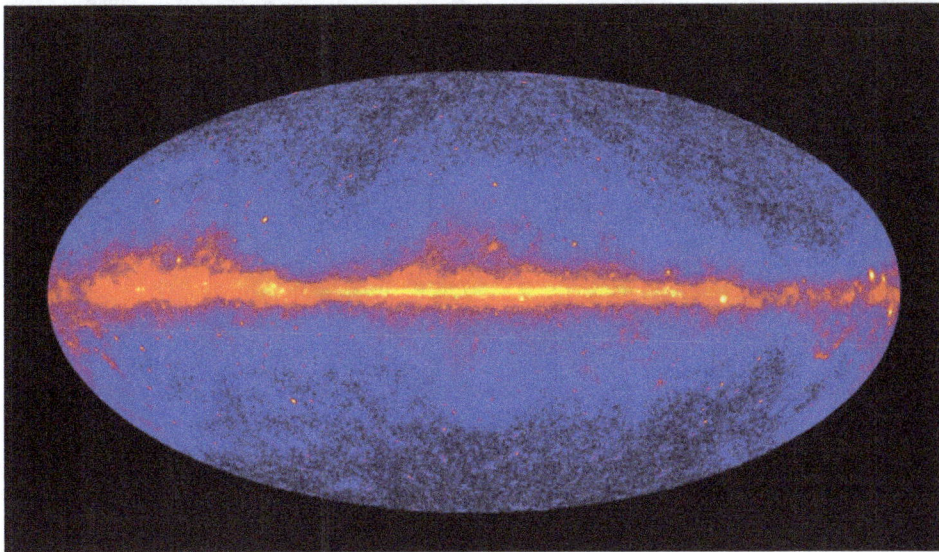

Figure 5.14: The Fermi Space Telescope's five-year map of the sky in gamma rays. Image credit: NASA/DOE/Fermi LAT Collaboration.

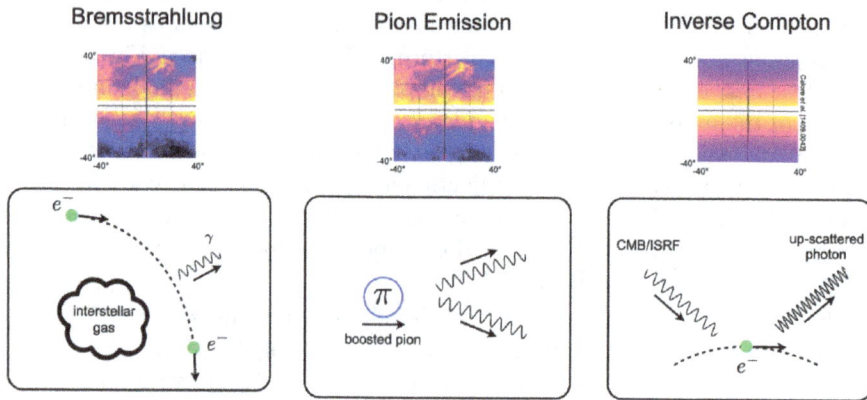

Figure 5.15: Illustrations showing the major sources of gamma rays in cosmic rays. The inset gamma ray heat maps are from Ref. [CCW15].

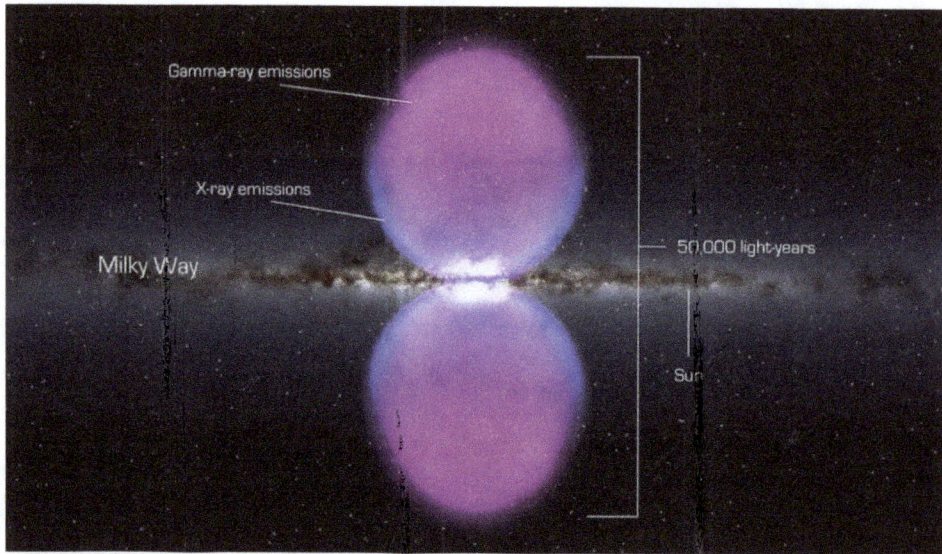

Figure 5.16: An illustration showing the Fermi Bubbles above and below the center of the galactic plane. The core of these bubbles emit gamma rays, causing the plane's edges to glow in x-rays. From top to bottom they are a combined 50,000 ly high. Image credit: NASA Goddard.

Fermi Bubbles, and, of course, a potential dark matter signature. A natural assumption is to imagine dark matter being most dense near the galactic core, the "deepest" part of the gravitational well formed by the halo of dark matter. A statistical modelling procedure is used to constrain the plausible range of flux to attribute to each template. There are a number of assumptions to be considered, including the spatial distributions and number of templates to be used in any analysis.

In the galactic center, the raw gamma-ray maps and the residual maps are determined after subtracting the best-fit galactic diffuse model, the template for interstellar hydrogen gas and the interaction of charged particles with that gas (the "20 cm template"), point sources, and extragalactic isotropic template. What remains in these residual and gamma-ray maps is a measure of the photons-per-square-centimeter, per second, per steradian of sky. One famous result of this kind of analysis is shown in Fig. 5.17 [Day+16]. In the right column of images there appears to be a residual, significant, and clear spatially extended excess, peaking in an energy range of $[1-3]$ GeV. This excess extends in space from the center out about $10°$ in the sky, or 5000 ly. It constitutes about 10% of the original unsubtracted flux and has high statistical significance.

A great deal of effort has been put forth by many groups, collaborations, and institutions creating a better understanding of this excess of gamma rays at the galactic center. For example, in the same publication that first established the existence of a possible excess, the spatial morphology and the energy spectrum of the structure was established. Superficially, as well as through the employment of template models, dark matter was a plausible explanation for this observation. The distribution can be well-explained, for example, by an NFW dark matter density profile with a $\gamma = 1.28$ spectral parameter (Fig. 5.18). The excess can then be interpreted, based on the NFW profile, in terms of the dark matter masses and velocity-averaged interaction cross-sections that would be required to explain the excess under the hypothesis of a $2 \to 2$ annihilation process that preferentially produces pairs, either of tau leptons or bottom quarks (Fig. 5.19) [CCW15]. This same work showed that the evidence for excess emission is robust against variations in models of the galactic diffuse emission (using the GALPROP framework).

Dark matter, however, may not be the only explanation. For example, a GeV-level excess could also be explained by a population of unresolved sources that are just below the Fermi Telescope's detection threshold. One example of such a population of sources is millisecond pulsars. These are rapidly spinning neutron stars, rotationally accelerated at the moment of their parent-star's collapse, with periods at the millisecond level. Pulsars emit regular bursts of radiation from the perspective of earth-bound observers, because they have a strong magnetic field that rotates as the star rotates. If one is located "down the beam" of the magnetic field as it sweeps across the sky, one will observe pulses of radio frequency or other kinds of electromagnetic radiation.

The magnetic environments of such pulsars are extremely strong and variable, making this an excellent recipe for generating natural particle accelerators that can, in turn, generate high-energy photons through synchrotron, bremsstrahlung, or other emission processes. Since stars cluster near the center of the galaxy, a population of relic millisecond pulsars could have emerged naturally in that same region, the corpses of the main sequence stars that preceded them. It is then possible to simulate a population of such sources that

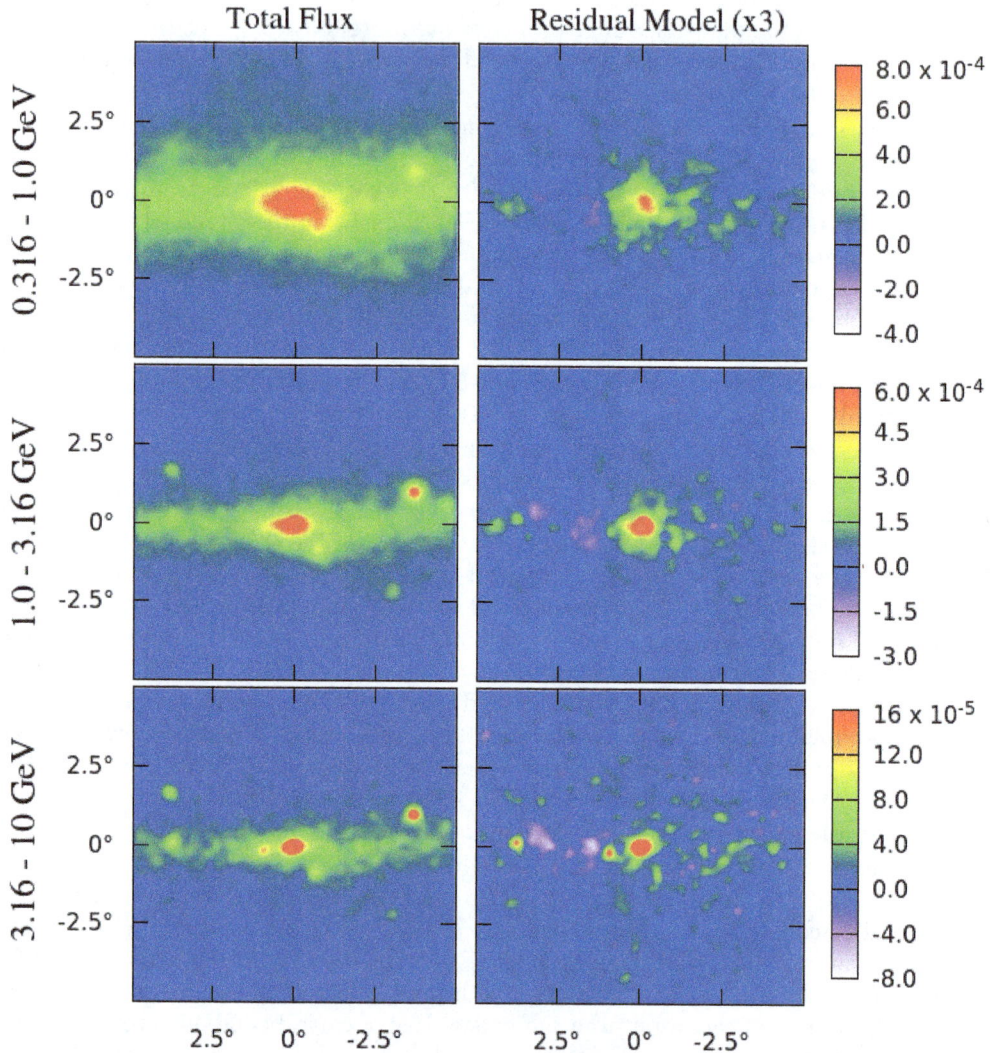

Figure 5.17: (Left Column) Heat maps showing the unsubstracted gamma ray signal from the galactic center binned from top to bottom in different energy ranges. (Right column) The residual map after subtracting the templates generated from other sources in the galaxy, leaving behind any additional contribution that might be unique to the galactic center. Images from Ref. [Day+16].

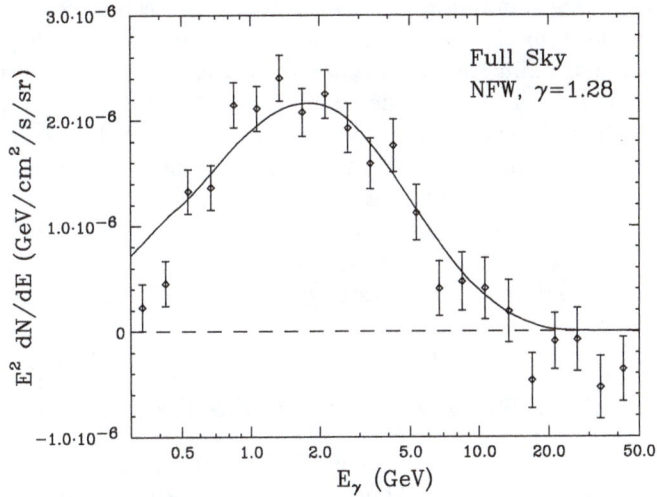

Figure 5.18: The energy spectrum of the galactic center gamma ray residual signal vs. the energy of the gamma ray, compared with an NFW dark matter profile. Figure from Ref. [Day+16].

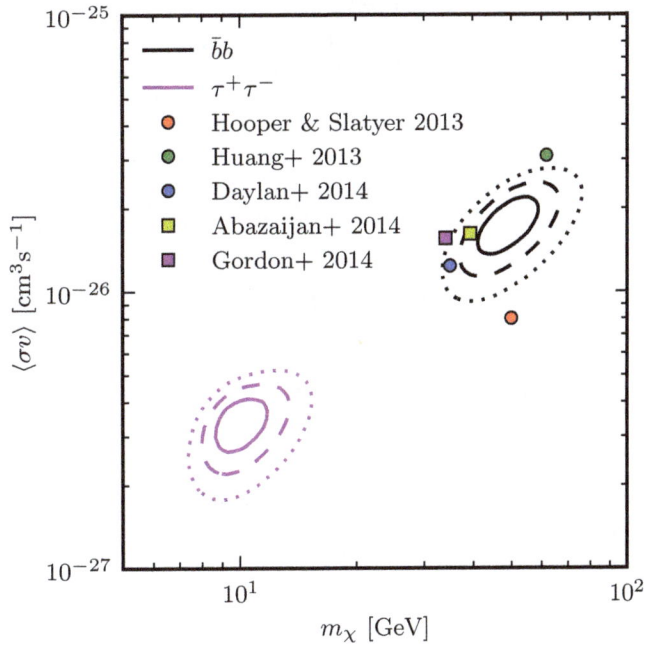

Figure 5.19: Interpretation of the gamma ray excess at the galactic center as dark matter (NFW profile) annihilates to SM particles in a $2 \to 2$ process. Figure from Ref. [CCW15].

traces the stellar population and apply image processing techniques to distinguish the pulsar-induced excess from other excesses that may be present. This is possible because the pulsars are, in fact, point sources; point sources too low in energy, and too small in size to be resolved by the Fermi Telescope, and, so, would appear as a "haze" concentrated near the galactic center. In contrast, the dark matter is truly a diffuse, if dense, gas that also collects near the galactic center. These two subpopulations can be subtracted from one another, as has been demonstrated by several recent efforts. There is some evidence that the excess at the galactic center may be explained through standard astrophysical means without needing to involve interacting dark matter. However, this remains a fertile and fervent area of investigation. The origin(s) of the galactic center excess are far from definitively determined.

Looking Beyond the Milky Way: Dwarf Satellite Galaxies

If the GeV excess in the galactic center is truly due to dark matter, then we should expect to see signals from other locations. There is nothing special about the galactic center except insofar as it is a place where dark matter has collected more densely over billions of years. There should be other places in the universe where this is true, including in our own galactic neighborhood.

Dwarf galaxies—small populations of stars that are more cloud-like in appearance due to their much smaller populations of stars—are a good choice as they are expected to be dark-matter dominated. Dozens of these galaxies have been discovered as satellites of the Milky Way (Fig. 5.20). Why would they be a promising laboratory for dark matter

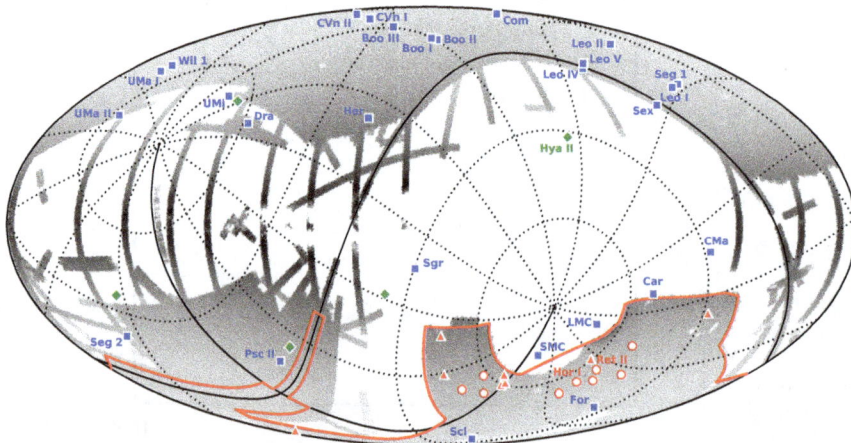

Figure 5.20: A sky map indicating the location of small (dwarf) satellite galaxies gravitationally bound to the Milky Way. The galactic plane spans the center of this map. Figure from Ref. [Drl+15].

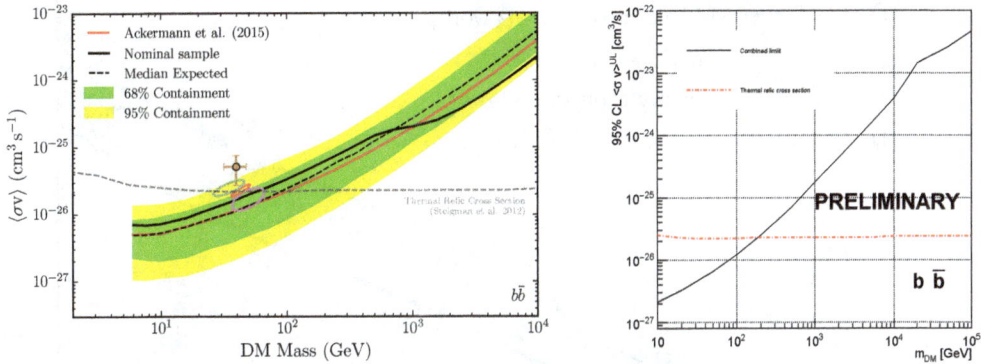

(a) Fermi-LAT and DES Collaborations

(b) Fermi-LAT, HAWC, HESS, MAGIC, and VERITAS Collaborations

Figure 5.21: Upper limits on the velocity-averaged dark matter cross-section vs. dark matter mass. (Left) Published results from 2016 using forty-five dwarf spheroidal galaxies. (Right) Preliminary results in 2019 from twenty dwarf spheroidal galaxies, but using more instrumental data on each. Both are interpretations assuming that $\chi\chi \to b\bar{b}$ dominates the annihilation process. Figures from (left) Ref. [Alb+17] and (right) Ref. [Oak+21].

discovery? They clearly represent places where gravitational wells are present, prompting the in-fall of stars. However, these gravitational wells are not as large as the ones that, for example, seeded the Milky Way or Andromeda galaxies. Stars become trapped in these wells, but they never themselves attract enough stars to form the kinds of dense structures we see in larger spiral galaxies. The result is expected to be a cloud-like galaxy, small in size, whose stellar-to-dark matter ratio is severely tilted toward the dark matter that originally seeded the gravitational well. As such, dwarf galaxies make excellent candidates for an annihilation search.

Observing dwarf galaxies is more challenging than observing the Milky Way as they occupy a small part of the sky and thus require good resolution. Their small angular size also makes them a fainter source of gamma rays than more nearby and extended structures. Six years of Fermi Telescope data has been used to search for γ ray emission from forty-five satellite galaxies included in the search. While not yet at the level of sensitivity to definitively observe or exclude a signal consistent with expectations from the galactic center interpretation, these independent observations are nevertheless becoming sensitive to thermal weak-scale dark matter (Fig. 5.21).

Looking Beyond the Milky Way: Other Galaxies

An obvious target for gamma ray searches would also be other galaxies like the Milky Way. The dwarf galaxies are small, but comparatively close. Other galaxies are all much farther

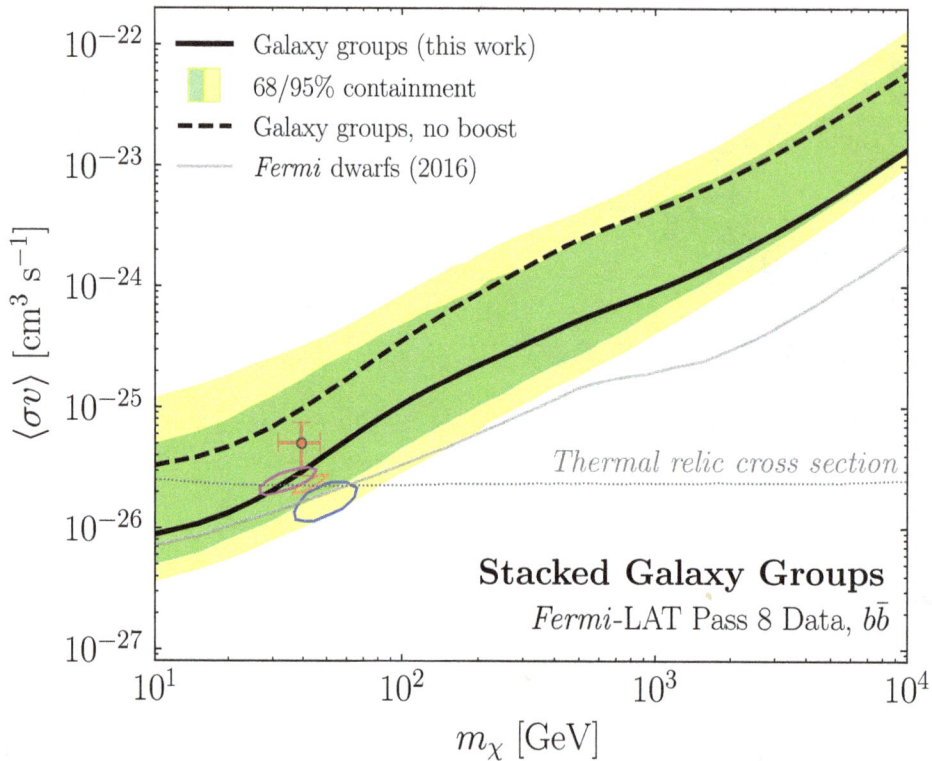

Figure 5.22: Constraints on the velocity-averaged annihilation cross-section of dark matter vs. dark matter mass, obtained from stacking observations of gamma rays from other galaxies. Figure from Ref. [Lis+18].

away, the closest being the Andromeda Galaxy at a distance of 2.5 Mly. (In comparison, the Large Magellenic Cloud is only 160 kly from the Milky Way.)

While harder to observe, it is possible to make similar gamma ray observations of galaxies and combine observations to draw conclusions about the flux of gamma rays from these sources. Figure 5.22 illustrates results from interpreting a stack of such observations [Lis+18], with the expectation from a thermal dark matter relic and weak-scale annihilation cross-section compared. These observations are indeed more statistically challenged than the dwarf spheroidal observations, but still yield constraints approaching one of the most interesting regions of parameter space (approaching the expected interaction rates of a thermal relic dark matter candidate).

Gamma/X-ray Ray Spectral Lines: the Galactic Center

We have, up to now, concentrated on the notion of a diffuse source of gamma rays (or other particles) that would signal dark matter decay or annihilation. However, this approach

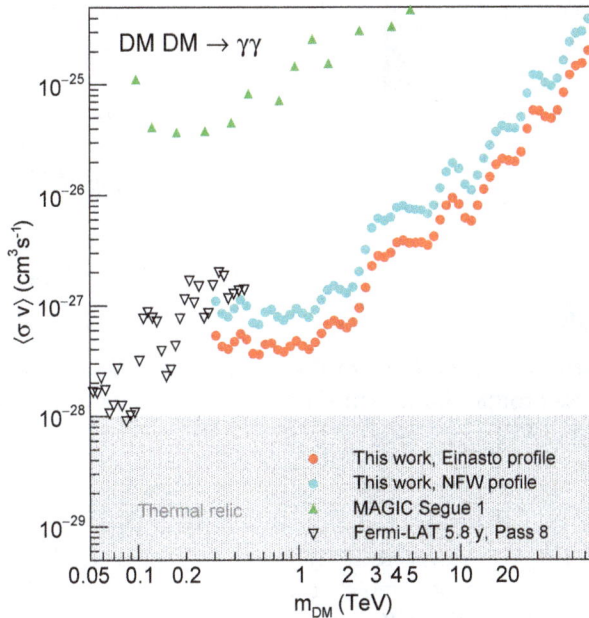

Figure 5.23: Gamma ray energy spectrum from the HESS telescope ("this work") using ten years of data and two different assumptions about the dark matter density profile. Constraints on dark matter are also shown from the MAGIC Telescope and the Fermi-LAT instrument. Gamma ray observations target the galactic center. No gamma ray excesses were observed, and the upper limits shown in the plot were derived from this non-observation. Figure from Ref. [Abd+18b].

neglects one of the most powerful tools in particle physics: the possibility of a resonance or threshold effect in, for example, annihilation. Such effects sharpen the energy spectra of the final-state particles. In doing so, they create a signal strongly peaking in energy (in frequency space) which is then much easier to distinguish from a continuous and featureless background spectrum. Let's study this approach for the possibility of detecting similar effects in dark matter interactions.

Consider gamma rays from the galactic center. If the $\chi\chi \to \gamma\gamma$ process proceeds via some intermediate resonance, the gamma rays will emerge with definite energies shaped by an interaction cross-section that maximizes, locally, around that resonant energy. A good example of a long-duration search for gamma ray spectral line excesses above expectation is shown in Fig. 5.23. The HESS telescope used ten years of data, observed no such excesses, and set upper limits on the cross-section and mass of dark matter based on that non-observation. For dark matter candidates with masses in excess of 1 TeV, annihilation constraints are set by gamma ray telescopes or neutrino telescopes.

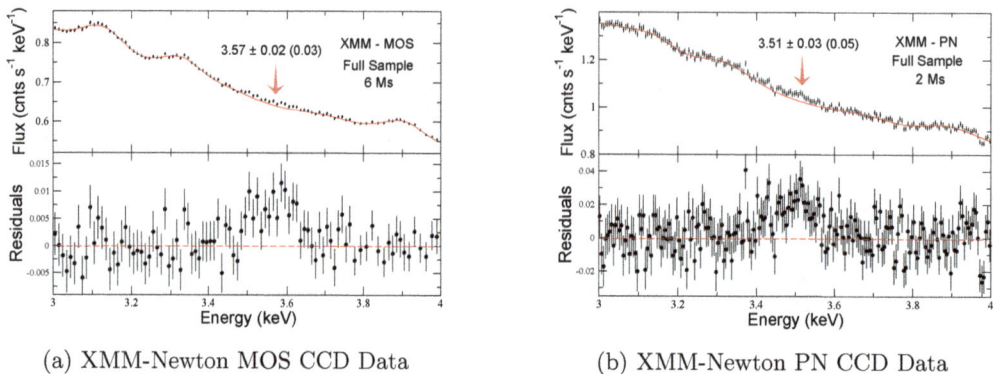

(a) XMM-Newton MOS CCD Data   (b) XMM-Newton PN CCD Data

Figure 5.24: XMM-Newton observations of the x-ray spectrum from the galactic center. Indications of an excess around 3.5 keV are indicated in two instrumental observations. Figures from Ref. [Bul+14].

Dark matter decay can also produce spectral lines indicative of the two-body decay of a heavy resonance. However, favored decay modes in various dark matter models are to pairs of heavy quarks or leptons, which then, in turn, decay. This washes out the resonance. Nevertheless, it would still leave some imprint on the resultant final-state particle spectrum. Limits on the decay of heavy dark matter are based primarily on data from cosmic ray telescopes and detectors.

There has been at least one x-ray spectral line excess that led to significant investigation: the 3.5 keV line. In 2014, a stacked-cluster observation detected an unidentified line near 3.55 keV at a $4-5\sigma$ significance (compared to the null hypothesis) using the instrumentation aboard the XMM-Newton satellite (Fig. 5.24) [Bul+14][Boy+14][Bul+16]. The signal was also detected in Chandra X-ray Telescope observations of the Perseus cluster. Significant follow-up work was done by XMM-Newton, Chandra, and the SUZAKU Telescope. They found additional indications of an excess.

So far, this same excess has been detected in the galactic center, in the cosmic x-ray background, and in the Draco dwarf galaxy (albeit faintly). The signal has NOT been found in the Andromeda Galaxy (using the Chandra Telescope), in stacked galaxies using XMM-Newton and Chandra, or in dwarf galaxies with XMM-Newton (which was able, from this non-observation, to set stringent limits).

The simplest dark matter-based explanation is a 7 keV sterile neutrino, a heavy neutrino in addition to the SM neutrinos which, due to its lack of significant interactions, makes an excellent dark matter candidate. This simple dark matter decay is a predictive model: the signal should be proportional to the amount of dark matter in the source. It is in tension with some of the null results mentioned above, where observations should have been expected.

There is an ongoing and robust debate as to possible contamination from potassium and chlorine plasma lines (due to atomic processes, these can mimic x-ray lines). There are several known x-ray lines close to 3.5 keV whose strength can be sensitive to plasma temperature. Charge-exchange processes, where the electron shells of a neutral atom and a negative ion come into contact, and an electron transfers from the atom to the ion, are another possible culprit. The transfer would emit a high energy photon and can be x-ray in nature. A summary of tests and critiques is available from Ref. [DRS20].

Is there a resolution to this interesting puzzle? Energy resolution at the eV-scale (0.1%-level) could isolate the 3.5 keV line from any known atomic lines and the expected Doppler shift of such lines (from the velocity of dark matter in the galaxy). The Hitomi Telescope would have had the required resolution, but failed a few days after launch. The Micro-X sounding rocket program has the potential to provide the needed resolution, but rocket-based observations are of short duration and we will have to wait to see if these can provide the statistics needed to understand this problem.

5.5 Mechanisms for the Collider Production of Dark Matter Interactions

The anticipated means by which SM particles can be collided to produce dark matter are simply the time-reverse versions of the Feynman diagrams in Fig. 5.1. Since the exact mathematical theory relating dark and normal matter is unknown, predictions of collider interactions are inexact. Instead, a common approach at colliders is to identify a set of potential dark matter candidates in a specific model (or features of dark matter candidates that might lead to signatures in collider data), and then search for the absence of observable particles in collisions consistent with the target model. In the absence of a detection, one returns to the motivating model(s) and uses the non-observation to constrain the assumptions of the model.

The advantage of a collider-based approach is the ability to discover and measure the properties of previously unobserved particles using a very high rate of experiment repetition (many independent collisions of matter), as well a high degree of control of the initial conditions of the experiments. The discovery of a new, electrically neutral, rarely interacting particle is not necessarily a discovery of dark matter (one has to establish that this new particle class is extant in the universe exterior to the collider) but can be used to constrain the pattern or degree of possible interactions between SM matter and other kinds of matter.

We will highlight the efforts to hunt for dark matter using colliders by investigating two major approaches: the use of missing momentum and energy as a signature of an undetectable final-state particle and the effort to identify possible dark force-carrying particles (e.g., dark photons) present in collider interactions.

Particle Colliders

We will briefly review the major features of particle accelerators and colliders. These are relevant to establish the strengths of this approach: control over energy and intensity in the initial reaction that might generate dark matter and the ability to observe all, or nearly all, of the final-state products of a reaction, generally including energy and momentum. We will focus our discussion on modern particle accelerator elements.

The first element is the particle that is to be accelerated and collided. These have to carry electric charge, given that accelerators rely on electromagnetic fields to perform both longitudinal acceleration (along the intended direction of travel) and transverse acceleration (to focus particles into a dense area around the longitudinal direction). Standard particle types for modern accelerators are electrons (and positrons), protons (and anti-protons), and heavy ions (such as lead or xenon). Electrons are readily available for removal with modest energy application to any atomic source. Photoemission is the most standard technique for liberating large numbers of electrons from an atomic source [HOS08], as full ionization of the more loosely bound atomic electrons requires only energies at the eV level. Protons are easily obtained by ionizing hydrogen gas; even a small gas cylinder is sufficient to supply an accelerator like the Large Hadron Collider (LHC) for $\mathcal{O}(10^5)$ years [Hes15]. Heavy ions like lead require more effort, but the principle is the same. A small volume of very pure lead-208 is heated so that the lead near the surface is liberated as a vapor. This vapor can then be ionized into a plasma for use in the accelerator [OLu13].

The reasons for using different particles can vary according to the needs of the experimental community, but usually arise from one of a few possibilities:

- there is a need to know all momentum vectors for the initial state or to know precisely what particles participated in the initial reaction, in which case the use of a fundamental particle (electron or positron) is preferable over a composite particle like a proton (where the quarks and gluons in the proton do the colliding but one cannot control which constituents actually interact, nor their exact longitudinal momentum at the time of collision); or

- there is a need to minimize energy loss during acceleration due to radiation from the particles, a process that is worse for low-mass particles (this then makes protons preferable over electrons, for example).

Acceleration is usually achieved by a series of stages, typically beginning with a linear acceleration process to take the early plasma from the above methods and accelerate it (and "bunch" the particles) so that it is ready for subsequent stages.

Accelerators generally take one of two forms: fully linear or circular (where early stages may be linear), as illustrated in Fig. 5.25. In the latter approach, particles are accelerated to high energy along a mostly straight path. This avoids, as much as possible, unnecessary bending of the trajectory of the beam particle, a factor that can minimize radiative energy loss from the beam. Linear accelerators of the highest energy require proportionally more length to achieve additional acceleration, requiring significant land usage. Circular accelerators reduce the use of larger land areas by permitting the beams to be accelerated repeatedly to higher and higher energy in the same length of accelerator path. This requires bending the beam into a circular path, a fact that leads to enhanced radiation and thus energy loss from the beam. In general, power transported away by radiative energy loss from the beam of particles bent in an arc of radius R (known as *synchrotron radiation*) is

$$P_{synchrotron} = \frac{q^2 c}{6\pi\varepsilon_0} \frac{(\beta\gamma)^4}{R^2}, \tag{5.42}$$

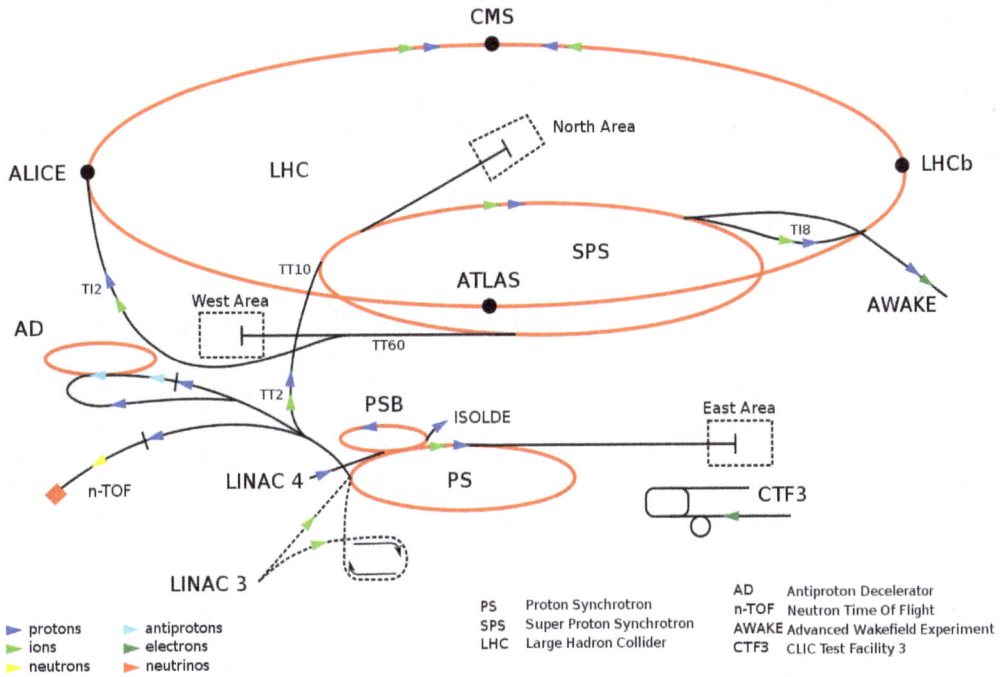

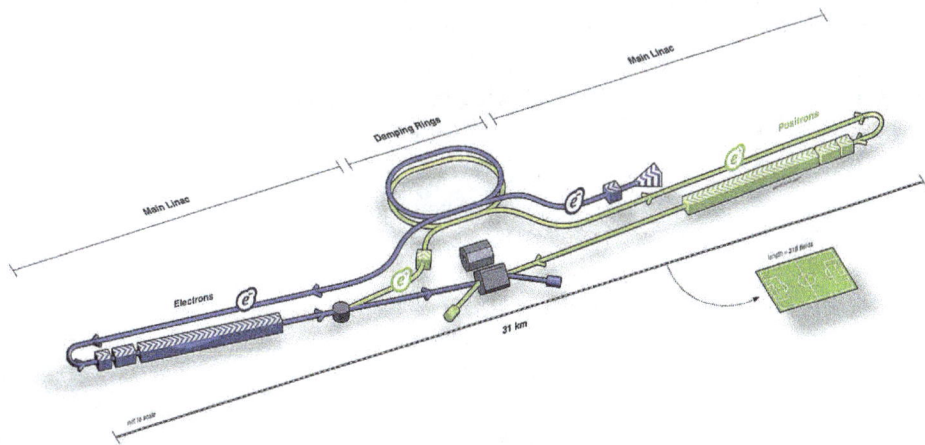

Figure 5.25: Common modern designs for circular (top, CERN accelerator complex) and linear (bottom, International Linear Collider conceptual design) particle accelerators. The LHC accelerator ring (top) is 26 km in circumference. A baseline ILC design would have a total length of about 30 km. Figures from Refs. [Wik20c; Wik23b].

where q is the particle's charge (e.g., $|q| = e$ for electrons and protons) and β and γ are the usual relativistic factors. Since $\beta\gamma = (pc)/(mc^2) = p/(mc)$ is just the ratio of the particle's momentum to its mass, we can see that, for two equal-energy beams of electrons and protons, the power radiated by the electron beam will be much greater.

We can insert the total energy into the above equation and isolate the relationship with mass using $p/(mc) = \sqrt{E^2 - m^2c^4}/(mc) = c\sqrt{E^2/(m^2c^4) - 1}$. For the special case that $E \gg mc^2$ we can approximate this as $c\sqrt{E^2/(m^2c^4) - 1} \approx E/(mc)$. In that case, the relative radiative energy loss by electrons, compared to protons, goes as $P^e_{synchrotron}/P^p_{synchrotron} \propto (m_p/m_e)^4$, a stunningly large dependence on the mass ratio. In this example, all things being equal between the beams except the mass of the particles, the radiative loss by the electron beam will be 10^{13} times greater than from the proton beam.

The above discussion leads to two conclusions regarding scaling the energy of an accelerator. If one is to go to as high an energy as possible (to, say, access very large dark matter masses), one should either use protons (to avoid synchroton radiation penalties) or one should build a linear, or the maximum-possible-radius circular, electron-based accelerator. On the other hand, if one is attempting to probe low-mass dark matter scenarios (with potentially very small couplings to normal matter) then intensity, not energy, is the goal. In that case, one can operate at smaller accelerator energies where either beam choice is practicable given radiation and size considerations.

The energy of an accelerator is determined by at least three factors: the energy region of interest to experimental physicists, the strength of the accelerating structures (electric fields to increase longitudinal velocity and magnetic fields to steer and focus particles), and civil engineering constraints (land area available and the cost of tunneling, which ultimately constrain the length and/or bending radius of the accelerator).

The intensity of an accelerator beam (cf. Ref. [HM06]) is determined by a number of factors that are summarized in the following equation (which assumes that transverse to the longitudinal axis, z, each bunch in the beam has a Gaussian number density profile):

$$\mathcal{L} = N_b f \frac{n_1 n_2}{2\pi\sqrt{\sigma_{1,x}^2 + \sigma_{2,x}^2}\sqrt{\sigma_{1,y}^2 + \sigma_{2,y}^2}}. \tag{5.43}$$

Here, $\mathcal{L}$ is the instantaneous luminosity of the accelerator (interactions per unit area, per unit time, typically in $cm^{-2}s^{-1}$); N_b are the number of particle bunches in the collider; f is the frequency of bunches (and therefore, collisions); n_i are the number of particles in each bunch in beam i ($i = 1, 2$); and $\sigma_{i,x}$ and $\sigma_{i,y}$ are the Gaussian widths of the bunches in the x and y directions in beam i at the interaction point, respectively. Here, "bunches" refers to the grouping of particles in the accelerator structure. For example, at the LHC a typical proton bunch contains about 10^{11} protons and has a transverse width in either dimension at the level of $\mathcal{O}(10)$ microns. The $N_b \approx 2800$ bunches are spaced by 25 ns, yielding a collision frequency of 40 MHz. Thus, we can estimate the instantaneous luminosity of an LHC-like collider to be at the level of 10^{34} $cm^{-2}s^{-1}$, consistent with typical modern operating conditions for that machine.

Production Cross Sections and Decay Branching Fractions

We discussed scattering cross-sections in, for example, Section 4.3, when we considered the probability of an interaction of initial-state particles producing a particular (or set of) final-state outcomes. In collider physics, the language is identical. We speak of the beam particles (initial state) colliding and producing some process with a certain probability. This is again determined by the cross-section for a process, which is usually determined from theory or experiment. This cross-section is again denoted σ, for instance $\sigma(pp \to H^0 + X)$ which denotes the probability that two protons interact and produce a Higgs particle plus any other set of particles, X.

The Higgs particle is unstable and decays to other SM particles, which we discuss below. The probability of decay is quantified in a *branching fraction*, which is the rate at which an unstable particle will transition to a particular final state. This is denoted by $\mathcal{B}$.

In any collider experiment, one can combine the rate of particle collisions, per unit cross-section, per unit time, with the cross-section of a target process and any branching fractions required to arrive at a particular final state. The number of events from a specific production and decay process ("signal") is provided experimentally by

$$N_{signal} \equiv N_{obs} - N_{bkg} = \mathcal{L}\sigma\mathcal{B}\alpha\varepsilon \tag{5.44}$$

where $\mathcal{L}$ is the collider's integrated luminosity (the total number of collisions per unit area, $\mathcal{L} \equiv \int L dt$, where L is the rate of collisons measured in $cm^{-2}s^{-1}$), α is the *acceptance* of the experimental apparatus (the probability that the final-state particles will travel within the active region of the detector), ε is the efficiency with which the final state can be reconstructed assuming all particles are within the acceptance of the detector, N_{bkg} is the number of background events that confound the observation of the signal, and N_{obs} is the number of events observed in the data set. Backgrounds inevitably must be subtracted to isolate the signal-like events, if that background expectation cannot be reduced to zero by experimental means.

Background processes will also be produced with particular cross-sections and branching fractions, which, again, must be estimated from theory or measured from experiment. For a set of background processes $pp \to X_{bkg}$, the total background expectation will be given by Eqn. 5.44, but using the cross-section(s) and branching fraction(s) related to the background process and eliminating the background term.

Uncertainty in the predictions of the expected number of signal and background will arise from a range of sources, including the luminosity measurement, cross-section estimates, branching fraction estimates, and uncertainties on the acceptance and efficiency of the experiment. All of these factors much be measured and considered when making predictions for the outcome of experiments.

Let us focus on proton–proton collisions. There are three major outcomes when protons collide: they can both scatter elastically ($pp \to pp$), or one can be inelastically scattered ($pp \to p + \sum_i x_i$, where x_i label i partons from the shattered proton), or both can be inelastically scattered ($pp \to \sum_i x_i$). The interaction of partons from each proton involved in the collision generally implies a double-inelastic scatter, the total rate of which at a given collision energy is generally known as the *inelastic cross-section* of the proton. This has been measured at the LHC (cf. Ref. [Aab+16]) and is known to be (78.1 ± 2.9) mb at a center-of-mass (CM) energy of 13 TeV.

Higgs Production at a Proton–Proton Collider

The rate of production of particles by the collider is determined by quantum field theory. For example, applying the SM to the LHC allows for predictions of the rate at which gluon-gluon, gluon-quark, or quark-quark interactions at LHC energies will lead to final states involving quarks, leptons, force-carrying particles ($W^{\pm}$, Z^0, gluons, and photons, and the Higgs particle (H^0). At a proton–proton CM energy of 13 TeV, the cross-section for producing a Higgs boson (with a mass of 125 GeV) via gluon-gluon fusion (ggF) is 48.52 pb [Ana+16]. The simplest diagram that facilitates this production mechanism is illustrated in Fig. 5.26a. Including all known production mechanisms, the total Higgs production cross-section under those same conditions is about 55.54 pb (ggF represents 87% of the total production probability at the LHC for these energy conditions) [Flo+17].

In the context of all inelastic proton–proton collisions, the probability to produce a single Higgs particle is $\sigma_H/\sigma_{inelastic} \approx 7 \times 10^{-10}$ per collision, or about one such Higgs boson produced in every billion proton–proton inelastic collisions at the LHC. All of those more common interactions that produce particles other than Higgs bosons can also produce patterns of final-state particles that resemble Higgs decay, and are thus also a background to the study of the Higgs. Rejecting the other non-Higgs-bearing collisions in favor of selecting those with at least one Higgs boson is the art of modern collider physics data reduction ("trigger") systems as well as refinement of the data after collection. We will return to this issue below.

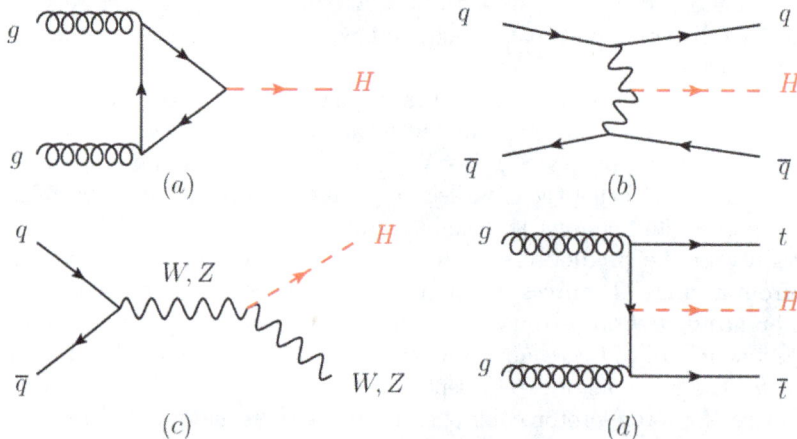

Figure 5.26: Feynman diagrams for the four major Higgs particle production mechanisms at the LHC. The processes are (a) gluon fusion, (b) vector-boson fusion (VBF), (c) vector-boson-associated (VBA) production, and (d) top-associated production. Figures from Ref. [Par22]

Final State	Higgs Branching Ratio
$b\bar{b}$	0.5824
WW^*	0.2137
gg	0.08187
$\tau^+\tau^-$	0.06272
$c\bar{c}$	0.02891
ZZ^*	0.02619

Table 5.1: Branching ratio (fractions of total possible ways of decaying) calculations for the Higgs boson at a mass of $m_H = 125$ GeV [Flo+17]. Values are shown only for final states with a contribution above the 1% level, though there are important decays with branching fractions below that level. Each of the above predictions has theoretical uncertainties which are not shown but are typically at the 1—10% level.

The rarity of Higgs production is compensated, to a significant degree, by the intensity of the collider. At a typical luminosity of 10^{34} cm^{-2}s^{-1} (which translates, in particle physics cross-section units, to 0.01 pb/s), the LHC is capable of producing about 0.5 Higgs bosons every second, or one every two seconds. Given that a typical data-taking campaign for a modern collider like the LHC lasts for about three-quarters or more of any given calendar year, the LHC is capable of producing $\mathcal{O}(10^7)$ Higgs bosons per year. However, a single experiment is not likely to actually see the complete decay products of the Higgs boson (a concept known as the *acceptance* of the detector), nor to then be able to identify that decay (even when all decay products enter the detector) with 100% efficiency. In addition, the need to reject a large background necessarily then reduces the experimental efficiency significantly (since backgrounds will generally outnumber Higgs bosons at the level of a million to one). Typical reconstruction efficiencies are at the $\mathcal{O}(1-10)$% level for standard analysis approaches at the LHC. We will explore a limited aspect of this in the next section as we discuss some techniques to hunt for dark matter at the LHC.

The largest predicted branching fractions for the decay of a Higgs are given in Table 5.1. An estimate of the production of a pair of bottom quarks proceeding through a single intermediate Higgs particle is obtained by multiplying the production cross-section of the Higgs by the branching fraction of the desired final state. In this example, $\sigma(pp \to H^0 \to b\bar{b}) = \sigma(pp \to H^0) \times \mathcal{B}(H^0 \to b\bar{b}) = 32.45$ pb.

Each of the Higgs particle's decay modes, which leads to the ultimate signature in a particle detector, is challenging to identify above the collider background. This is due to the fact that the LHC can produce these same final states in other ways at much higher rates. For example, the $b\bar{b}$, gg, or $c\bar{c}$ final states will be marked by the production of pairs of particle *jets* (collimated but spatially distributed sprays of subatomic particles originating from the progenitor quark or gluon). Within the fiducial region of a typical large LHC multi-purpose detector, the cross-section for di-jet production at $\sqrt{s} = 13$ TeV is about 340 nb [ATL24], still many thousands of times higher than the Higgs production and decay rates. Similarly, for WW^* the generic production rate under the same conditions is about 130 pb, higher than the Higgs production and decay rate. This pattern continues for other prevalent decays. The major effort in searching for and reconstructing Higgs candidates is then on identifying differences in the kinematic patterns of the final-state particles that allow for distinction between the signal and background processes.

We will investigate these challenges somewhat more deeply in the context of collider searches for particles that are not easily directly detected. In particular, we will delve into some of the specifics of the Higgs search and backgrounds using the possible coupling of the Higgs particle to a dark-matter-like final state. We will also highlight efforts to detect particles that appear alongside dark matter candidates in dark sector theories. We begin with the search for "missing" or "invisible" signatures at a collider like the LHC and close the collider discussion with a search for a possible force-carrying dark-sector particle, a search that can be carried out at colliders with a wide rage of energies.

Missing Momentum and Energy Approaches

Dark matter is expected to have very small interaction cross-sections (and thus interaction couplings) with normal matter. As a result, once produced, the dark matter should traverse a material system (such as a particle detector) with almost zero chance of interacting with the detector. For example, consider a detector like the Compact Muon Solenoid (CMS) instrument at the LHC. CMS is a nearly cylindrical particle detector 21 m long and 15 m in diameter, giving it a cylindrical volume of about $15,000$ m^3. It has a mass of about 14,000 tonnes. Its average density is thus about 1800 kg/m^3.

Direct detection experiments suggest that the interaction cross-section between dark matter and a single nucleon (proton or neutron) is not much greater than 10^{-47} cm^2 for a WIMP mass of 20 GeV. Based on this, we can estimate the probability that a single dark matter particle, passing through CMS, will interact with one of its components. We will assume the most optimistic scenario: that the dark matter can interact with any nucleus it encounters and that any such interaction leads to a 100% detectable electronic signature inside CMS. If this scenario for detecting dark matter directly looks grim, then any more realistic scenario will be even more grim.

We can estimate the "average atom" from which CMS is composed by using its density as our guide. Given the atomic densities of known elements, this number is closest to that of the following materials: beryllium, magnesium, phosphorous, or cesium.

For this calculation, to pack the most nucleon targets into our detector, we will assume the atomic mass of cesium (^{133}Cs, with mass 133 u) is our target, yielding 55 protons and 78 neutrons, for a total of 133 nucleons.

How many such atomic targets would lie along a straight path from the center of the detector and travelling radially outward to the edge? For a distance of 7.5 m, this would give us $\rho_{CMS} \times r_{CMS} = 1.38 \times 10^4$ kg/m^2. Dividing this by the mass of each cesium atom gives us the number of such atoms-per-unit area, yielding 6.3×10^{28} atoms/m^2.

The probability of a single dark matter–nucleus interaction is going to be the number of atoms per unit area multiplied by the number of nucleons per atom and the interaction cross-section with dark matter, yielding $p_{interaction} = (\rho_{CMS} \times r_{CMS}/m_{Cs}) \times 133 \times \sigma_{\chi-N} \approx 10^{-20}$.

This is an exceedingly small probability for a single dark matter–nucleon interaction anywhere along the radius of the detector. Let's see if the nature of the collider (high-intensity collisions) can offset this small number. The LHC delivers something like one billion collisions each second. It operates for about nine months each year, and will have operated for about thirty years when it concludes its program in the 2030s. Assuming every collision produces a dark matter particle, the chance of seeing one dark matter interaction any time in thirty years of operations is 1%. Another way to say this is that an experimental physicist, under the most unrealistic and optimistic conditions, could hope to see one such interaction somewhere in the CMS data if the LHC ran for about 3000 years.

It is obviously impractical to use such a particle detector to directly observe a novel interaction from a particle dark matter candidate. Instead, the approach is to avoid the penalty one pays by direct detection (the multiplication by the dark matter–nucleon cross-section) and instead look for events where there is an *absence of information* in a region of the detector. This could indicate that something with high energy and/or high mass passed through a region of the experiment, but left no trace of itself. This is known as a "missing energy" or "missing momentum" approach.

We can understand the basics of this approach using the application of the conservation of energy and momentum. Let us assume a collider in which we know precisely the initial conditions of particles in each beam (which we shall label 1 and 2, where, by construction, the energy of beam 1 will always be equal to, or greater than, the energy of beam 2). In that case, in the laboratory reference frame (the frame in which the collider detector experiment is at rest with respect to an external observer and thus the frame in which energy and momentum are actually measured) the total energy and momentum will be

$$E_{total} = E_1 + E_2 \tag{5.45}$$
$$\vec{p}_{total} = \vec{p}_1 + \vec{p}_2. \tag{5.46}$$

We can also write the same equations in the CM frame of the collider, in which particles in the two beams have equal but opposite momentum:

$$E'_{total} = E'_1 + E'_2 \tag{5.47}$$
$$\vec{p}'_{total} = \vec{p}'_1 + \vec{p}'_2 \equiv 0. \tag{5.48}$$

There are designs where the two reference frames are the same, and where they are different. For example, in a *symmetric collider* composed of beam particles with equal energies in both beams, the laboratory and CM frames are identical. Since this isn't universally the case and depends on collider design, we make the distinction for now and note later if the equivalence of the frames is a valuable constraint. Note also that we make a distinction between *beam particles* and *the participants in the fundamental collisions* themselves. For example, in a collider whose beam particles are non-composite objects (e.g., electrons or positrons), there is no distinction between these two things. Since electrons contain no known substructure, one expects the whole electron to participate in any collisions with the particles in the other beam.

In a proton collider, however, it is generally not the whole proton that participates in a fundamental collision. Rather, individual quarks and gluons make the collisions, resulting in the production of new particles and the dissolution of the parent proton. This is important since, in this latter case, the *parton* (gluon or quark) that actually participates in the collision carries an unknown fraction, x_1, of parent proton energy, E_1, meaning the actual energy with which it enters the collision is $x_1 E_1$, not just E_1.

Particle Reconstruction and Missing Particles

With these points in mind, let us proceed to consider a situation in which a collision of two beam particles results in the production of at least one undetectable final-state particle. The total energy and momentum of these products will be denoted E_{inv} and $\vec{p}_{inv}$, respectively, in the laboratory frame. Thus we can see that the total energy is composed of the sum of observable (visible) and unobservable (invisible) energy, $E_{total} = E_{vis} + E_{inv}$.

Note that, at this stage, we assume that all observable particles are, in fact, observed. In a real detector, this is not the case. Collider detectors (Fig. 5.27) necessarily have a region through which the accelerator enters and exits, creating a hole in angular space

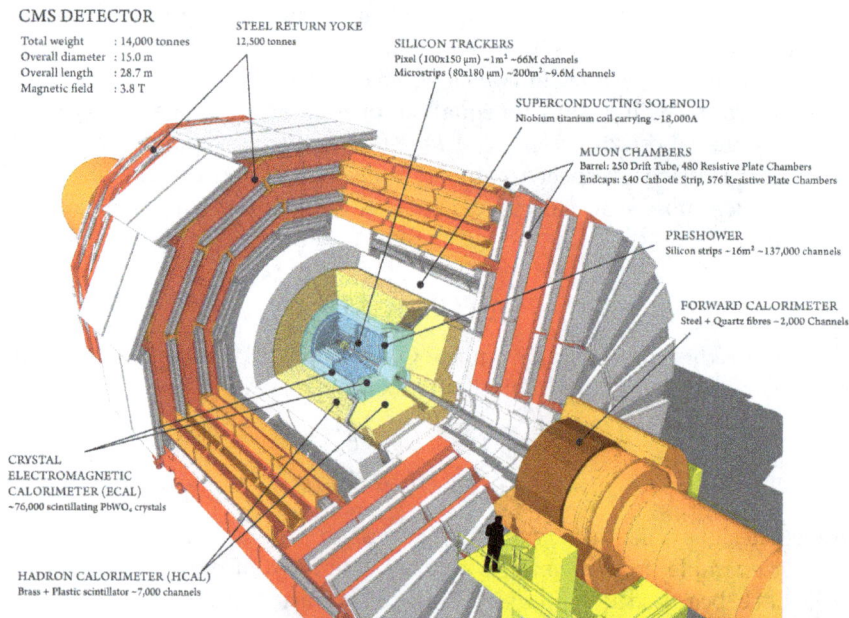

Figure 5.27: A schematic of the CMS detector, showing the components and general composition of the detector from the center (where the colliding beams intersect) to the outermost extent of the experiment. Figure from Ref. [Col23].

where observations are nearly impossible. Experimental designs strive to avoid gaps in the fiducial volume of the detector. These can, nevertheless, occur as there are unavoidable seams between active detector elements or segments of the detector, as well as places where services like electricity, water, dry air, and cryogenics must enter and exit the internal systems of the detector.

Any realistic detector naturally has particle detection inefficiencies (a natural inability to reliably detect 100% of all observable particles passing through its active volume). That efficiency can vary with time, due both to aging effects and to upgrades. We will neglect these realities for now, but return to them later when considering the challenge to the experimentalist in assessing the limits of missing energy detection.

A typical collider detector has two major features: the ability to see charged particle ionization and the ability to detect energy deposition. Charged particle ionization provides information about the direction of travel of these particles. Subjecting the particles to a magnetic field while they produce ionization will result in observing the bent trajectories of those particles, from which the momentum of the charged particles can be inferred.

Charged particles can lose energy in material through a variety of mechanisms [Par22]. These vary in degree with the energy of the particle. Charged particles can ionize atoms. If a charged particle exceeds the speed of light in a material (which is always less than the speed of light in empty space, the fastest anything is known to travel), they can emit Čerenkov radiation (photons) that reduces the kinetic energy of the original particle. A charged particle that is deflected by an interaction with matter (through the action of nuclear electric field, for example) can emit radiation in response, known as *bremsstrahlung* or "braking radiation." Photons can Compton scatter from electrons in material, altering their wavelength (and thus their energy). Any particle detector experiment must account for all such possible energy losses when designing an instrument in order to select the best combination of technologies and achieve the necessary momentum and energy measurements.

Some detectors employ Čerenkov or transition-radiation detection schemes that provide independent information about charged particles and lead to the ability, combined with momentum information, to extract the mass of the charged particle and, hence, the identity of that particle. Electrically neutral particles (such as neutrons and photons) do not produce ionization, so to detect them they must be stopped in material in such a way that they deposit all of their energy in the medium. Photons can be readily stopped using materials like lead. By interleaving the dense material with an active medium, one can identify the shower of particles that are radiated as the photon interacts with material.

Heavier particles require more intervening material or more distance to stop, so calorimeter systems often contain an inner system for stopping lighter particles (electrons and photons) and an outer system for stopping the heavier particles (protons, neutrons, and long-lived but unstable quark matter such as pions and kaons). The performance of a calorimeter is often described using the concept of *radiation length*, the average distance a particle needs to travel in material (of atomic mass A and atomic number Z) before reducing its kinetic energy to $1/e$ of the original value. This varies by particle species and how they lose energy in material.

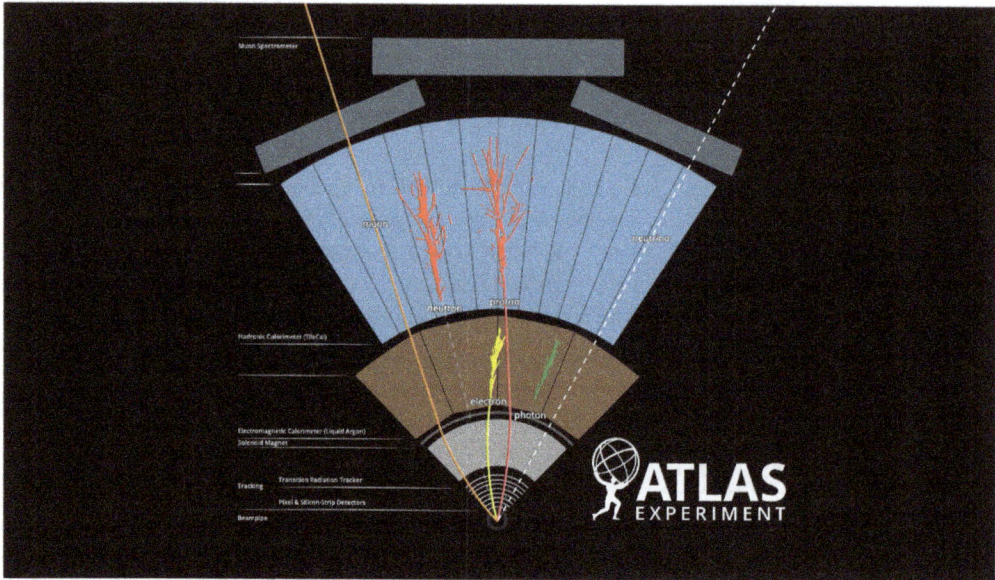

Figure 5.28: An illustration depicting a slice of the ATLAS detector, showing how different long-lived particle species traverse elements of the detector and can thus be observed. Figure from Ref. [Meh21].

Muons, which are heavier cousins of the electron, often traverse the entirety of calorimeter systems and continue beyond. This is due to the fact that high-energy muons lose only a small fraction of their energy to ionization and are not as easily deflected as electrons during interactions, so their energy loss due to effects like bremsstrahlung is much lower. Since high-energy muons can penetrate through particle detectors, this means that it is not possible to collect and observe all of the muon's energy, only its momentum. This necessitates dedicated systems in the outermost regions of a detector that make one final ionization measurement for any particle that survives passage through the calorimeter. A depiction illustrating different particle types and the means of each one's detection in the ATLAS experiment at the LHC is shown in Fig. 5.28.

Collider experiments will typically emphasize precision and accuracy in the reconstruction of energy and charged particle information (their trajectories, or *tracks*, in the detector). For detectors where the calorimeters have been designed to stop both light and heavy particles, and segmented and instrumented as a level to sufficiently ensue a precise energy measurement, it is more typical to rely on the calorimetric information to define particle energies. However, experiments may choose to emphasize precision tracking and reduce the coverage of calorimeter systems (for example, employing an electromagnetic calorimeter but foregoing a dedicated hadronic calorimeter that would stop higher-mass particles). In that case, employing the momentum information, combined with measurement of the mass (through Čerenkov detector approaches) can provide a superior energy measurement.

The process of combining information from across detector systems to identify specific particle categories is known as "particle identification" or "particle reconstruction." We provide an example of this using electrons, photons, and muons as classes with very different behaviors in the kinds of detector systems we have outlined earlier.

An electron is distinguishable by its energy deposits in the charged particle tracking system, where it will also bend in a magnetic field. The path of the electron in the tracking system should point into the region of the calorimeter where it interacts next. There, the electron should experience a total or near-total deposition of its kinetic energy in the inner calorimeter system. Electrons should leave little or no energy in the outer calorimeter, although it is possible for some leakage to occur. In addition, the outermost muon detection systems should contain no energy deposits along the extrapolated flight trajectory of the electron. These behaviors are illustrated in Fig. 5.28. This example provides the basic framework for particle reconstruction: a pattern of behaviors in multiple systems (positive detection in the tracking and inner calorimeter systems, negative detection in the outer calorimeter and muon system). Physicists will algorithmically combine patterns of information from these systems to determine if a given track and calorimeter energy deposit could be from an electron or something else.

Photons are distinguished by a lack of interactions in the tracking system [2] and a total or near-total deposit of all of their kinetic energy into the inner-most calorimeter system. Little or no activity should be present in the outer calorimeter or in the muon system. In contrast to both electrons and photons, muons will leave energy in the tracking system, bend in the magnetic field while doing so, leave very little kinetic energy behind in either of the calorimeter systems, and then leave energy deposits in the muon system as it exits the entire detector.

Particle identification requires the use of information from many systems. As a result, the uncertainty on particle identification is the combination of uncertainties from tracking, calorimetry, and potentially from the muon system. Let us imagine, for a moment, an experiment that can be done simply by looking at information from the calorimeter for all particle categories, regardless of the identity of those particles. In that case, any experimental quantity $\mathcal{F}$ derived from the use of this information will depend solely on the total uncertainty from using calorimeter energy information, $\mathcal{F}(\sigma(E_{cal})_i)$ where i labels the distinct energy deposits. If one employs more systems in the same analysis, the derived experimental quantity must depend on uncertainties from measurements in all of those systems, $\mathcal{F}(\sigma(\vec{p}_{track})_i, \sigma(E_{cal})_j, \sigma(\vec{p}_{muon})_i, \ldots)$. For a young experiment, these uncertainties may be large and there may be disadvantages to relying on fine-grained

---

[2]A caveat to the statement about photons lacking interactions in charged particle tracking systems arises quite often in high-energy particle colliders: photons can convert to electron-positron pairs when passing through the material of the tracking system. If this happens, the original photon ceases to exist and in its place an electron and a positron finish traversing the tracking system. These can be identified by first finding electron and positron candidates using electron identification techniques (energy in the tracking system, bending trajectories in a magnetic field, energy deposits in the inner-most calorimeter system) and then looking for pairs of electrons and positrons that seem to share a common origin point in space. These co-originating pairs are indicative of photon conversions and are then grouped together into photon candidates. Nearly 50% of high-energy photons in detectors like ATLAS and CMS can convert to electron-positron pairs and must be identified this way.

particle categorization. A more mature experiment will have more experience measuring and controlling these uncertainties and can opt for more complexity in defining an analysis strategy.

Modern experiments will also layer their definitions when they categorize tracks and energy deposits into identifiable particle classes. A hierarchy of categories will be developed, generally with the easiest-to-classify categories first and the hardest-to-classify last. Tracks and energy deposits used in earlier parts of the hierarchy are removed from consideration from later parts to avoid double-counting the same information in two different categories. For example, physicists want to avoid using the same calorimeter energy deposit in an electron and then later for a photon. This would prevent the ability to impose energy and momentum conservation on a particle collision.

As an example of a hierarchy, consider a typical approach used by experiments like ATLAS and CMS. Muons are often the most recognizable particles in such experiments as they traverse the entire detector. If a track in the inner tracking system, energy deposits in the calorimeter systems, and a trajectory in the muon system are combined to make a muon candidate, those elements are removed from consideration for making other particle identification choices. Photons and electrons tend to then be the next easiest-to-define particles, owing to their *absence* of signatures in the most distant parts of the calorimeter and in a muon system. Tau leptons are often next. They decay to electrons and muons about 30% of the time (and so would be classified as electron or muon candidates), but to hadrons (pions and kaons) the remainder of the time. This latter class of decays resemble quark and gluon particle jets. However, algorithms have been developed that can detect the differences. So, at this stage it is typical for taus to be selected into their own class.

We have mentioned jets before as a collimated arrangement of particles that originate from the quark and gluon radiation that results from a progenitor quark or gluon. However, in an experiment it is not possible to know which particles are supposed to go together into a single jet. Rather, an algorithmic definition of a jet is required. Different approaches to gathering energy clusters or tracks ("objects") into *jet candidates* have strengths and weaknesses, and map back onto the "true" jets with a varying degree of accuracy.

A review of jet algorithms is available in Ref. [Sal10]. They generally appear in two categories: cone algorithms and sequential recombination approaches. The former impose the hypothesis that a jet should be cone-shaped in some geometric space and then group objects that are within the cone definition. The latter approach begins by ranking objects using kinematic information and employing the first-ranked object as a seed to build the first jet candidate. Lower-ranked objects are added to the jet candidate so long as they are deemed to be close enough to the seed in the kinematic space. The kinematic measure and the cut-off are two parameters of this approach that need to be adjusted to suit the experimental needs. A jet candidate is complete when the next object on the list is beyond the cut-off. That object is then used as a seed for the next jet candidate. The benefit of sequential recombination is that every object is ultimately used to build a jet, while the cone algorithmic approach may leave objects unused because they lie outside the boundary of any previously identified cone.

Jets tend to be the last particle candidates defined because they tend to use most or all of the particles available to them, so any tracks or energy deposits in the earlier categories must first be removed from consideration for jet-finding. Jets tend to be constructed from energy deposits first. A refinement of the jet definition might then replace energy deposits in a jet with the tracks that produced those deposits, as momentum is often better measured than energy in a modern detector system. In some cases, jets will be entirely built from tracks.

Finally, any charged particle tracks that have not been used in any other category, and which tend to be very low in momentum ("soft"), are then placed in their own category. We will refer to this as the "soft tracks" class.

Constructing Missing Energy and Momentum

Let us consider a case where the calorimeter system is sufficient to stop both light and heavy particles (excepting muons) and provide excellent energy resolution. In this case it is typical to define the visible energy as equivalent to the calorimeter energy, $E_{vis} = E_{calo}$. However, in the case where at least one muon is present, this is not exact and it is better to define $E_{vis} = E_{calo} + E_\mu$. Since muons are not stopped by the calorimeter, their energy must be inferred from their momentum. Absent the use of Čerenkov radiation to infer particle mass, it is typical to assume that any particle that ionizes *after* the calorimeter must be a muon. In that case, $E_\mu^2 = p_\mu^2 + m_\mu^2$. For high-energy colliders like the LHC, $m_\mu \ll |p_\mu|$ and so, $E_\mu \approx |p_\mu|$.

Thus it is that, hypothetically, one can infer the missing (invisible) energy and momentum by solving the following equations:

$$E_1 + E_2 = E_{vis} + E_\mu + E_{inv} \tag{5.49}$$

$$\vec{p}_1 + \vec{p}_2 = \sum_i \vec{p}_{i,vis} + \sum_j \vec{p}_{j,\mu} + \vec{p}_{inv} \tag{5.50}$$

Missing momentum can be inferred as $\vec{p}_{inv} = \vec{p}_1 + \vec{p}_2 - \sum_i \vec{p}_{i,vis} - \sum_j \vec{p}_{j,\mu}$, and a similar equation can be written for missing energy. In the special case where $\vec{p}_1 = \vec{p}_2$, this reduces to $\vec{p}_{inv} = -\sum_i \vec{p}_{i,vis} - \sum_j \vec{p}_{j,\mu}$, or just the negative of the vector sum of everything that is visible to the detector. We note that it is possible to combine missing energy and missing momentum information (separately determined from calorimeter and tracking system approaches) to infer the *missing mass* from a particle collision, $m_{inv}^2 = E_{inv}^2 - p_{inv}^2$.

The above equations are experimentally useful when all momentum components of the colliding beam particles can be controlled. Since the beams collide along a single axis, typically denoted as the z axis of a cylindrical coordinate system (with ρ or r being in the radial coordinate direction, perpendicular to z and the angle ϕ denoting the azimuthal angle of r about z), the momentum of the beam particles transverse to the z direction is slight. Generally, collider physicists approximate the radial motion of the beam particles to be approximately zero. In a collider where both beam particles are non-composite, the momentum along the z axis is entirely determined by the properties of the accelerator. However, in a collider with composite beam particles it is not those particles that participate in the interactions, but the constituents inside the beam particles. Thus the momentum along z is not known collision-by-collision, as the partons each carry a fraction, x_i, of the parent beam particle's energy whose value is unknown at the time of collision.

In the case of proton or heavy ion colliders, it is traditional then to abandon the use of conservation of momentum along the z direction since the initial-state momentum cannot be known. Experimentalists then deal only in the directions *transverse* to the z axis. The initial state is known, in this case, to be $\vec{p}_{T,1} + \vec{p}_{T,2} \approx 0$, where the T denotes the vector components transverse to z. These are referred to as *transverse momentum*. It is then possible to rewrite the above equations, ignoring the z component of the momentum, to obtain

$$\vec{p}_T^{\,\text{inv}} = -\sum_i \vec{p}_{T,i}^{\,\text{vis}} - \sum_j \vec{p}_{T,j}^{\,\mu}. \tag{5.51}$$

For a high-enough energy collider, where the masses of the final-state particles are small compared to the energy and momenta involved after the collision, one can then define $E_T^{inv} \equiv |\vec{p}_T^{\,\text{inv}}| \dots$ a *transverse missing energy*.

Transverse energy can be given a direction using the transverse momentum. This is achieved by asserting that

$$\vec{E}_T \equiv E_T \hat{p}_T \tag{5.52}$$

where $\hat{p}_T$ is a unit vector of length, 1, that points in the transverse direction of the momentum. A few vector algebraic steps are sufficient to reveal the relationship between total energy, transverse energy, and the mass, m, of the decaying particle:

$$\vec{E}_T = E_T \hat{p}_T \tag{5.53}$$

$$= E \frac{p_T}{p} \frac{\vec{p}_T}{p_T} \tag{5.54}$$

$$= \frac{E}{\sqrt{E^2 - m^2}} \vec{p}_T. \tag{5.55}$$

This allows us to write the total energy in terms of the transverse energy as follows: $E = E_T p/p_T$. This will be useful in a later discussion.

The above definition is extremely broad and can be used by an experiment that wants to solely source information from a single system, such as its calorimeter. A more mature definition of missing transverse momentum, using fine-grained particle classification, would be given by

$$\vec{p}_{inv,T} = -\sum_j \vec{p}_{T,i}^{\,\mu} - \sum_i \vec{p}_{T,j}^{\,\gamma} - \sum_k \vec{p}_{T,k}^{\,e} - \sum_l \vec{p}_{T,l}^{\,\tau} - \sum_m \vec{p}_{T,m}^{\,\text{jets}} - \sum_n \vec{p}_{T,n}^{\,\text{soft tracks}}. \tag{5.56}$$

We take this opportunity to comment on a few confounding issues that face collider detector experiments. First, their resolution on missing energy and momentum will be limited by the least precise component that makes measurements of visible energy. Typically, the ability to resolve energy deposits in calorimeters is worse than the ability to resolve momentum in a charged particle tracking system. The subject of calorimeters and the various effects that can lead to uncertainty on the energy measurement is well-documented elsewhere (cf. [LW19]). A standard technique to improve energy reconstruction in modern detectors is to, where possible, use the track information in lieu of the calorimeter information (e.g., energy flow or particle flow approaches, Refs. [Bus+95; Sir+17]). Nevertheless, since the direct detection of the source of invisible energy is not possible in such collider experiments, the detection mechanism will always rely on a sum of other information where the contribution of uncertainties from nearly all aspects of the detector is almost maximal. Nevertheless, progress in this area has been remarkable in recent decades.

A second confounding effect is that there are real, invisible particles that are already known to us and produced in colliders: neutrinos. These particles are readily produced in electroweak interactions such as the decay of Z and W bosons as well as the prompt decay of many hadrons (e.g., bottom and charm mesons). Any claim of a detection of excess missing energy in a collider signature must account carefully for the effects that real neutrinos, also potentially present in the collisions, play in producing real missing energy.

A third confounding effect is fake missing energy. This is due to at least a few effects, including particles lost in the inactive regions of the detector (visible but unseen) and inefficiencies in reconstructing particles in active regions of the detector (for example, track reconstruction is not 100% efficient). This can be studied in collisions where the final state of the collision is easily reconstructed and a full accounting of all particles is possible, but where, nevertheless, non-zero missing energy is present. For example, at the LHC it is possible to reconstruct the Z bosons produced by $pp \to ZZ$ using $Z \to e^+e^-$ and/or $Z \to \mu^+\mu^-$. Reconstructing electrons and muons is fairly straight-forward, and there are no real sources of true missing energy in this process, such a neutrinos. In this case, it should be that $\sum_i \vec{p}_{T,i}$ for the final-state electrons and muons should be exactly zero. However, these approaches will often observe deviations from zero in the real measurement due to misreconstruction of energy and momentum associated with the electrons and muons. That results in the appearance of a missing energy or momentum independent of real invisible final-state particles. This effect degrades the precision and accuracy with which actual missing energy can be determined.

Nonetheless, searches for invisible final states are completely possible and routinely performed at collider experiments. We focus on a single example here as a case study of the potential for this approach although there are many others that we will mention briefly.

Invisible Decay of the Higgs Particle

Our case study is the use of the Higgs boson to search for undetectable particles, hypothesized to compose dark matter, whose mass is $m_\chi < (1/2)m_H$. If there is a coupling between the Higgs particle and dark matter, then pair production of such dark matter could be possible from Higgs decay; for example, $H^0 \to \chi\chi$. Assuming all background processes can be distinguished from this potential signal process, it is important to establish any "invisible Higgs" signature above any expectations from the SM.

Does the SM predict invisible Higgs particle decay? In fact, it does. For example, consider the decay of the Higgs particle to Z boson final states, $H^0 \to Z^0 Z^{*0}$. Due to the mass of the Z^0 being greater than half the Higgs boson mass, pair production of this final state requires that at least one of the Z bosons be *virtual*, possessing of a mass other than the nominal Z pole mass. This is allowed in quantum mechanics, but suppresses the decay in the process, as the favored decay occurs when both final-state particles are real. The SM prediction of this process is shown in Table 5.1: about 2.6% of all Higgs particle decays results in a pair of Z bosons. The Z boson has a significant branching fraction to neutrinos, $Z^0 \to \nu\bar{\nu}$, which is 0.2000. Thus, from these two facts we can conclude that the invisible decay of the H^0 in the SM is represented by a total branching fraction of $\mathcal{B}(H^0 \to \text{invisible})_{SM} = \mathcal{B}(H^0 \to Z^0 Z^{*0}) \times (\mathcal{B}(Z^0 \to \nu\bar{\nu}))^2 = 0.001048$, or 0.1%. Any attempt to measure the non-SM contribution to the invisible Higgs particle decay process will ultimately be limited by this expectation. Of course, if a search for $H^0 \to \text{invisible}$ is performed and determines the rate to be 10%, we can then assume that there is a significant excess above the SM expectation.

The above discussion assumed that an H^0 is produced but misconstrued as a possible non-SM invisible decay by virtue of a SM process ($H^0 \to Z^0 Z^{*0}$) that yields a final-state composed entirely of real missing energy sources (neutrinos). There are other ways to obtain a signature when the final-state is only partially (or not at all) composed of neutrinos. For example, $H^0 \to W^\pm W^{*\mp}$ is another significant Higgs particle decay (Table 5.1) and each $W^\pm$ boson can decay to either $e^\pm \nu_e$, $\mu^\pm \nu_\mu$, or $\tau^\pm \nu_\tau$. These *leptonic* W decays do contain real neutrinos but also electrically charged leptons that can be missed or misreconstructed in the detector. The latter effect (missing the lepton entirely) is an acceptance effect, while misreconstructing or misidentifying the lepton (implying it left traces of itself in the detector but these did not pass algorithms designed to find them) is an efficiency effect. An analysis of the data that requires no electrically charged leptons be present in the final state (intended to veto the $H^0 \to W^\pm W^{*\mp}$ topology) would nevertheless pick up events where the leptons are lost by a combination of acceptance and efficiency effects. This results in a subset of SM Higgs decay events that appear to have a large missing energy, faking the invisible decay of the H^0.

This is a serious issue, as the rate of $H^0 \to W^\pm W^{*\mp}$ is about 21% and W bosons produce leptons and neutrinos in about 33% of their decays. Such large rate processes have the potential to fake the invisible Higgs signature. In addition, collisions that produce no Higgs bosons at all could be misreconstructed as invisible Higgs boson decays.

A good example of this is the production of a lone Z boson that then decays entirely to neutrinos (real missing energy), but, due to the missing energy resolution, it gets misconstrued as a $H^0 \to \text{invisible}$ candidate. In cases like this, the job of the experimental physicist is to constrain such confounding backgrounds as much as possible through careful use of the data (e.g., using $Z \to \mu^+ \mu^-$ reconstruction to estimate the rate at which $Z \to \text{invisible}$ can fake the Higgs signature) and simulation.

Exercise 14

If the rate of individual muon (electron) reconstruction in an LHC detector is 98% (95%), estimate the probability that $H^0 \to W^{\pm}W^{*\mp}$ will fake the $H^0 \to$ invisible signature (where the denominator of the probability is any Higgs decay). Only consider the W branching fractions directly to the final state involving either the electron or the muon.

Both the ATLAS and CMS Experiments at the LHC have pursued searches for invisible Higgs decay. They reduce the potential backgrounds for this search by using rare, but distinctive, reconstructed Higgs production mechanisms. The main approaches have been the production of a Higgs along with a Z boson ("vector-boson-associated [VBA] production"), vector-boson fusion (VBF) production of the Higgs, and the Higgs produced along with top quarks ("top-associated production", or ttH production). VBA production proceeds as $pp \to Z^0 H^0$, where the Z^0 can be fully reconstructed (e.g., $Z^0 \to \mu^+\mu^-$ or e^+e^-) leaving only the decay of the other particle (a Higgs boson) to be determined. VBF proceeds when a quark and an antiquark from each proton in the collision radiate Z or W bosons, which subsequently fuse into the Higgs. These processes are illustrated using Feynman diagrams in Fig. 5.26.

The generic concept of missing energy is universally useful in these searches. However, the related concept — missing mass — is applicable for hypotheses where a parent resonance decays entirely (or nearly entirely) to invisible final states. We will derive, under a basic situation, how the missing mass for $H \to$ invisible can be determined from the visible properties of the event. Since this is a hadron collider, we can only derive this in the plane transverse to the beam axis (z). Only in that plane do we know with high certainty the initial energy and momentum state before the collision.

First, we should recognize the most general definition of mass in terms of energy and momentum. This is given by the relativistic mass-energy-momentum relationship from special relativity (written using natural units), $m^2 = E^2 - \vec{p} \cdot \vec{p}$. In the directions transverse to the z axis, we can write a similar equation: $m_T^2 = E_T^2 - \vec{p}_T \cdot \vec{p}_T$.

Let us consider the specific case of $pp \to Z^0 H^0$ with a fully reconstructed Z^0 candidate ($Z^0 \to \mu^+\mu^-$). The fundamental parton-level interaction that leads to this final state is, for example, $q\bar{q}' \to Z^{*0} \to Z^0 H^0$, where a quark from one proton and its anti-quark partner from the second proton annihilate into a virtual Z^0 boson. The virtual boson becomes a real boson by radiating energy, here in the form of H^0 emission. The final-state Z^0 and H^0 then decay. This is an *electroweak process*, as it involves the direct coupling of quarks to one of the weak bosons.

The proton–proton collision will not only result in the Z^0 and H^0 bosons. While one of the individual parton interactions would certainly produce this final state, additional interactions within the protons and the scattering of the proton "remnants" (the remains of the protons after inelastic collision) will also be observable in the detector. In addition, the strong nuclear interaction will play a significant role in the collisions even if the central process is electroweak in nature. For example, one or both of the initial-state quarks can radiate one or more gluons before they annihilate. This results in a particle jet from one or more gluons, which can also enter the detector.

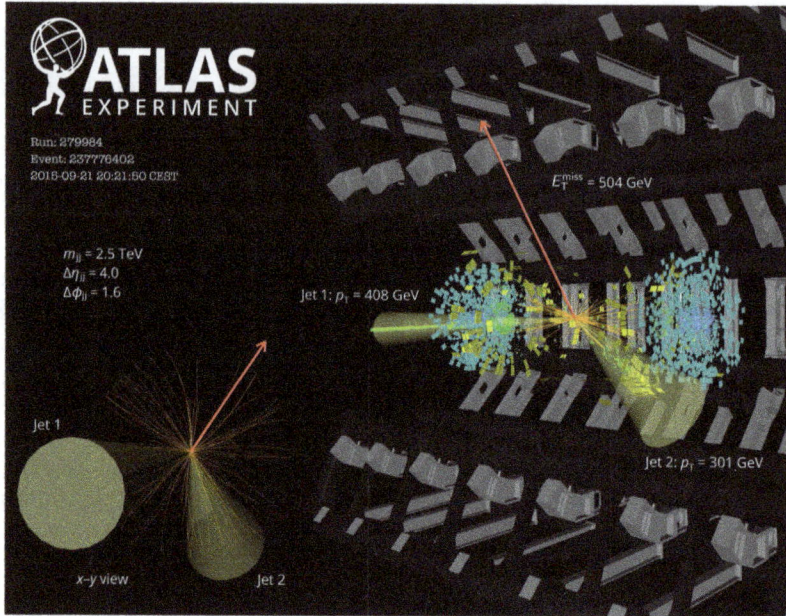

Figure 5.29: A reconstruction of a proton–proton collision in the candidate VBF final-state, including a large momentum imbalance that is consistent with invisible Higgs boson decay. Two jets from the VBF process are visible at low angles with respect to the long axis of the detector (the accelerator beam line). The red arrow indicates the reconstructed direction of the missing momentum. Image from Ref. [Aad+22].

An example of a complex collision event consistent with VBF production of the Higgs particle is shown in Fig. 5.29, and a red arrow indicates the probable direction of the missing momentum. In any process like this we expect a minimum set of challenging phenomena to be considered:

- The invisible decay of the Higgs to either neutrinos or dark matter determined via missing energy;
- The decay of potential accompanying particles, like a Z^0 boson, to observable final-state particles;
- Zero or more particle jets resulting from initial-state radiation by the progenitor partons (quarks) before they annihilate;
- The remnant(s) of the protons after dissolution in an inelastic collision at high energy;
- Additional interactions from other protons also present in the proton bunches that participate in the collision (a phenomenon known as "in-time pile-up" or, for short, "pile-up")[3].

---

[3] "Out-of-time" pileup occurs in high-intensity colliders where the time between collisions is less than the time it takes to clear the previous collision's energy from the detector. When this happens, energy

The missing transverse energy and momentum will then consist not only of the muons from the Z^0 decay, but a potential multitude of other energy deposits from other particles associated with the primary parton interaction. Pile-up can be reduced by eliminating from consideration electrically charged particles originating from spatial points along the z axis that are not the same as the spatial origin of the muons. Energy clusters without charged particles, however, have limited or no pointing information back to their origin. In this case, experiments typically utilize weighting procedures to re-scale the energy clusters that enter the missing energy calculation (cf., Ref. [Aab+18b; CMS19]).

The missing energy alone may not provide sufficient information to reject backgrounds. For the same collision energy between the partons, it may be possible to produce $W^{\pm}Z^0$, Z^0Z^0, or Z^0H^0 final states. The invisible decay of the Higgs or Z^0, or the decay of a W or Z to particles that go undetected and fake missing energy, can look very much the same in only the missing energy distribution. The missing energy will capture both the mass-energy and kinetic energy of the decaying parent, which, in all three of these cases, would be nearly identical since the original collision energy is the same. However, one can combine missing energy magnitude and direction along with the momentum and energy of the $Z^0 \to \mu^+\mu^-$ decay, to improve the background rejection.

We can now proceed to write down the *transverse mass* using the missing-energy and di-lepton system. Examples of the application of missing transverse energy and mass in real experiments is shown in Fig. 5.30. As with any relativistic quantity, one can define transverse mass from the Einstein special relativity mass-energy-momentum relationship,

$$m^2 = \left(\sum_i E_i\right)^2 - \left(\sum_i \vec{p}_i\right)^2 \tag{5.57}$$

where in our case $i = 1, 2$, labeling the di-lepton system (Z^0 decay) and the Higgs candidate (missing energy magnitude and direction), respectively. For now, let us work with all momentum components. We can try to write the above in terms of total quantities (such as the total mass of a decaying particle, (i) and longitudinal (z) and transverse momentum quantities:

$$
\begin{aligned}
m^2 &= (E_{\mu\mu} + E_{\ inv})^2 - (\vec{p}_{\mu\mu} + \vec{p}_{\ inv})^2 & \text{(5.58)} \\
&= E^2_{\mu\mu} + 2E_{\mu\mu}E_{\ inv} + (E_{\ inv})^2 - p^2_{\mu\mu} - 2\vec{p}_{\mu\mu} \cdot \vec{p}_{\ inv} - (p_{\ inv})^2 & \text{(5.59)} \\
&= (E^2_{\mu\mu} - p^2_{\mu\mu}) + 2E_{\mu\mu}E_{\ inv} + \left[(E_{\ inv})^2 - (p_{\ inv})^2\right] - 2\vec{p}_{\mu\mu} \cdot \vec{p}_{\ inv} & \text{(5.60)} \\
&= m^2_{\mu\mu} + (m_{\ inv})^2 + 2E_{\mu\mu}E_{\ inv} - 2\vec{p}_{\mu\mu} \cdot \vec{p}_{\ inv} & \text{(5.61)} \\
&= m^2_{\mu\mu} + (m_{\ inv})^2 + 2E_{\mu\mu}E_{\ inv} - 2\vec{p}_{\mu\mu} \cdot \vec{p}_{\ inv} & \text{(5.62)}
\end{aligned}
$$

At this step, it is useful to recall that total energy for a decaying particle can be rewritten in terms of the particle's transverse energy, mass, and momentum. Doing so will allow

---

from the previous collision is still present, in part, when the readout of the next collision begins. This requires the subtraction of that energy from the next event. This is typically handled by electronics or algorithmic approaches (cf., Ref. [Aad+23b]).

us to then separate the longitudinal and transverse components of momentum, isolating them for the final step:

$$
\begin{aligned}
m^2 &= m_{\mu\mu}^2 + (m_{\ inv})^2 + 2E_{\mu\mu}E_{\ inv} - 2\vec{p}_{\mu\mu} \cdot \vec{p}_{\ inv} & (5.63)\\
&= m_{\mu\mu}^2 + (m_{\ inv})^2 & (5.64)
\end{aligned}
$$

$$
+2\left[E_{T,\mu\mu}\frac{p_{\mu\mu}}{p_{T,\mu\mu}}E_{\ inv,T}\frac{p^{inv}}{p_T^{inv}} - \vec{p}_{T,\mu\mu} \cdot \vec{p}_{\ inv,T} - \vec{p}_{z,\mu\mu} \cdot \vec{p}_{\ inv,z} \right] \quad (5.65)
$$

$$
= m_{\mu\mu}^2 + (m_{\ inv})^2 \quad (5.66)
$$

$$
+2\left[E_{T,\mu\mu}\frac{\sqrt{p_{T,\mu\mu}^2 + p_{z,\mu\mu}^2}}{p_{T,\mu\mu}}E_{\ inv,T}\frac{\sqrt{(p^{T,inv})^2 + (p^{z,inv})^2}}{p_T^{inv}} \right. \quad (5.67)
$$

$$
\left. -\vec{p}_{T,\mu\mu} \cdot \vec{p}_{\ inv,T} - \vec{p}_{z,\mu\mu} \cdot \vec{p}_{\ inv,z} \right] \quad (5.68)
$$

$$
= m_{\mu\mu}^2 + (m_{\ inv})^2 \quad (5.69)
$$

$$
+2\left[E_{T,\mu\mu}\sqrt{1 + (p_{z,\mu\mu}/p_{T,\mu\mu})^2}E_{\ inv,T}\sqrt{1 + (p^{z,inv}/p_T^{inv})^2} \right. \quad (5.70)
$$

$$
\left. -\vec{p}_{T,\mu\mu} \cdot \vec{p}_{\ inv,T} - \vec{p}_{z,\mu\mu} \cdot \vec{p}_{\ inv,z} \right] \quad (5.71)
$$

In the case that we only have access to the transverse components of the final state, we can compute the transverse mass merely by asserting that all z components are zero. This leads to the definition

$$
m_T^2 = m_{\mu\mu}^2 + (m_{\ inv})^2 + 2\left[E_{T,\mu\mu}E_{\ inv,T} - \vec{p}_{T,\mu\mu} \cdot \vec{p}_{\ inv,T} \right]. \quad (5.72)
$$

In the special case that the decaying particles have small masses compared to their total energy and momentum, we find that

$$
\begin{aligned}
m_T^2 &\approx 2\left[E_{T,\mu\mu}E_{\ inv,T} - E_{T,\mu\mu}E_{\ inv,T}\cos\phi_{\mu\mu,inv} \right] & (5.73)\\
&= 2E_{T,\mu\mu}E_{\ inv,T}(1 - \cos\phi_{\mu\mu,inv}) & (5.74)
\end{aligned}
$$

where $\cos\phi_{i,j}$ is the cosine of the azimuthal angular separation between the di-lepton and invisible momentum directions.

So far, no evidence has been observed for the non-standard-model invisible decay of the Higgs particle. This is determined by applying Eqn. 5.44 in various ways to the data from experimental searches for invisible Higgs particle decay. For example, one can merely add up all events observed and expected after developing the analysis (a simple "counting experiment" on all events, also known as a "one-bin analysis"). Or, one can use distributions of key discriminating variables (cf. Fig. 5.30) and apply the interpretation equation in each bin of the distribution, executing multiple counting experiments for each such distribution (with an overall constraint that requires the total number of events in all bins to be compatible with the observed total number of events, etc.).

The latter approach has the benefit of accounting for the fact that some regions of the discriminating variables can be richer in signal than others, and can thus emphasize the weight of regions expected to be more sensitive to a given signal (or background).

Background-rich regions of variables are often referred to as "control regions" because they are the regions where processes other than the signal process are expected to dominate, and thus offer assessments of the analysis away from a region containing potential signal events. This allows one to assess the reliability of the analysis techniques before proceeding to apply them to the subset of the data were signal may (or may not) be observed.

Equation 5.44 also provides a road map for the experimentalist regarding the effects of assumptions on their interpretation of the data. First, the data itself will be subject to statistical uncertainties that must be included in any interpretation. Uncertainties inherent in the colliding beam process will affect $\mathcal{L}$. For example, the measurement of instantaneous luminosity is done by one or more experimental means, usually yielding uncertainties in a modern particle collider at the $1-5\%$ level. These are irreducible for the experimentalist because they signify the fundamental limitation in knowledge of the total number of collisions.

Assumptions about the production rate of particles like the Higgs boson appear in σ_H, limited by theoretical uncertainties. If the branching fraction is also unknown, it is subject to similar uncertainties. In the present case of the production of a Higgs particle, the theoretical uncertainties can range from 1—20% depending on the specific production cross-section and final-state of the Higgs particle decay.

The acceptance and efficiency of the detector can be constrained both from independent experimental tests and simulation of the detector itself. For example, consider collisions where two particles are produced back-to-back and one is absolutely detected. Conservation of momentum allows an experimental scientist to constrain the flight direction of the second. These conditions provide useful tests of the ability of the detector to see the second particle. Simulation and data together, using cases like the one described here, can be used to constrain the systematic uncertainties due to detector modeling, including the acceptance of the detector and the efficiency of reconstruction choices and techniques. Finally, the background prediction relies on the simulation of non-signal processes whose properties have been constrained using independent data or on control regions, enriched in particular background processes, to extrapolate the background expectation in the signal region of the data.

In the case of invisible decay of the Higgs, the production cross-section for the Higgs bosons is generally well-established from both experiment and theory. The backgrounds are modeled and additionally constrained and tested using control regions in the data and simulation. The acceptance and efficiency are similarly constrained using simulation and real data. The essential unknown is the branching fraction for $H \to \chi\chi$, which is then the target for interpreting the measurement. In simple terms, one is then left to test the compatibility of a given $\mathcal{B}_{inv}$ with the actual observed events and the expected background. Any $\mathcal{B}_{inv}$ that would lead to significantly more events than observed, even with background present, can be excluded with some level of statistical confidence. The interpretations usually select a target level of confidence and then state the constraints on the hypothesis given that benchmark.

An example interpretation is shown in Fig. 5.31 [Tum+23]. The most model-independent part of the interpretation process is shown in Fig. 5.31a, which is the constraint on invisible Higgs branching fraction given no evidence for the process in the

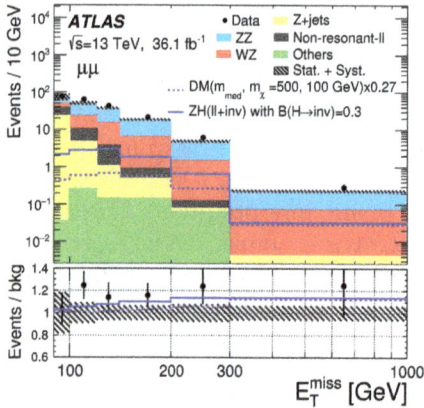

(a) ATLAS Experiment, missing transverse energy

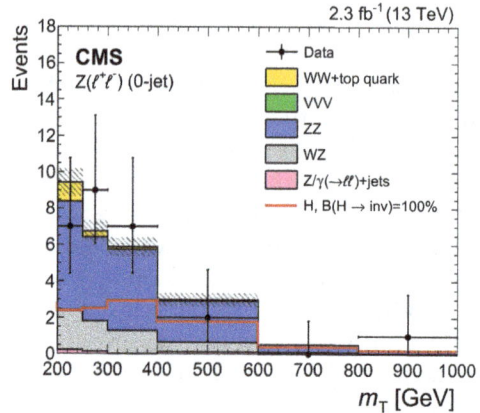

(b) CMS Experiment, transverse mass

Figure 5.30: Missing transverse energy distribution [Aab+18a] (left) and total transverse mass of the candidate ZH system [Kha+17] (right) in searches for invisible Higgs. The solid-filled distributions are expectations from a variety of background sources. The unfilled histograms indicate models of invisible Higgs decay. The black points are data.

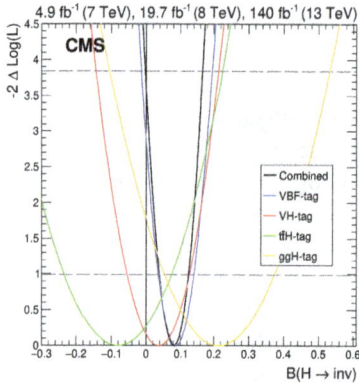

(a) CMS Experiment, likelihood contours for $\mathcal{B}_{inv}$

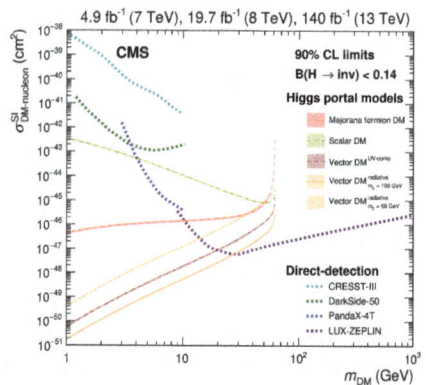

(b) CMS Experiment, dark matter cross-section constraints

Figure 5.31: CMS Collaboration interpretation of their searches for $H^0 \to$ invisible. The left figure shows likelihood contours, where the minima indicate the branching fraction hypothesis most compatible with the data, individually for each search channel and for the combination of all channels. The right figure shows a further interpretation that uses a dark matter theoretical framework ("Higgs portal models" that assume how the Higgs and dark matter interact) to convert the experimental observation into a statement about the dark matter nuclear interaction cross-section. Figures from Ref. [Tum+23].

data. We can see this in the comparison of data and background expectations like those shown in Fig. 5.30b. The data within uncertainties agrees well with the background distribution. The background-only situation is considered to be the *null hypothesis* and assumes that fluctuations in the background alone can explain the data. The alternative hypothesis is that the data contains both background and a signal contribution, where an example of $\mathcal{B}_{inv} = 100\%$ is shown in the figure. Certainly, that case is ruled out even without applying the mathematics of Poisson statistics. Such an outcome (background fluctuating high enough to be compatible with the observed data) is very likely given the relevant counting statistics and associated uncertainties. Thus the data can be satisfactorily explained by the null hypothesis and disfavors the alternative hypothesis.

The CMS analysis uses all available data collected from 2010 to 2018. It also combined multiple ways of producing the Higgs boson; i.e., in association with a vector boson, top quarks, or with jets. Nevertheless, this is a challenging measurement given the large uncertainties typically associated with looking for excesses in data using missing energy processes. The observed branching fraction with its uncertainties is $\mathcal{B}_{inv} = (8 \pm 4)\%$. This is a positive result, but one that is also highly compatible with the null hypothesis (a zero branching fraction). Thus the data are interpreted as constraining the branching fraction to be be $\mathcal{B}_{inv} < 15\%$ at the 95% confidence level.

It is worth noting that the collaboration expected, based on prior knowledge and the absence of a signal, to have set a slightly better limit on the invisible branching fraction of $\mathcal{B}_{inv}^{expected} < 8\%$. However, they note that the expectation and the observation are still very statistically compatible given the uncertainties.

The more model-dependent part of the experimental interpretation is shown in Fig. 5.31b. Its reliability is self-illustrating. This figure represents an effort to make a statement on the cross-section of the dark matter and nucleus interaction so that a comparison to direct detection experiments can be made. However, as the exact mechanism for coupling a Higgs to dark matter is not known (and may not even exist), this interpretation is confounded with a significant degree of theoretical uncertainty.

An observation of dark matter interactions by a direct detection experiment may tell us something about the Higgs coupling to dark matter (or not), but a non-observation of Higgs decay to dark matter does not necessarily tell us anything about the dark matter and nucleus interactions for interactions that don't involve the Higgs. Alternatively, a strong positive observation of invisible Higgs decay, perhaps along with non-SM processes, would provide valuable information for theoretical physicists and might inform the community as to the correct theoretical framework for dark matter. That, in turn, might be used to constrain future observing prospects at direct detection experiments. For now, we don't have positive and reproducible observations from any experiment.

Using the Higgs portal models that are considered, absent specific knowledge of the correct parameters in these models, it is apparent that different choices yield very different expected constraints on the nuclear interaction cross sections. These predictions, and the constraints that result from them, span almost ten orders of magnitude at the lowest considered dark matter masses.

Direct detection experiments (Chapter 4) suffer far less from this interpretation problem because astrophysics constrains, to a far better degree, the rate at which we *should* have expected a SM–dark matter interaction to have been possible. At a collider (a facility not subject to the relic density population of dark matter and our planetary velocity through the galaxy's halo) we lose that valuable constraint and must rely on a model to tell us how probable such an interaction could have been (thus allowing us to say whether we should have seen it or not).

A similar analysis from the ATLAS Collaboration [Aad+23a] expected to find $\mathcal{B}_{inv} < 7.7\%$ in the absence of signal and at the 95% confidence level. They observed $\mathcal{B}_{inv} < 10.7\%$ and while, again, this implied a slight excess in the data that worsened the observed upper limit, it is nonetheless compatible with the null hypothesis (that the SM is a reliable description of the observed data). The observed branching fraction itself was $\mathcal{B}_{inv} = (4 \pm 4)\%$, illustrating the compatibility with the SM. Since the total uncertainty on both the ATLAS and CMS observations are highly compatible (4%), it is easy to perform a naive combination of the two results to obtain a rough estimate of the complete and current picture from both experiments. We assume that the errors are fully uncorrelated between the two experiments [4] and merely perform an error-weighted average of the two branching fractions, where

$$\mathcal{B}_{inv}^{combined} = \frac{\sum_i \frac{\mathcal{B}_{i,inv}}{\Delta \mathcal{B}_{i,inv}^2}}{\sum_i \frac{1}{\Delta \mathcal{B}_{i,inv}^2}}, \qquad (5.75)$$

where $\Delta \mathcal{B}_{i,inv}$ is the reported uncertainty on the branching fraction from experiment i, and

$$\frac{1}{\left(\Delta \mathcal{B}_{inv}^{combined}\right)^2} = \sum_i \frac{1}{\Delta \mathcal{B}_{i,inv}^2}. \qquad (5.76)$$

Solving for the combined branching fraction and its uncertainty, we obtain $\mathcal{B}_{inv}^{combined} = (6 \pm 3)\%$. Even a naive combination is estimated to still be within two standard deviations of the null hypothesis. The precision era of the LHC, scheduled to begin in the late 2020s, will be an interesting phase during which to pursue this analysis. The main limiting factors of these measurements are related to the reconstruction of particles in the detectors, all effect from which influence the precision and accuracy of missing energy.

An increasingly mature understanding of the detectors and the collider, improved efforts at reducing the contributions of pile-up collisions to missing energy calculations, and long-planned upgrades to experimental facilities, should all have a positive impact on the ultimate reach of this search strategy.

---

[4]The assumption of fully uncorrelated uncertainties between ATLAS and CMS is only completely correct for the statistical uncertainties on their observed number of candidates in the data since they are independent experiments and do not observe the same proton–proton collisions. However, the uncorrelated error assumption breaks down when considering other sources of uncertainty. Both experiments rely on the LHC to report luminosity uncertainties, and those are 100% in common between the two experiments (though they account for a small part of the total uncertainty in the final measurement). More serious is the fact that both experiments rely on the same theoretical calculations for production cross-sections of the Higgs, so these are also completely in common. Both collaborations have very different detectors, but rely on the same underlying simulation software to model their experiments, so it can be argued that there is a common uncertainty lurking at the heart of their acceptance and efficiency measurements. However, this naive approximation will suffice for a rough conclusion to be drawn.

Projections of sensitivity for this search in the final era of the LHC (cf. Ref. [Dai+19]) suggest that, in the absence of non-standard-model signals, the combinations of ATLAS and CMS results could yield a direct observation limit of $\mathcal{B}_{inv}^{HL-LHC} < 2.5\%$ at the 95% confidence level. This projection, while well-motivated at the time, was conservative, assuming only two search channels per experiment (VH and vector boson fusion) and no improvements in the analysis techniques used in 2019.

We have chosen a very narrow thread in the subject of direct observation of invisible decays at a particle collider. The purpose was to illustrate approaches and challenges. There are, obviously, many other ways in which one can search for such a signature (whether motivated by a theoretical model or not). At this time, there is no compelling evidence that such processes exist at any particle collider. Interpreting that in the context of constraints on the mass and coupling of SM matter to dark matter is extremely challenging and highly model-dependent.

Dark Gauge Boson Searches

In the previous discussion on missing-energy-based dark matter searches, we emphasized a key problem in translating non-observation of a signal into constraints that allow comparison of null results to those from direct detection experiments. That problem is the unknown structure of the dark sector, particularly the means by which the dark sector interacts with the SM. This problem is precisely addressed by the subject we cover here: dark gauge boson searches.

It is assumed that the lessons of quantum field theory and symmetry (group theory) that resulted in the success of the SM can be translated into the dark sector. We note that this is an assumption, but, as no clear successor to quantum field theory has been identified, this presently defines the most reliable boundary condition on a dark sector theory. We touched on the idea of an *effective field theory* in Section 4.3. We expand on that a little more here and explore other interaction terms (not just those affecting $\chi N \to \chi N$ scattering) that would allow us to predict, and interpret, searches for new gauge bosons.

We focus here on the search for a complement to the SM photon, a *dark photon*. The idea of a new spin-1 (vector) gauge boson that couples significantly in a *hidden sector* (a gauge theory describing a group of particles and their interactions where particles in that sector frequently interact with one another, but are "secluded" from interactions with the SM fermions and bosons) apart from the SM, was articulated in detail in Ref. [Hol86]. A key concept here is *kinematic mixing* between the dark and SM photon, which then manifests as a diminished coupling to electric charge and allows for the possibility that the dark photon has been in plain sight this whole time, its effects diminished by the seclusion of its sector from ours.

We can motivate this concept by starting from the term in the SM Lagrangian that describes the motion of the gauge field itself. This term is known as the *kinetic term*. The kinetic term for photons after breaking electroweak symmetry is given by

$$\mathcal{L}_\gamma = -\frac{1}{4}F_{\mu\nu}F^{\mu\nu} \tag{5.77}$$

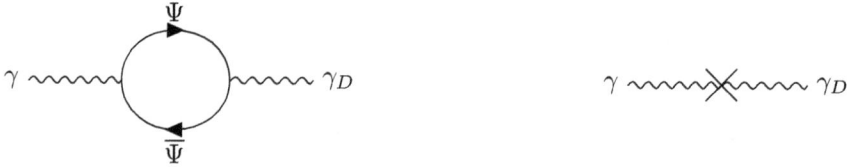

Figure 5.32: An example of the kinematic mixing of a SM and dark photon via a heavy fermionic state Ψ in a full high-energy field theory (left) and represented in a lower energy effective field theory (right).

where $F_{\mu\nu}$ is the *field tensor* (with $\mu, \nu = 0, 1, 2, 3$) describing the electromagnetic interaction. This tensor can be written as four unique equations that describe the electric and magnetic forces, classically known as Maxwell's equations. Imagine that there is a secluded sector—a *dark sector*—with its own dark electromagnetism. Mathematically, this means that the dark sector consists of at least one $U(1)$ symmetry group akin to the one in the SM that, after electroweak symmetry breaking, represents the electromagnetic interaction [5]. This is denoted $U(1)'_D$ and the generator of this group is a spin-1 gauge boson whose free-field kinetic term is similar to the photon's in the SM Lagrangian density,

$$\mathcal{L}' = -\frac{1}{4}F'_{\mu\nu}F'^{\mu\nu}. \tag{5.78}$$

The particle that results from this symmetry group is denoted γ_D, the *dark photon*. In the above notation the primes denote quantities in the dark sector.

Many dark sector models build from the simple assumption of a single additional symmetry group to describe interactions among the dark matter particle species. These models, though simple, take many forms and are well-explored in Refs. [Ark+09; HWX12]. Even a more complex model, with additional particles, can manifest at low energies (those currently accessible to experiments) as a simplified sector consisting of a singular force and a stable dark matter particle. For example, consider a more general theory of the dark sector in which there exist heavy fermions Ψ that can couple to both the dark photon and the SM photon (by being charged under both gauge groups). Quantum loops involving these heavy fermions can induce a mixing of the SM photon and the dark photon (Fig. 5.32). It is easier, at low energy, to write an effective field theory that captures this effect, but does not need the details of how it emerges. We can simply add a term to the SM Lagrangian density that permits the kinematic mixing of the two photons. This effect needs to be suppressed to keep it small (otherwise it would already have been noticed by experiment). This effectively captures the fact that the additional fermions are very heavy and the quantum loop would be suppressed. These fermions could be candidates for dark matter or simply extra fermions in a dark sector that contains lighter dark matter candidates.

---

[5] See Section 1.3 for a brief overview of the symmetry groups in the SM, with additional discussion of $U(1)$ symmetry groups and their implications in Section 6.2.

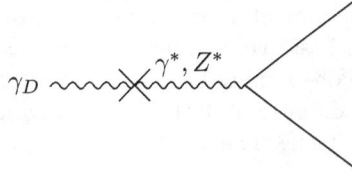

Figure 5.33: An example of the kinematic mixing inducing a decay of the dark photon into pairs of SM particles.

The effective field theory Lagrangian density containing both the SM and dark electromagnetic interactions is then:

$$\mathcal{L}_{SM+DARK} = -\frac{1}{4}F_{\mu\nu}F^{\mu\nu} - \frac{1}{4}F'_{\mu\nu}F'^{\mu\nu} + \frac{1}{2}\epsilon F_{\mu\nu}F'^{\mu\nu}, \qquad (5.79)$$

where ϵ is the kinetic mixing parameter (whose value is less than 1) and summarizes the degree to which the two fields can interchange. Effectively, this mixing manifests in an experiment as if particles are millicharged under the new interaction, with an effective electromagnetic charge $e\epsilon$ that appears suppressed relative to ordinary electric charge because any production of a dark photon must first proceed by emission of a SM photon followed by a kinetic mixing to the dark photon state. Unlike the photon, the dark photon is not necessarily constrained to be massless.

We note that additional terms can, of course, be added to the Lagrangian density for the dark sector. For example, one would expect an interaction term that couples the dark matter fermions to the dark photon. If $m_{\gamma_D} \geq 2m_\chi$, then it is possible for the dark photon to decay to dark matter. This leads us back to the direct detection problem, and for a collider experiment we are returned to the missing energy scenario. However, if $m_{\gamma_D} < 2m_\chi$, then this decay process is not available. Only kinematic mixing to the SM photon would lead to directly observable signatures. We focus on this case.

The case of dark photon mixing already leads to a rich phenomenology that can be probed using accelerators and particle detectors. The simplest case is a photon appearing where it should not and then interacting as a photon in a calorimeter or other similar device. However, one consequence of the dark photon models is that, when the dark photon mixes back into a SM photon, that photon need not be real, with $m_\gamma = 0$. Instead, it can be virtual with $m_{\gamma^*} > 0$, or it can even be a virtual Z boson, allowing for decays to pairs of SM particles [FGL20; PRV08]. This is illustrated in Fig. 5.33. In fact, any particle-antiparticle pair, $f\bar{f}$, is possible so long as $m_{\gamma_D} > 2m_f$. For example, the decay width for a dark photon into any lepton pair $\ell^+\ell^-$ is

$$\Gamma(\gamma_D \to \ell^+\ell^-) = \frac{1}{3}\alpha_{EM}\epsilon^2 m_{\gamma_D}\sqrt{1 - \frac{4m_\ell^2}{m_{\gamma_D}^2}}\left(1 + \frac{2m_\ell^2}{m_{\gamma_D}^2}\right). \qquad (5.80)$$

for $m_{\gamma_D} > 2m_\ell$.

We can immediately begin to appreciate the phenomenological implications of such an hypothesis. For example, any process that leads to photon radiation with $E_\gamma > m_{\gamma_D}$ can also lead to dark photon production at a rate proportional to ϵ. There is also no guarantee that the massive dark photon is stable, and many models assume that, once produced, it is possible for it to decay before mixing back to the SM photon. How, experimentally, can one then probe for the existence of dark photon mixing (and decay) and, thus, the existence of the dark photon itself? The key to this in a particle collider context will be based on two principles:

- An environment that leads to a high rate of photon production so that the rarity of dark photon mixing can be overcome; and
- A detector that looks for interactions away from the collision point, in a place where particles directly from the collider should not be detected.

The experiment could look for a photon appearing where there should not have been one. Alternatively, the detector could look for missing energy signatures (akin to the discussion in Section 5.5), decay products from a non-SM particle, or displaced decay positions (vertices) that indicate that a photon traveled from its origin, mixed into a dark photon, and then resulted in a decay far from the origin.

Fixed-Target Dark Photon Searches

The heart of this idea is that, at best, the SM matter fields are very weakly charged under the dark gauge group. A dark photon can then pass through SM matter without being absorbed, emerging in another location far from where SM photons could plausibly have traveled. If the dark photon remains in that state, it is challenging to detect it directly. However, there is a small chance of the dark photon mixing back to the photon. The combined probability of this complete sequence of events is proportional to ϵ^2, so this double-mixing process is suppressed. However, the signature will be distinctive: a photon seeming to pass through an impenetrable wall[6], whose time of arrival on the other side of the wall is causally linked to the original process that should have created it. Since, in a particle collider, we can know the time at which the original photons were produced (owing to the precise time structure of such particle beams), we can look for coincidences between (a) a particle collider beam structure hitting its target and (b) a photon emerging far from the beam in a place where photons cannot normally travel.

The "target" in the above discussion could be a counter-circulating bunch of particles in a second beamline (for example, another bunch of electrons or protons) or a stationary atomic or nuclear target (a "fixed target" into which the beam is steered). There are many examples of such experiments reviewed in Ref. [FD20]. We will narrowly focus on beam dump experiments (a fixed target) and specifically on the E137 experiment as an example of this style of investigation. The E137 experiment was originally designed to search for long-lived but non-interacting particles[7] that are produced when a beam of particles strikes a solid target, traverse a long region beyond the target, and then decay (or mix into SM particles), and are detected.

---

[6]This concept of "light shining through a wall" is at the heart of other searches for dark matter, which we highlight in Section 6.4.

[7]Specifically, E137 was designed to look for axions. We discuss these in Chapter 6.

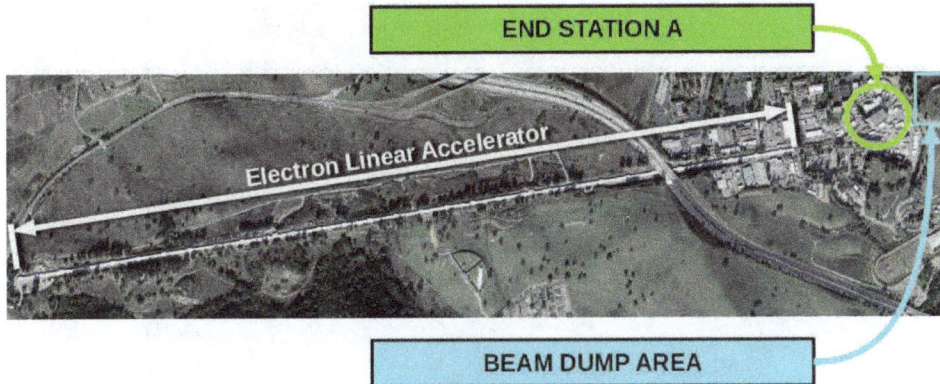

Figure 5.34: An annotated aerial view of the SLAC National Accelerator Laboratory, showing the electron linear accelerator, historic End Station A, and the beam dump area beyond the end stations. Photograph is in the public domain and from Ref. [Wik20e].

The E137 experiment involved an electron accelerator at the SLAC National Accelerator Laboratory (known at the time as the Stanford Linear Accelerator Center). The linear accelerator delivers bunches of electrons to various beamlines, including those that enter the historic "end station" area. Specifically, E137 utilized the beamline to End Station A. Electrons (with 20 GeV of energy) were delivered onto aluminum targets in that hall, but not all electrons in a bunch interacted with the target. The remaining high-energy electrons needed to be safely disposed, or "dumped." This was accomplished by transporting the unused electrons in vacuum to a target farther downstream. There the remaining electrons interacted and the products of those interactions were monitored.

An aerial view of the laboratory and relevant experimental areas is shown in Fig. 5.34. E137 was set up 209 m beyond the "Beam Dump East" section after End Station A. The beam dump itself is followed by 179 m of earth beneath a hill on the site. This was determined to be sufficient to stop all beam-dump products except neutrinos. The E137 detector consisted of a calorimeter that was eight radiation lengths in thickness, sufficient to stop nearly all electromagnetically interacting products from the beam dump. Each unit of the calorimeter consisted of a dense material to induce particles to interact and shower, as well as a scintillator to measure energy deposition and a wire chamber to provide very good angular resolution. We note that the proportional wire chambers in this experiment represent an early use of charge-coupled-device (CCD) technology, which now is commonly used in solid state electronics in commercial cameras and personal devices. We note this because CCDs play an increasing role in a new generation of dark matter search experiments that we will summarize in Chapter 7.

The E137 detector was set to trigger in two ways: either on a deposit of total energy exceeding 400 MeV (equivalent to an electromagnetic shower induced by electrons or photons) or on a series of consecutive energy deposits in the same vertical region of neighboring detectors (equivalent to a high-energy muon punching through the detector,

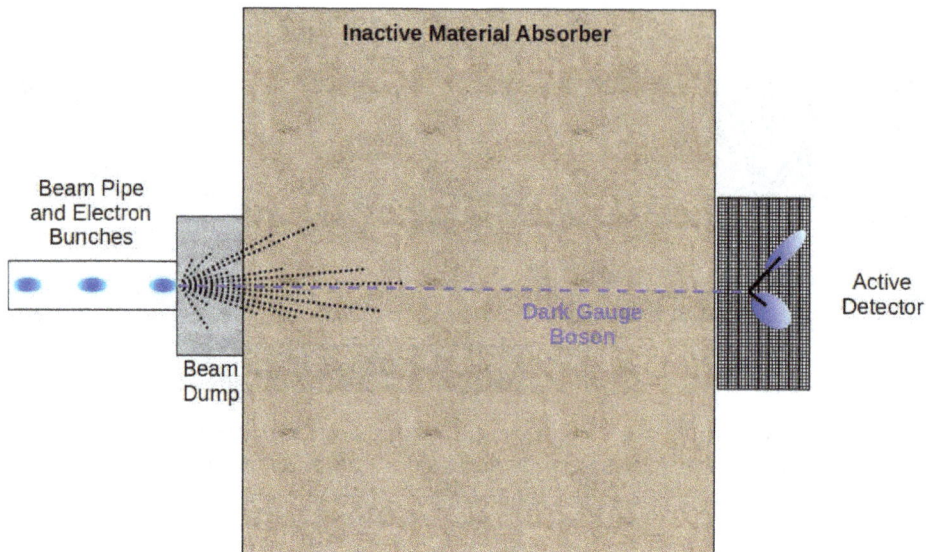

Figure 5.35: An illustration of a typical dark photon beam-dump experiment. The beam travels to a region of dense material designed to stop it; a region of inactive material lies beyond to insure that all particles except neutrinos are stopped. A dark photon is produced from photon radiation at the beam dump, followed by kinematic mixing to the dark photon state. That state traverses the inactive region and then mixes back into the SM photon, which decays in the active detector region.

but depositing little energy in the process). The data analysis of interest to dark photon searches involved searching for pairs of particles being detected in E137, originating from the decay of a common heavier particle.

The concept of dark sectors was developed long after E137 concluded data-taking and the results re-analyzed in the context of dark gauge bosons. The basic physics model used to interpret the data is shown in Fig. 5.35. This image depicts a photon first mixing into a dark photon, traversing a long region of inactive, but dense material, and then interacting in the far active detector region. In this example, the dark photon decays into an electron-positron pair and leaves two energy deposits in the detector.

The major backgrounds in this approach derive from two sources. The first is cosmic ray muons. These randomly pass through the active particle detector. Occasionally, one would expect a random cosmic ray (or a particle produced from a cosmic ray shower) to be coincident with a time window just after the accelerator beam particles striking the beam dump. This would make random background particles appear as though they arose from the beam dump process, but they would have nothing to do with the accelerator. Cosmic rays were determined to dominate over other effects during nominal running conditions. They were capable of producing a higher energy spectrum of particles inside E137 than observed from the other significant source of background.

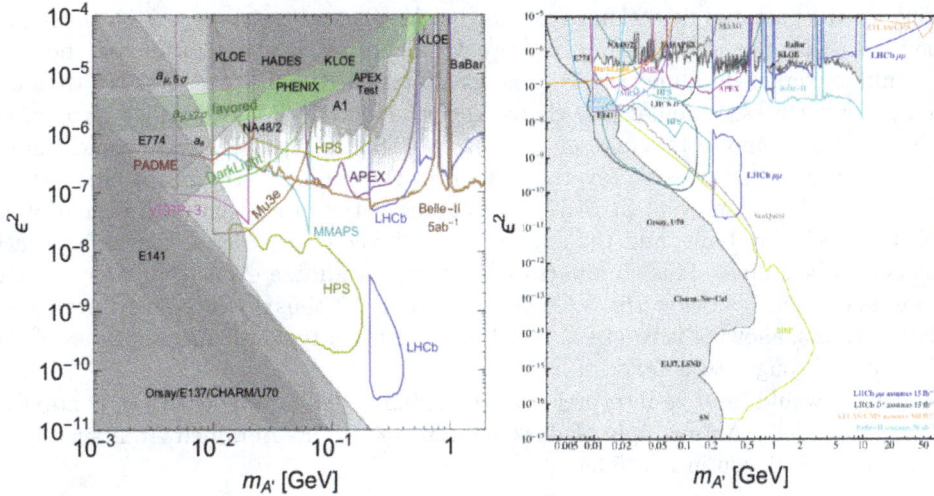

Figure 5.36: An example of exclusion limits on dark photon properties (ϵ^2 vs. dark photon mass) derived from a range of experimental sources (solid, filled regions) or projected for ongoing and future experiments (unfilled regions). The left figure is a narrower region of parameter space than the right. Figure from Ref. [FD20].

The subleading source of background is known as "skyshine" and is related to the accelerator beam, not to cosmic rays. Beam particles interact, both inside the beam dump and earlier (if the vacuum inside the beam pipe is not perfect and there are stray gas molecules or other ions in the pipe), and produce pions. Pions at the energies available from the E137 configuration were able to travel many tens of meters away from the beam dump; this included traveling vertically above the hill (toward the sky). The pion can then interact with an air molecule or decay, producing secondary particles that can rain down on the E137 detector.

Experimentalists were able to study the effect of skyshine on the E137 detector by intentionally inserting and removing an aluminum target into the beam in End Station A. This allowed them to estimate the degree to which skyshine influences the observations at the E137 instrument and led to the installation of additional shielding along the beamline to reduce the effect. Skyshine was observed to mostly produce low-energy deposits in the E137 detector—dominated by muons.

The combined data samples from two runs of the E137 experiment yielded almost 30 C of dumped charge, and, since each electron has a fixed fundamental charge of 1.602×10^{-19} C, this equates to about 2×10^{20} electrons-on-target. This is a significant number of opportunities for a dark photon to have been produced (Fig. 5.36), even if the effective electromagnetic interaction strength is suppressed by the mechanisms outlined above.

This illustrates one of the clear advantages of particle accelerator-based approaches: the sheer number of opportunities experimentalists can have to repeat a measurement, allowing for the potential detection of extremely rare phenomena. Focusing on the region in their data where at least 3 GeV of energy was deposited in the calorimeter, they use the aluminum target data and the low-energy no-target data to predict fewer than 0.2 skyshine events for every Coulomb of charge deposited on the target. This leaves only the cosmic ray background in this high-energy region of the E137 data. These could readily be eliminated or reduced using several strategies: the timing of the beam bump event relative to the detection event in E137; the characteristics of the shower information in the E137 detector systems; and the degree to which the energy deposition points back to the actual beamline. This eliminates all but one candidate event that was rejected because its energy was below the 3 GeV threshold. E137 thus observed zero promising candidate events, allowing it to use a model interpretation to constrain properties of the particles it had sought to observe.

Physicists would later re-interpret this (and similar) experiments in the context of dark sector models. An example of dark photon constraints obtained from such data interpretations is shown in Fig. 5.36.

Colliding Beam Dark Photon Searches

A search for dark photon production and (displaced) decay can also be carried out using detectors at particle colliders (Fig. 5.37), where two beams are accelerated and collided into one another. A range of collider processes can result in the production of real SM photons. Some of these may mix into dark photons, travel from the interaction point of the beams, and either (a) remain undetected as missing energy or (b) subsequently decay (via kinematic mixing) into SM particles. The latter process leads to either prompt or displaced

(a) Belle-II Detector   (b) BaBar Detector

Figure 5.37: The Belle-II (left) and BaBar (right) multipurpose particle detector experiments. Each was sited at an electron-positron collider facility. The BaBar image is credited to the SLAC National Accelerator Laboratory, and the Belle-II image is from Ref. [Wik24d].

production of easily identified particles like electrons and muons. Many experiments have searched for these signatures (for an incomplete selection, see Refs. [Aai+20; Abl+17; Ada+15; Aga+14; Ana+15; Ana+18; Aub+09; Lee+14; Lee+17])). We illustrate an alternative to the dark photon decay signature discussed above, focusing instead on a search by the BaBar Collaboration for a missing energy dark photon signature.

The BaBar and Belle Collaborations operated, and the Belle-II Collaboration now operates, detectors at low-energy e^+e^- circular collider facilities. Belle, and now Belle-II, uses the KEKB accelerator in Japan (upgraded now to SuperKEKB) while the BaBar Collaboration used the PEP-II accelerator at SLAC. Both accelerators were designed to achieve high luminosities (at the level of 10^{33}—10^{34} cm^{-2}s^{-1} for PEP-II and KEKB and now 10^{34}—10^{35} cm^{-2}s^{-1} at Super-KEKB) at center-of-mass collision energies around 10.6 GeV. Both programs changed the collider energies to study a range of physics at low energy, though most of the data was concentrated at $\sqrt{s} = 10.58$ GeV where the machines can operate at a resonance in $e^+e^- \to b\bar{b}$ production known as the $\Upsilon(4S)$ resonance. The detectors were optimized to study bottom quark matter but could perform a wide range of investigations of low-energy and rare phenomena.

One example is to search for $e^+e^- \to \gamma\gamma_D$ where the dark photon is unobserved while the other SM photon is readily detected using the calorimeter. The dark photon goes undetected either because it does not mix back into the SM photon or Z, does not decay to SM particles by doing so, or primarily decays to dark matter because $m_{\gamma_D} > 2m_\chi$. The search for invisible dark photon production is thus able to constrain scenarios with very light dark matter particles, as well as kinematic mixing between the dark and SM photons.

An electron-positron collider is, in contrast to a proton–proton collider, one in which it is possible to strongly constrain the entire momentum and energy of the initial state. This is because electrons are fundamental particles, not known to be made from any other subatomic structure. All energy that goes into accelerating the electrons and positrons in the beams also goes into the production of final-state particles. It is possible, therefore, to fully constrain the collision to know what energy and momentum is expected in the final state, and to compute a full missing momentum vector as a result. In particular, it is possible to fully calculate the CM frame of the colliding beam particles.

In the case of the invisible dark photon search, momentum conservation in the collider CM frame requires that the dark photon travel in a direction opposite the SM photon. This allows for the suppression of backgrounds caused by well-established SM processes. We can highlight the strategies for this kind of search using a result from the BaBar Collaboration [Lee+17]. In this case, the event topology was expected to be a single photon and little or no other activity in the detector. Especially, by using the reconstructed SM photon energy and position in the calorimeter, it was possible to construct its CM energy-momentum vector, $P_\gamma^* = (E_\gamma^*, \vec{p}_\gamma^*)$, and require that there was specifically little or no activity in the direction $-\hat{p}_\gamma^*$ corresponding to the direction the γ_D should have traveled. Here, a $*$ denotes CM-frame quantities.

The major background processes that confound the search for the dark photon signature come from $e^+e^- \to \gamma\gamma$ ("di-photon" background) and $e^+e^- \to e^+e^-\gamma$ ("radiative Bhabha" background, as the process $e^+e^- \to e^+e^-$ is known as "Bhabha scattering"). The former process produces high-energy photons whose energy is typically half the CM energy in the rest frame of the colliding electron and positron. The latter process tends to produce lower-energy photons as the final-state photon originates via radiation from either the electron or positron in the final state.

In both cases, the missing energy arises from the loss of at least one final-state particle. For example, the second of the photons in the di-photon events may interact with an inactive region of the calorimeter (in-between active detector elements) or entirely miss the calorimeter at angles closer to the beamline where detector coverage tends to decline rapidly. The electron and positron in the radiative Bhabha background events may be very low in angle relative to the original collision axis, escaping detection completely. This leaves a single photon in the detector, plus missing energy and momentum, faking the dark photon signature.

The invariant mass of the hypothetical dark photon can be inferred from the properties of the final-state SM photon and the four-momentum of the beam electron and positron. The invariant mass of a system described by a four-vector, $P \equiv (E, \vec{p})$, is obtained by squaring the four-vector, which restores the special-relativistic mass-energy-momentum equivalence, $P^2 = E^2 - p^2 \equiv m^2$. Since this quantity is Lorentz-invariant, we can compute it using four-vectors defined in any frame of reference and still obtain the same quantity. It is typical to choose a reference frame in which the mathematics simplifies quickly.

In the case of a missing energy search like the one discussed here, it is often convenient to select the collider CM frame. In that frame the momenta of the two beam particles are explicitly equal in magnitude and opposite in direction, $\vec{p}_+^* = -\vec{p}_-^*$, where the $+$ $(-)$ refers to the positron (electron). We can explicitly choose these momenta to lie along a single axis, which we will denote z. In addition, even for colliders at this energy scale the mass of the electron (and thus the positron) can be neglected compared to the momentum of the beam particles, making them effectively massless. For example, in the CM frame of the PEP-II collider, the energy of either beam particle will be about 5 GeV. The momentum computed from this energy using (a) either the nominal electron mass or (b) $m_e = 0$ results in values that differ by only about a part in a billion. This justifies the massless-electron assumption.

Thus, in the CM frame, $\vec{p}_+^* = \hat{z}p_+^* \approx \hat{z}E_+^*$ using the massless-electron approximation. In this frame, the total collider energy is given simply by $\sqrt{s} = E_+^* + E_-^*$. Conservation of energy and momentum from the initial to the final state requires that:

$$P_+^* + P_-^* = P_\gamma^* + P_{\gamma_D}^*. \tag{5.81}$$

The energy and momentum of the photon candidate are determined experimentally from the energy deposit in the calorimeter and the angular position of this deposit. The beam energy and momenta are determined by the geometry of the collider and measurement of the beam properties from the accelerator. This allows us to rewrite the above equation in terms of the dark photon four-momentum (unknown) as it is related to known experimental quantities,

$$P_{\gamma_D}^* = P_+^* + P_-^* - P_\gamma^* \tag{5.82}$$

$$m_{\gamma_D}^2 \equiv (P_{\gamma_D}^*)^2 = \left[(E_+^* + E_-^* - E_\gamma^*), (\vec{p}_+^* + \vec{p}_-^* - \vec{p}_\gamma^*) \right]^2. \tag{5.83}$$

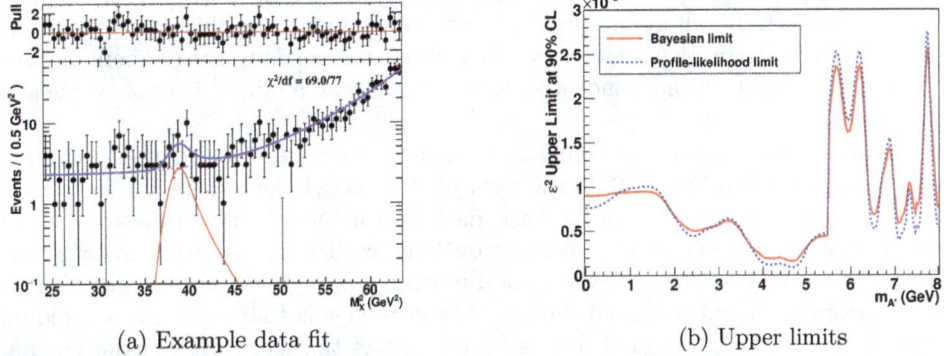

(a) Example data fit   (b) Upper limits

Figure 5.38: Examples of modeling the invariant mass spectrum in an invisible dark photon search (left) and the obtained upper limits on the rate of $e^+e^- \to \gamma_D\gamma$ from that search. Figures from Ref. [Lee+17].

We recognize that the momentum vectors of the beam particles sum to zero in the CM frame and substitute with $\sqrt{s}$ where possible,

$$m_{\gamma_D}^2 = \left[(\sqrt{s} - E_\gamma^*), (-\vec{p}_\gamma^*)\right]^2 \tag{5.84}$$

$$= (\sqrt{s} - E_\gamma^*)^2 - (-\vec{p}_\gamma^*)^2 \tag{5.85}$$

$$= s - 2\sqrt{s}E_\gamma^* + (E_\gamma^*)^2 - (E_\gamma^*)^2 \tag{5.86}$$

$$= s - 2\sqrt{s}E_\gamma^*. \tag{5.87}$$

Since the mass of any hypothetically invisible particle can be inferred from experimental quantities, it is possible to construct the mass event-by-event (Fig. 5.38a) and search for deviations in that spectrum from what would be expected from the background-only hypothesis. Signal events are modelled using mass distributions determined from simulation of a dark photon signal. Background is modelled by using real data events that have been selected by a machine-learning algorithm trained to discriminate simulated signal from real backgrounds. In this case, the analysis uses a small portion of the available data, small enough that any signal would be insignificant (compared to what would be present in the remainder of the data), treats that data as pure background, and trains the machine learning algorithm to distinguish events of that type from the simulated signal events. Those training data are never used again in the analysis. The output of the algorithm is then used in the remainder of the data to explicitly select background-compatible events, from which the m_χ background shape is determined.

The measurement described here does not definitively observe a deviation from the background-only hypothesis. However, there are small excesses in the data for dark photon mass hypotheses above 5 GeV, as shown in Fig. 5.38b. The data are interpreted using two statistical approaches and obtain consistent results. The most significant deviation from background occurs at $m_\chi = 6.21$ GeV but, taking into account the fact that the experiment looked at a wide range of masses and that statistical fluctuations faking signals is possible, the global significance of the observation is below the 3σ threshold for claiming evidence of an effect. It will be interesting to watch for updates to this kind of analysis from the current program, Belle-II, as it will collect much more data using a more modern detector and thus will have independent statistical power and a different set of systematic uncertainties.

At the present time there are no compelling confirmed excesses observed by experiments. However, as indicated in Fig. 5.36, in the dark photon model alone there is a significant portion of parameter space (even in the basic dark photon model presented above) left to explore. Unlike the missing energy searches outlined earlier, the ability to reconstruct all or most of the final state of the dark photon process provides additional distinctions between backgrounds and potential signals. The challenge is that the masses explored are generally low (below or around the GeV level), a fact that necessitates experimental design with attention to energy and momentum resolution to allow the distinction of a resonant decay from other background processes (say, from normal SM photon conversion in material to e^+e^- final states). A new generation of experiments, some designed as beam-dump-style projects and others using colliding beams of leptons (electrons and positrons), are running or proposed to run in the next couple of decades [Ale+16; Cli+22].

It is important to keep in mind that a discovery in this search effort does not necessarily give experimental physicists any direct information about dark matter itself[8]. The discovery of a new boson or a new symmetry group will only given direct information about that aspect of the theory. It may help to constrain the gauge boson sector of dark sector theories (and, in the case of invisible dark photons, might tell us something about the minimal lifetime of the dark photon or about the mass of the dark matter), but it is not guaranteed to constrain the properties of the new dark fermions in such theories. Complementarity between approaches and experiments remains crucial to make broader statements about a possible dark sector.

5.6   Perspectives on Annihilation and Production Methodologies

We have concentrated in this chapter on approaches to expand the possible number of places in which to search for dark matter (e.g., in astrophysical locations far from Earth or by using particle accelerators). While these methods expand the opportunity to study dark matter, they introduce a number of new uncertainties, particularly on the exact model by which one expects the dark matter to interact with itself or with the SM.

Indirect detection approaches open up a set of astrophysical messengers for use in studying the nature of dark matter. Apart from the theoretical drawback mentioned above, the main additional challenge is to disentangle possible dark matter signals from nominal astrophysical processes, some of which have yet to be completely understood. Indirect detection has provided some mysterious excesses. So far, none of these excesses can point conclusively to dark matter as the best explanation. We will need increased astronomical

---

[8]We are grateful to Professor Steven Robertson (IPP Scientist, University of Alberta) especially for the points raised in the concluding discussion about dark photon searches and the implication of a positive discovery of such a phenomenon.

observing platforms and enhanced sensitivity across a portfolio of complementary approaches to understand the origins of these excesses.

Colliders allow for the probing of dark matter production or force-carrying particles that may interact with dark matter. We presently suffer from a lack of knowledge of the fundamental model of interaction that leads to dark matter production (or decay). Colliders are attractive, in that they allow for great control over the frequency and timing of collisions (and thus of observing experiments). In the absence of a signal they can provide constraints on a range of models that contain dark matter or dark gauge boson candidates. Ultimately, it may be that dark matter simply doesn't interact much or at all with SM particles, in which case the constraints don't tell us much. Alternatively, a positive detection of an invisible particle at a collider would provide a roadmap for other experiments to then hunt for evidence of the same particle in the surrounding cosmos. We again emphasize the need for complementarity between approaches.

In an "all-of-the-above" approach to dark matter searches, the combination of direct, indirect, and collider experiments will one day be crucial to unraveling the nature of this part of the universe. We encourage ambition in experiments, novel uses of current and old data, and attention to the relationships that necessarily must exist between what is observed "out there" in the cosmos and what is observed "in there" using controlled collider experiments.

CHAPTER 6

Non-Particle Dark Matter and Methodologies

The previous chapters have focused intensely on an implicit hypothesis: that dark matter behaves in a particle-like manner. It could scatter elastically like a billiard ball in a non-relativistic collision. It could decay or annihilate in a way akin to the matter of the SM (e.g., described by a quantum field theory). Tied to this hypothesis is a range of masses for our dark matter candidates, going from the keV or GeV level up through the multi-TeV level. However, we already know from earlier considerations (Section 3.5) that there is no *a priori* lower limit on the mass other than that imposed by Fermi-Dirac or Bose-Einstein statistics and the observation of halo formation. This would permit fermion (boson) dark matter to have masses as low as $\mathcal{O}(10)$ eV/c^2 ($\sim 10^{-22}$ eV/c^2) (see section 3.5). There is a vast space of possibilities below the keV–MeV level.

In this chapter, we will explore an example of a possibility. What if dark matter candidates with masses below the keV energy scale have wavelengths that cannot induce a momentum transfer in traditional particle scattering settings? Dark matter becomes expressly wave-like in this mass regime, and a completely different set of approaches is needed to probe such candidates.

6.1 Considerations from the de Broglie Wavelength

We have used the de Broglie wavelength earlier (Section 4.3) to consider the scale of dark matter interactions and whether they might or might not be coherent. However, we can also use this same approach to determine the relationship between the wavelength and the dark matter momentum transfer to an atomic system. When such a transfer becomes too small, the use of traditional elastic or inelastic scattering approaches to observe dark matter candidates will not apply.

For example, atomic methods of detecting dark matter rely on the momentum transfer to the system being sufficient to break atomic bonds (ionize or change atoms in some detectable way). The energy scale of such atomic signatures can go as low as about 1 eV, which is to say that such bonds can be broken so long as the change in the kinetic energy

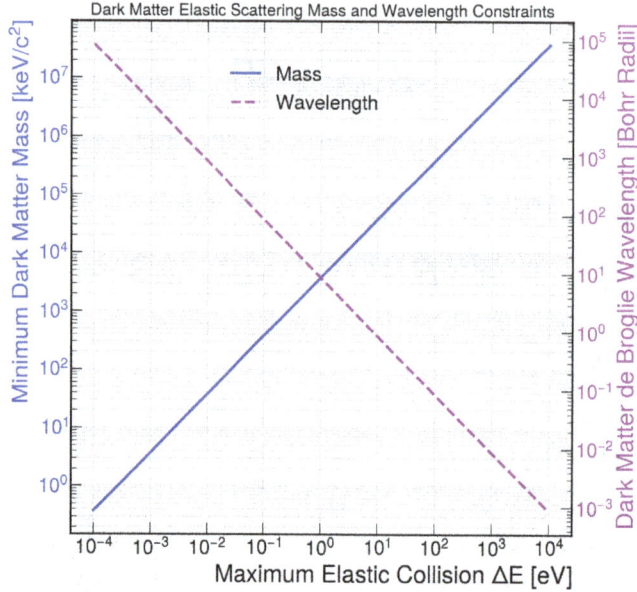

Figure 6.1: For a given maximal energy transfer to a target system (ΔE) in a non-relativistic elastic collision, the minimal dark matter mass (and corresponding de Broglie wavelength) that can be probed by the interaction is shown.

of a struck electron is *at least* $\Delta E \approx 1$ eV. The direct use of even-more-weakly bound systems, such as Cooper pairs (see section 4.7 for the indirect use of Cooper pairs to probe ionized nuclei), would permit energy transfers as small as about 1 meV to also break weak bonds.

In an elastic scatter, the amount of energy transferred to a target system (an electron or a Cooper pair) will be equal to the amount of kinetic energy lost by the dark matter in the process, $\Delta E = \frac{1}{2}m_\chi v_i^2 - \frac{1}{2}m_\chi v_f^2$. Let us assume all kinetic energy is transferred from the dark matter to the electron—the maximal case—but that this transfer is just enough to induce a measurable effect. This will be our limiting case for estimating the minimum mass that can achieve this. This maximal transfer case corresponds to $v_f = 0$ and $v_i = 220$ km/s. A graphical relationship between minimum dark matter mass, maximum energy transfer to the target, and de Broglie wavelength is shown in Fig. 6.1 (valid only under the conditions stated above).

In the case of the energy required to induce an atomic-scale effect ($\mathcal{O}(1)$ eV), we can then solve for the dark matter mass to find that $m_\chi \approx 7m_e$. This places the minimum dark matter mass scale at the level of $\mathcal{O}(1)$ MeV/c^2. From the de Broglie wavelength relationship with non-relativistic momentum, $p = h/\lambda = m_\chi v$, we find that $\lambda_\chi \approx 0.5$ nm, or about 10 Bohr radii. This is already in excess of the atomic size scale, and a lower mass dark matter candidate would lead to even longer wavelengths.

If we consider the case of breaking Cooper pairs by depositing energy at the level of 1 meV, then we find that the minimum dark matter mass we can probe is reduced to $\mathcal{O}(1)$ keV and a de Broglie wavelength at the level of 10,000 Bohr radii. Cooper pairs have wavelengths that can span macroscopic scales, so this is plausible. However, lower masses than this will require even larger-scale quantum phenomena to maintain the consistency of the dark matter and target size scales.

We conclude from this estimate that the energy scale at which atomic dark matter elastic scatters become infeasible as a detection mechanism, corresponds to the wavelength at which dark matter, relative to the target size, becomes more wave-like than particle-like. A recent review of scattering and low-mass dark matter candidates is available in Ref. [Zur24].

Let's explore one framework that predicts precisely this kind of dark matter candidate. If real, it would potentially resolve a long-standing problem in the SM. However, candidates that do this are not required to exist only to solve this problem; there could, in principle, be a whole class of particles like this one with a range of properties that could be detectable at current or future experiments. This particle is called the *axion*. We emphasize this framework because there is presently a revolution ongoing in building suites of experiments to probe many possible scenarios within this candidate space.

6.2   The Strong CP Problem and the Axion

The axion, as a particle concept, originates from an effort to solve a problem seemingly unrelated to dark matter: the *strong CP problem*. This problem is rooted in an observation of nature and the difficulty of explaining it using the SM. We begin with the neutron. It is composed of three valence quarks—two down quarks and one up quark—bound together by a sea of gluons. This gluon sea represents a strong binding energy field that keeps the valence quarks locked into the net-colorless neutron state. By virtue of the strong interaction and quantum mechanics, there are virtual quark-anti-quark pairs constantly coming into and going out of existence. These *sea quarks*, along with the gluons, make significant and undeniable contributions to the structure of baryons like the neutron.

It is tempting, therefore, to make a first-principles calculation of the properties of the neutron. The SM certainly permits this. One observable property of the neutron is its net electric charge, and, more specifically, its electric dipole moment (EDM). While the net charge of the neutron is observed to be essentially zero, one must keep in mind that the neutron has physical extent (about 1 fm), and contains electrically charged objects—quarks—with spatial separations at the scale of, or smaller than, 1 fm.

Spatial separation of charges, we learn in introductory physics, minimally leads to a small *dipole* electric field and thus a non-zero EDM. A dipole is simply a net positive and a net negative electric charge of magnitude q, separated by any distance, d. The larger the charge, or the larger the separation, the larger the EDM. A non-zero EDM makes the system susceptible to manipulation by external electric fields.

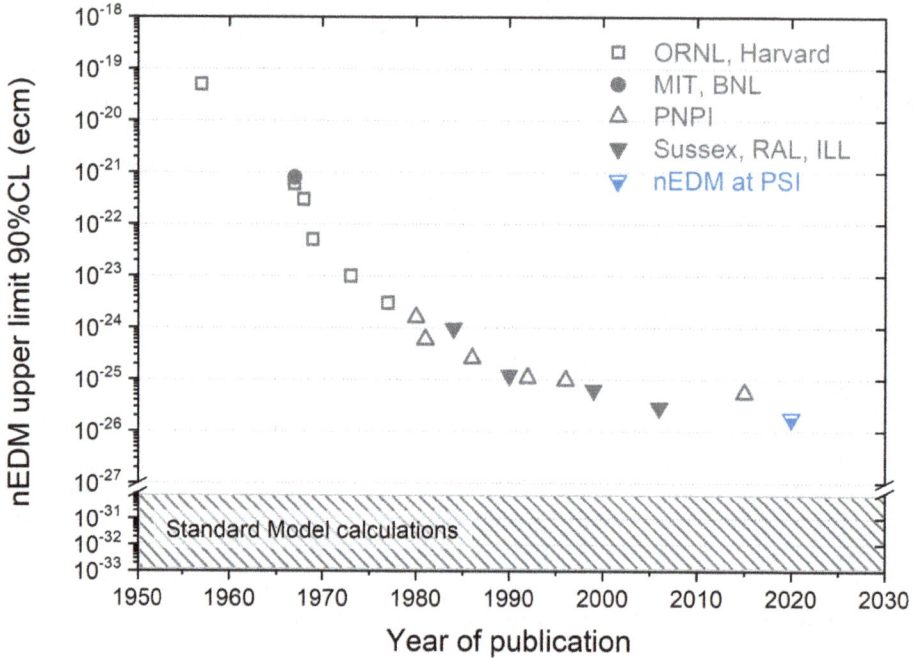

Figure 6.2: A history of measurements of the neutron EDM. The SM calculations shown here assume the only source of CP violation that could induce a neutron EDM is from the electroweak sector. Figure from Ref. [Wik22b] and the most recent measurement is from Ref. [Abe+20b].

One can subject the neutron to experiment to evaluate, observationally, its EDM (Fig. 6.2). The EDM of the neutron is currently known to be $d_n = (0.0 \pm 1.1_{stat.} \pm 0.2_{syst.}) \times 10^{-26}$ e · cm (where e is the fundamental electric charge) [Abe+20a]. This can be converted to an upper limit on the neutron EDM, $d_n < 1.8 \times 10^{-26}$ e · cm at the 90% confidence level.

We can perform a naive back-of-the-envelope estimate of the neutron's EDM. The naive calculation gives us the "best case scenario" for the neutron EDM. The up quark has a charge of $+(2/3)e$, while each down quark has $-(1/3)e$. The pair of down quarks forms the negative-charged part of the dipole, $-q = -(2/3)e$, and is separated by about $r = 1$ fm, on average, from the up quark. This valence configuration would be the core element of the neutron's electric dipole. Using the classic electric dipole formula, we find:

$$d_n = qr \approx 10^{-16} \text{ e} \cdot \text{cm}. \qquad (6.1)$$

Clearly, the simplistic model described above is ruled out. But other theoretical features of the SM similarly imply that the neutron EDM could have been this large ... but appears experimentally to be highly suppressed in nature.

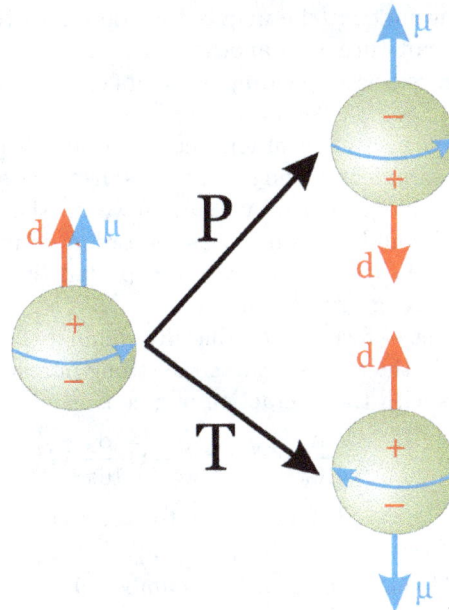

Figure 6.3: A visual example of how an electric dipole moment is a parity-symmetry-violation and a time-symmetry-violating phenomenon. The EDM $\vec{d}$ is reversed by swapping all coordinate axes via $\vec{r} \to -\vec{r}$. The magnetic dipole moment $\vec{\mu}$ is reversed under time-reversal, $t \to -t$. Under the combined transformation, PT, the system is also not left invariant. This implies CP must also be violated to preserve CPT as a symmetry of the system. Figure from Ref. [Wik20d].

The strong interaction as implemented in quantum chromodynamics naturally contains terms that violate fundamental discrete symmetries like parity inversion (P), charge conjugation (C), or time reversal (T) (see Section 1.3). The electroweak interaction also naturally provides for violations of these symmetries, each by themselves (C, P, or T), or in pairs. All known self-consistent and physical quantum field theories require that CPT be conserved. What is remarkable is that violations of P and T symmetries have been long observed in electroweak interactions, but not in strong interactions. If P and T are separately violated, then the combination CP must *also* be violated to preserve CPT as a whole.

Our back-of-the-envelope calculation (in-line with much more detailed estimates, c.f. [Dar00]) suggests the neutron EDM should be very large. Non-zero EDMs naturally violate both parity-inversion symmetry and the combination of parity-inversion and time-reversal symmetry, as illustrated in Fig. 6.3. However, non-observations of CP violation in the strong interaction would again confirm that the EDM is very small.

The SM calculations in Fig. 6.2 are made using the only known source of CP violation: the electroweak sector [Dar00]. This is due to a single known CP-violating phase, δ_{EW}, in the quark flavor mixing matrix (the matrix that describes the degree to which one quark flavor, such as u, can turn into another one, such as s, in weak interactions). Assuming only this phase, estimates yield upper limits on the neutron EDM at the level of 10^{-31} e · cm.

The gap between current measurement and the upper bounds predicted by electroweak theory is vast, with measurement having achieved sensitivities that are six orders of magnitude above this scenario provided by δ_{EW}. However, the naive calculation would suggest the neutron EDM could be as many as ten orders of magnitude *above* present experimental sensitivities. Where, then, are the expected (but unobserved) large CP violating effects from the strong interaction?

One resolution to this set of facts is to artificially eliminate or reduce the contribution of the CP-violating terms in the SM as regards the strong interaction. The portion of the SM Lagrangian that deals with the interaction of gluons is

$$\mathcal{L}_{QCD,gluon} = \frac{\alpha_s}{16\pi^2} G^a_{\mu\nu} G^{a\mu\nu} - \overline{\Theta} \frac{\alpha_s}{16\pi^2} G^a_{\mu\nu} \tilde{G}^{a\mu\nu} \qquad (6.2)$$

where a is the color index ($a = 1, 2, 3$ for each of the color charges red, green, and blue); μ and ν are spacetime indices (e.g. $\mu = 0, 1, 2, 3$); $G^a_{\mu\nu}$ is the gluon field strength tensor; and $\tilde{G}$ is the dual of G, $\tilde{G} = \epsilon^{\mu\nu\lambda\rho} G^a_{\lambda\rho}/2$. The number $\overline{\Theta}$ is a CP-violating phase that can take any value between $-\pi \leq \overline{\Theta} \leq \pi$.

For CP-violation to be so suppressed in the strong interaction would require a CP-violating phase that is also very small, ($|\overline{\Theta}| < 10^{-10}$), eliminating the CP-violating term involving the dual of the gluon tensor. To have a parameter in a model that is anything other than $\mathcal{O}(1)$ in size is considered "unnatural" and is usually treated as a place in the model requiring a theoretical explanation for that parameter. This is an example of a *fine-tuning problem* in the SM, where parameters that one would hope to be $\mathcal{O}(1)$ turn out to be much smaller. One is then forced to explain such specific choices for parameters ...that is, to explain why the model is so fine-tuned to get the right answer in nature. In the present context, this is known as the "Strong CP Problem."

There are solutions to this problem, the most well-known of which is one proposed by Roberto Peccei and Helen Quinn [PQ77a; PQ77b; Wei78; Wil78]. It is a dynamical solution to the problem, achieved by adding an extra symmetry group to the SM, a global $U(1)_{PQ}$ symmetry.

The solution begins by adding a field, Φ_{PQ}, to the SM whose properties are symmetric under a global symmetry that represents a change to the phase of the field by a constant amount,

$$\Phi_{PQ} \rightarrow e^{i\alpha}\Phi_{PQ}, \qquad (6.3)$$

where α is a real constant. The fact that α does not depend on spacetime coordinates is what makes any invariance under such a change a global symmetry of the field. If two sequential phase transformations $U(\alpha_1) = exp(i\alpha_1)$ and $U(\alpha_2) = exp(i\alpha_2)$ are applied, their order does not matter, so these transformations are said to commute and are thus *Abelian*. We also see that the transformations are unitary, such that $U(\alpha)^\dagger U(\alpha) = I$, the identity. The mathematical group that describes all such unitary Abelian transformations of this type is known as U(1).

The implications of a field that obeys such a global symmetry are explored in many references (c.f., Ref. [HM84]). For example, if we require that the field's Lagrangian density be invariant under a global U(1) group then there must be a conserved current, and a conserved charge, in the theory. Any other field that couples to Φ_{PQ} must carry this conserved charge. Since the precise goal of this idea was to dynamically eliminate the CP violating term in the QCD Lagrangian, the field must couple, at minimum, to the gluons (through the $G\tilde{G}$ term). The specific details of PQ field interactions with quarks, leptons, or other gauge fields leads to various scenarios to be explored in this theory (see below).

The field, in key models, is usually implemented by adding a doublet, $\Phi_{PQ} = (\phi_1, \phi_2)$. The PQ field will be described by a Lagrangian density, from which the equations of motion of the field can be determined. In its simplest form, this might look like

$$\mathcal{L}_{PQ} \propto (\partial_\mu \Phi)^2 - \lambda |\Phi|^2. \tag{6.4}$$

This first term is related to the kinetic energy of the field and the second to the potential energy. The problem with this potential is one inherent in the earliest quantum field theories: it is quadratic, symmetric about zero, and takes its minimum at 0, guaranteeing massless field quanta. This represents the unbroken PQ symmetric theory and is shown in Fig. 6.4a.

The PQ symmetry must be broken, or we cannot eliminate the CP-violating term in the QCD Lagrangian by a dynamical method. There must then be an energy scale at which this symmetry breaks, denoted f_a. This quantity will be later identified as the decay constant of the axion. In the early universe, when the temperature dropped below this threshold, the PQ symmetry broke. The Lagrangian density of the broken symmetry can be written in its simplest form taking a cue from the Brout-Englert-Higgs mechanism that successfully explained the origin of fundamental mass in the SM. We replace the potential above with

$$V(\Phi_{PQ}) = \lambda |\Phi_{PQ}|^2 \longrightarrow \lambda \left(|\Phi_{PQ}|^2 - f_a^2 \right)^2. \tag{6.5}$$

This potential has a new minimum displaced from $\langle \Phi_{PQ} \rangle = 0$, as the field now has an expectation value $\langle \Phi_{PQ} \rangle = f_a$ related to the symmetry breaking energy scale itself. This is represented in Fig. 6.4b. The field will be expected to settle into the lowest point in the potential, which here is a circular ring of infinite and equally likely minima. The field doublet can be recharacterized in terms of two distinct components: a radial piece (the *radion*) and an azimuthal piece, the *axion*. Let us denote the axion field by $a(\vec{r}, t)$. At this point, the axion has not developed a singular vacuum expectation value as it can take all values in the circular minimum of the potential, corresponding to an angular parameter φ in ϕ_1, ϕ_2 field space that can take any value $\varphi = [0, 2\pi] f_a$. We can notate $\Theta = \varphi / f_a$, indicative of the role of this feature of the theory in ultimately eliminating or suppressing $\overline{\Theta}$.

The axion field will be invariant under the $U(1)_{PQ}$ global symmetry group, meaning that shifts in the field by an overall small constant (corresponding to an infinitesimal rotation, $exp(i\alpha) \approx 1 + i\alpha$, will leave the field invariant. This is equivalent to rotating the

(a) $T > f_a$

(b) $\Lambda_{QCD} < T < f_a$

(c) $T < \Lambda_{QCD}$

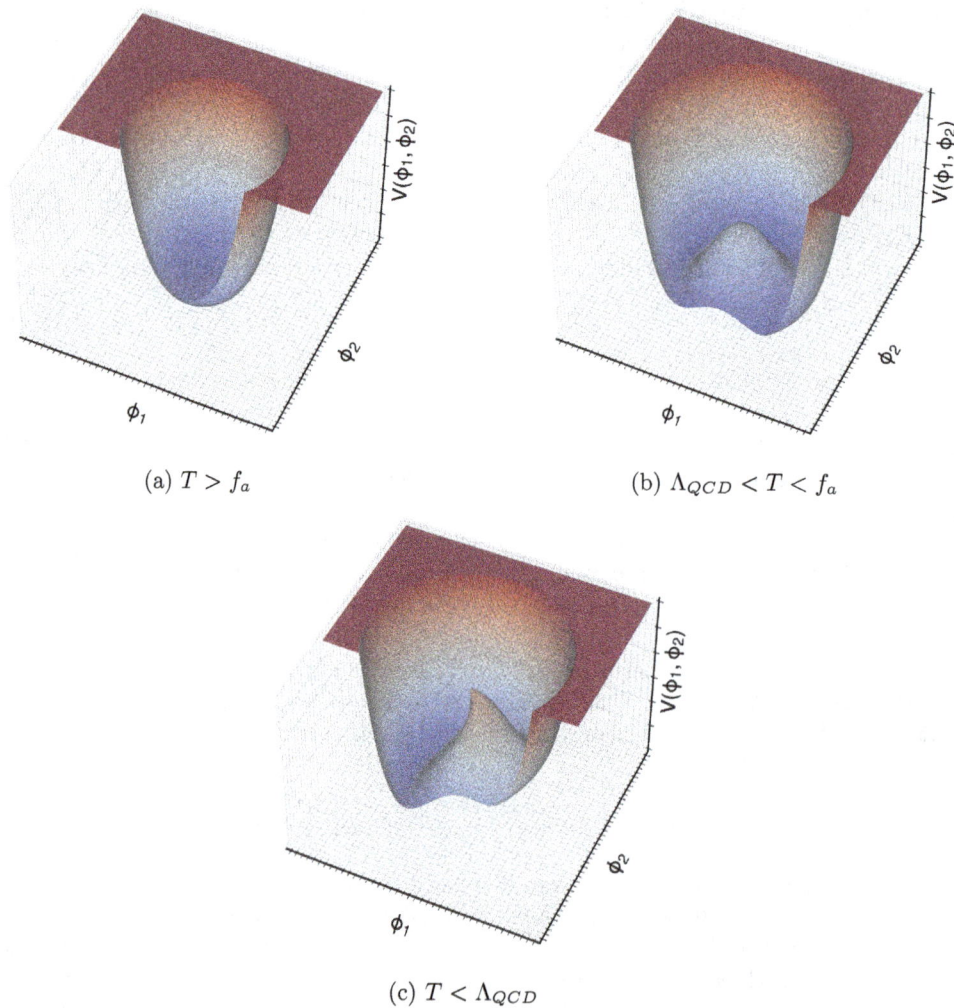

Figure 6.4: Graphical representation of the PQ field potential at various times in the early universe. The full 3-D surfaces are artificially cut away in the front to allow for viewing of the interior structure. From left-to-right: when the temperature is above the PQ symmetry breaking scale, f_a; when the temperature is below the PQ breaking scale, but above the QCD condensation threshold, Λ_{QCD}; and when the temperature is below Λ_{QCD}.

field in Fig. 6.4b anywhere in the minimum by an amount α. This, in turn, corresponds to shifting the field by

$$a(\vec{r},t) \to a(\vec{r},t) + \varphi. \tag{6.6}$$

Since this field will appear in the Lagrangian of the full theory, along with the SM components, it can be written with any such shift and not change the Lagrangian at all. This ultimately is what allows for the elimination of the CP-violating free parameter, $\overline{\Theta}$, by writing the interaction terms for the field and asserting that the expectation (minimum) value of the field occurs at:

$$\langle a(\vec{r},t) \rangle = -f_a \overline{\Theta}. \tag{6.7}$$

Minimal CP violation (neutron EDM) in the strong sector is achieved when $\overline{\Theta}$ is the value at which the axion potential achieves its global minimum. Thus, if

$$d_n \propto \frac{a(\vec{r},t)}{f_a} + \overline{\Theta}, \tag{6.8}$$

the neutron EDM can be suppressed if Eqn. 6.7 is true. This drives the EDM to zero as the result of a dynamical feature of the field. Since the EDM and strong CP violation are related, we expect this mechanism to also drive toward zero the CP-violating term of the QCD SM Lagrangian,

$$\mathcal{L}_{QCD,Axion} = (\partial_\mu a)^2 + \left(\frac{a(\vec{r},t)}{f_a} + \overline{\Theta} \right) \frac{\alpha_s}{32\pi^2} G^a_{\mu\nu} \tilde{G}^{a\mu\nu}. \tag{6.9}$$

Here, we finally encode the coupling of the axion to the $G\tilde{G}$ term of the QCD Lagrangian and show how the field (unaffected by a constant shift) dynamically suppresses the naturally large CP violation in the strong interaction.

However, at this stage we have not said anything about the axion mass. Since the axion and QCD are coupled to one another, when the temperature in the universe drops below Λ_{QCD}, we transition from free quarks to quark condensates (bound states of quarks that are only asymptotically free when in very close proximity to other quarks). The newly broken symmetry in QCD results in the formation of the pions. This phase transition also induces one more shift in the PQ field potential [Mar16; Pec08],

$$V(\Phi_{PQ}) \propto \lambda \left(|\Phi_{PQ}|^2 - f_a^2 \right)^2 \longrightarrow \lambda_1 \left(|\Phi_{PQ}|^2 - f_a^2 \right)^2 + \lambda_2 \left(1 - \cos(\varphi/f_a) \right). \tag{6.10}$$

This is illustrated in Fig. 6.4c. This phase transition breaks the symmetry one more time, creating a "bundt pan" potential[1] with the bottom unevenly tilted toward a singular minimum both for the radion and the axion. It is now that the axion develops a mass, and that mass can be quite small as it's related to $1/f_a$. Specifically, the *QCD axion* sketched here will have a mass that is well-prescribed by parameters of QCD and the PQ theory.

---

[1]We wish to credit the description of the doubly-broken PQ field potential's symmetry as a "bundt pan potential" to Christian Capanelli, communicated to us by Professor Katelin Schutz (McGill University). We were delighted with this description, as one of the authors (JC) is an avid baker. Bundt pans often have circular, but azimuthally irregular, bottoms, leading to the description applied by the proponent of this term.

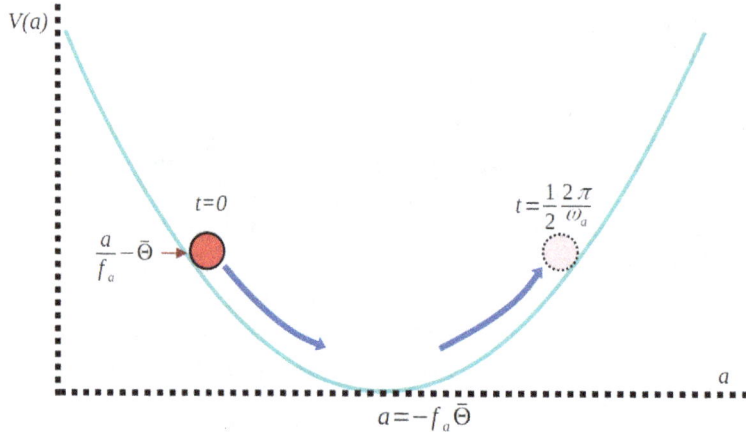

Figure 6.5: An illustration showing an overly simplistic model of the axion field potential and the behavior of the field at (and after) time $t = 0$, after PQ symmetry breaking and the QCD phase transition. The parameter $\bar{\Theta}$ is assumed to be located in some space x such that at this value $V(x)$ takes its minimum. The axion field oscillates in the potential about this minimum value, and without dissipation does so forever and is stable for the lifetime of the universe.

A Taylor expansion of the field about the minimum of $V(\Phi_{PQ})$ would produce a quadratic-order description of the field that is oscillatory with time, such that the field can take on non-zero values but does so very close to $V(a) = 0$, where Eqn. 6.7 holds. A simple model of this oscillation of the field around the minimum is shown in Fig. 6.5. Oscillation close to the minimum also naturally maintains the suppression of CP violation in the strong interaction.

The oscillation of the axion field occurs at a frequency, ω_a, that will, by the usual de Broglie relationships, correspond to the mass of a particle excitation of the field[2]. The field's time-dependence at leading order can be written as

$$a(t) \propto \frac{2\sqrt{\rho_a}}{m_a} \cos(m_a t). \tag{6.11}$$

where ρ_a is the axion density which, in a radiation-dominated universe, scales as $1/a^3$ where this a refers to the scale factor of the universe.

The axion field exists at all points in spacetime and so is present everywhere in the universe. Since it oscillates about its minimum, it has an excitation that is particle-like (in that it has properties like mass). The QCD axion mass would have to be related

---

[2]For the rest of this discussion, equations will generally assume that natural units are being employed. In this case, frequency and energy have the same units. This is based on Planck's relationship that $E = \hbar\omega$ such that when $\hbar \to 1$ the units of frequency are also those for energy (e.g., eV). Since, in natural units, mass and energy have the same units, frequency is completely interchangeable with mass (e.g., $m = \omega$)

to a number of SM parameters, including (but not limited to) the up and down quark masses, the pion mass and the pion decay constant [Gri+16]. The most generic sequence of events in the early universe that ultimately leads to a specific axion mass are the breaking of the PQ symmetry (energy scale unknown, resulting in an axion potential whose minimum is displaced from $\langle \Phi_{PQ} \rangle = 0$), the breaking of the EW symmetry (around a temperature/energy of 250 GeV, which results in an electroweak potential whose minimum is also displaced from zero, resulting in non-zero quark masses, etc.), and then the QCD phase transition (around 100 MeV) which results in the formation of quark bound states. It is that last transition that ultimately locks in the relationship between the axion mass and QCD strong dynamics [Mar16]. These relationships, inherent in the axion specifically designed to solve the strong CP problem, yields

$$m_a = 5.691 \left(\frac{10^9 \text{ GeV}}{f_a} \right) \text{meV}. \tag{6.12}$$

The QCD axion mass is only controlled by the PQ symmetry-breaking scale (f_a), whose value is unknown. Here it has been normalized to a value below the GUT scale (10^{16} GeV) under the assumption that the breaking of the $U(1)_{PQ}$ symmetry happens *below* the energy scale at which the strong, weak, and electromagnetic forces unite. The lack of a narrow constraint on the potential mass of the axion poses a problem for axion search experiments. While the potential range is narrower than the broader scope of generic dark matter candidates, it is still difficult to design dedicated experiments that hunt for axions or axion-like states. This is also made additionally challenging by the specific details of *when* the PQ symmetry broke (corresponding to the actual value of f_a).

The axion will have mass but is expected to carry no electric charge. If it is light enough, it should be very stable (we'll explore this below). These qualities make it an ideal candidate for dark matter. However, we can see that the mass is expected to be small (something, perhaps, at the level of the μeV–meV scales) unless $f_a \ll 10^9$ GeV. Based on the above arguments, any axion that solves the strong CP problem via the PQ mechanism will have a small mass. This will make it an exceptionally wave-like phenomenon that cannot be detected by the direct detection means we have described earlier (see Fig. 6.1).

Let us make a final comment on the axion before moving on to its other consequence, which will aid in establishing a detection mechanism. Let's assume, for a moment, that some other mechanism explains the smallness of strong CP violation. Nevertheless, there may still be at least one additional $U(1)$ global symmetry that is broken. Equation 6.12 only holds in the case of the axion that solves the strong CP problem because its properties are constrained by the QCD phase transition and having to dynamically cancel out $\overline{\Theta}$. In the case that an axion field does not play a role in solving the strong CP problem, but nevertheless exists, the universe could still be filled with axions . . . just not the ones constrained by Eqn. 6.12. These *axion-like particles*, or ALPs, would make excellent dark matter candidates . . . but we would have no natural constraint on their mass, which widens again our search window.

There are multiple generic classes of axions. Kim-Shifman-Vainshtein-Zakharov (KSVZ) axions involve a mechanism that introduces additional heavy quarks as well as the PQ field [SVZ80]. In that scenario, the PQ scalar couples to leptons. The Dine-Fischler-Srednicki-Zhitnitsky (DFSZ) axion mechanism introduces an additional Higgs field as well as the PQ field. This results in an axion that couples to quarks and leptons. Other higher-energy theories, such as string theories, generically include ALPs.

Axion dark matter behaves like a classical field, which we explore in later sections of this chapter. Let us assume the mass is somewhere in the range of μeV–meV. Employing again the de Broglie wavelength, assuming the axions move at 220 km/s relative to Earth, we find

$$\lambda_a = \frac{hc}{pc} \approx \frac{hc}{mvc} = \frac{hc}{m\beta c^2}, \tag{6.13}$$

where for $m_a \approx 1$ meV/c^2 yields $\lambda_a \approx 0.3$ m. That's the scale of radio waves. If the axion mass is even smaller than the meV scale, then this de Broglie wavelength grows larger. At the μeV scale, the axion wavelength is hundreds of meters; at the neV scale, it's hundreds of kilometers. In comparison, the traditional WIMP paradigm with its 100 GeV/c^2 mass scale leads to wavelengths of femtometers.

6.3   The Axion and Cosmology

Let's take a brief look at the axion in cosmology [Mar16; Sik08]. We will summarize the cosmological effects of when the PQ symmetry breaks and the implications of axion decay. After the universe comes into existence, and very quickly thereafter, it expands and cools. This rapidly brings it well below the Planck and then the GUT energy scales.

Inflation in the very early universe causes such a rapid spacetime expansion that different parts of the universe become causally disconnected. Any feature that is present in the universe before inflation will be expanded to scales corresponding to the entire visible universe, and thus would be "constant" from that perspective. On the other hand, any features that arise *after* inflation may differ from place to place in the visible universe as their effects could not have been spread everywhere and all at once prior to causal disconnection. We would observe the average of those features across the visible universe.

The precise time at which PQ symmetry breaks thus has interesting cosmological consequences [BFS20; Mar16]. For example, if PQ symmetry breaks before inflation, then it was possible for all parts of the universe to be at the same global minimum of $V(a)$ after symmetry breaking, leading to a single CP-violating phase throughout the visible universe. The drawback of this case is that the relic density of axions depends on two parameters: the CP-violating phase and the axion decay constant. The relic density of axion dark matter in this scenario goes like $\overline{\Theta}^2 f_a^2$ and thus depends on two unknowns, making mass constraints from cosmology extremely challenging.

However, should PQ symmetry break after inflation it is possible for different parts of the universe to have settled into different global minima of $V(\Phi_{PQ})$, meaning there won't be a singular universally applicable CP-violating phase. That average CP violating phase is determined by a sample of values drawn randomly from the range $[-\pi, \pi]$ and winds up being constrained, $\langle \overline{\Theta}^2 \rangle = \pi^2/3$. Thus the relic density of axions depends only on the axion mass.

Let us consider, in addition to the above, the effect that the competing dynamics of spacetime expansion and axion field behavior will contribute to the cosmos. The axion equation of motion is given by:

$$\ddot{a} + 3H(t)\dot{a} + m_a^2 a = 0, \tag{6.14}$$

where $H(t) \equiv H$ is the Hubble rate and a is the axion field. This equation is the same form as that for a damped harmonic oscillator. The term containing the Hubble rate plays the role of the damping effect.

This equation already implies two interesting epochs. The first is when $m_a \ll H$, in which case the damping term completely dominates and the system is said to be *over-damped*. During this period in the early universe, any oscillatory behavior of the axion field would be suppressed by the strength of the damping term.

The second epoch is when $m_a \gg H$. This situation results in the suppression of the damping effect in the equation of motion (the second term is smaller than the third), admitting oscillatory solutions like Eqn. 6.11. In this phase of the universe's evolution, the oscillation of the axion field is no longer suppressed by the expansion of spacetime, allowing for the relic density of axions with a mass m_a to be locked into the universe. If this transition occurs when the universe crosses into the radiation-dominated era, $H \approx 10^{-28}$ eV, this would imply that axions heavier than this value will compose the dark matter. This axion population density will scale just as the matter density, inversely proportional to the cube of the scale factor, which confirms that these axions will behave in a matter-like way.

Now we can turn to the CMB and what we know about relic densities to see how cosmology might constrain the properties of the axion if it is the dominant contributor to a cold dark matter population. If the PQ symmetry breaks before inflation [Bal+17; BHK08; Bor+16; WS10], asserting that $\Omega_a h^2 = \Omega_{CDM} h^2 \approx 0.12$ leads to a constraint that the axion mass be around the level of 6 μeV. However, this statement is highly dependent on where, exactly, in the PQ field potential the universe has landed in our visible universe, and how the field oscillations behave in the potential. In the case where the PQ symmetry breaks after inflation, the observed cold dark matter density implies an axion mass more at the level of 30 μeV or higher. Again, constraints here are very dependent on assumptions of the underlying model. What we can conclude from this, however, is that experiments targeting such constraints need to be able to probe axions that, at least, are at the μeV level ... but experiments also need to be prepared to go higher and lower.

The lifetime of the axion is a final key piece in the puzzle. If the axion cannot decay, then it must be completely stable once produced. However, models of the axion imply that it can have non-zero couplings to bosons and fermions, but those couplings are highly dependent on the assumptions of a given scenario. Nevertheless, the observable universe helps to constrain some of the possibilities.

The axion should decay to particles lighter than m_a. If the axion has a mass below $2m_e$, then it can only decay to neutrinos or photons. For now, let us set aside couplings to fermions and focus on the possibility of $a \rightarrow \gamma\gamma$. In the DFSZ or KSVZ scenarios, the partial width for this decay goes roughly as

$$\Gamma_{a\to\gamma\gamma} \propto g_{a\gamma\gamma}m_a^3 \approx 10^{-24}\text{s}^{-1}\left(\frac{m_a}{\text{eV}}\right)^5. \tag{6.15}$$

Since the decay rate is related to the axion mass, and since the lifetime of an unstable particle is given by $\tau = 1/\Gamma$ (in natural units), we can estimate the maximum possible axion mass assuming it must have a lifetime at least as small as the age of the universe, $t \approx 13.78$ Gyr. In that case, we find that $m_a \lesssim 20$ eV.

The previous discussions about using cosmology to try to confine the possible mass range of axions leads us to one strong conclusion. The concerns outlined earlier in this chapter are to be taken seriously; such a dark matter candidate would be so light as to evade particle-based methods of detection, requiring approaches sensitive to wave-like behaviors.

6.4   Detecting Axions

The axion field will add terms to the SM Lagrangian. We focus here on the terms that could involve interactions of the axion with other particles:

$$\mathcal{L}_{a,\gamma} = -\frac{1}{4}g_{a\gamma\gamma}aF_{\mu\nu}\tilde{F}^{\mu\nu} \text{ (axion-photon)} \tag{6.16}$$

$$\mathcal{L}_{a,n_{EDM}} = -\frac{i}{2}g_d a\overline{N}\sigma_{\mu\nu}\gamma_5 N F_{\mu\nu} \text{ (axion-neutron EDM)} \tag{6.17}$$

$$\mathcal{L}_{a,N} = g_{aNN}(\partial_\mu)\overline{N}\gamma^\mu\gamma_5 N \text{ (axion-axial nuclear moment)} \tag{6.18}$$

$$\mathcal{L}_{a,N} = g_{aee}(\partial_\mu)\overline{e}\gamma^\mu\gamma_5 e \text{ (axion-axial electron moment)}. \tag{6.19}$$

We re-emphasize a statement we made earlier in the context of the axion-photon coupling: all couplings listed here (e.g., $g_{a\gamma\gamma}$) are expected to be highly model-dependent. While additions to the SM Lagrangian can be written, not all may be realized even if axions exist. Experiments should be guided by a balance between designs that target specific scenarios and those that achieve a broad spectrum of sensitivity with less model-dependent choices.

The axion today behaves as a field, not a particle, as we have emphasized above. The picture we should have in mind is more like how we think of the electromagnetic field acting on charges. This is a classical picture of a field of force exerting changes on the state of motion of the objects with which the axion field can couple. Axion dark matter has a *collective*, not an individual, action on matter.

The most striking aspect of this is that the axion field contributes new terms to Maxwell's Equations, themselves the classical consequence of a quantum field theory of the electron and photon. In terms of the free-space formulations of Maxwell's Equations, the modifications appear as follows:

$$\vec{\nabla}\cdot\vec{E} = -g_{a\gamma\gamma}\vec{B}\cdot\vec{\nabla}a \tag{6.20}$$

$$\vec{\nabla}\cdot\vec{B} = 0 \tag{6.21}$$

$$\vec{\nabla} \times \vec{E} = -\frac{\partial \vec{B}}{\partial t} \tag{6.22}$$

$$\vec{\nabla} \times \vec{B} = \frac{\partial \vec{E}}{\partial t} - g_{a\gamma\gamma}\left(\vec{E} \times \vec{\nabla}a - \frac{\partial a}{\partial t}\vec{B}\right). \tag{6.23}$$

In the presence of a large static magnetic field, axion dark matter creates an "effective current" that generates oscillating magnetic fields in response,

$$\vec{\nabla} \times \vec{B} \approx \frac{\partial \vec{E}}{\partial t} + g_{a\gamma\gamma}\frac{\partial a}{\partial t}\vec{B} \tag{6.24}$$

$$\vec{J}_{eff} = g_{a\gamma\gamma}\sqrt{2\rho_{DM}}\vec{B}_0\cos(m_a t). \tag{6.25}$$

The mass of the axion will affect the wavelength of the collective field phenomenon. Much as in the case of detecting electromagnetic waves, one will have to consider the wavelength of the collective axion behavior compared to the size of the experimental apparatus, R_{exp}.

In the case that the wavelength is comparable to R_{exp}, we will be dealing with a resonance detectable in a cavity of that size. This is the "cavity regime" of axion searches and is pursued by experiments like the Axion Dark Matter eXperiment (ADMX) [Asz+10a]. Cavity searches focus on the observation of resonances in a radial electromagnetic field via

$$\vec{\nabla} \times \vec{B} \approx \frac{\partial \vec{E}}{\partial t} + g_{a\gamma\gamma}\frac{\partial a}{\partial t}\vec{B} \text{ (Cavity Regime)}. \tag{6.26}$$

In this case, both terms in the strong magnetic field condition contribute, and the oscillating axion field sets up a time-varying electromagnetic field in the cavity.

In the case that the wavelength is much greater than R_{exp}, we enter what is known as the "quasistatic regime" where the field oscillation is on a time/distance scale far greater than the apparatus. This regime is probed by proposed experiments like A Broadband/Resonant Approach to Cosmic Axion Detection with an Amplifying B-field Ring Apparatus (ABRACADABRA) [Oue+19] or Dark Matter Radio (DM Radio) [Sil+17]. The emphasis here is on

$$\vec{\nabla} \times \vec{B} \approx g_{a\gamma\gamma}\frac{\partial a}{\partial t}\vec{B} \text{ (Quasistatic Regime)}, \tag{6.27}$$

where the axion field generates an effective current that, in turn, creates a magnetic field where none should exist.

In the case of the wavelength being much smaller than R_{exp}, we enter the "radiation regime." Here the collective oscillation has a characteristic size much smaller than the apparatus. An experiment that probes this condition is the MAgnetized Disk and Mirror Axion eXperiment (MADMAX [Bru+19a]. In this case,

$$\vec{\nabla} \times \vec{B} \to 0 \text{ (Radiation Regime)} \tag{6.28}$$

probes the relationship between the oscillating axion field and a time-varying electric field.

What all of these regimes and approaches have in common is that we are measuring a frequency, not a mass or other particle-like properties. While that frequency relates to the axion mass, the frequency itself (e.g. manifesting in electromagnetic behavior that should not exist without the axion field) is the key observable. In a plot of power vs. frequency (much as in the CMB spectral analysis), one should observe a "noise floor" from the experiment—a fairly structureless "hum" of EM frequencies—and then a spike (a lot of power in a single frequency) where the axion effect appears. This spike will not be a delta function; rather, the axion frequency will have a line shape related to the halo velocity distribution. The axion field's oscillations will have a dispersion of velocities identical to the halo velocity dispersion, owing to the fact that the axion oscillations are not those of a massless field, but, rather, one with mass acquired through symmetry breaking. This in turn will lead to differing Doppler shifts for axions at different velocities relative to the earth, broadening the frequency peak. The precise shape of the frequency line will then be useful to determine the dispersion of axion velocities in the Milky Way dark matter halo.

Searching for Axions in the Cavity Regime

The term in the Lagrangian that relates the axion field to the photon results in a prediction: in a magnetic field, axions will convert into photons. An example of a cavity intended to observe this conversion is ADMX. ADMX uses a strong, static magnetic field to create a volume dense with photons. Axions entering this region can interact with a photon to produce a new photon of a specific frequency via the inverse Primakoff effect. This effect is, essentially, a re-writing of the axion decay $a \rightarrow \gamma\gamma$, where momentum and energy are conserved in the process due to the fact that the static magnetic field stores momentum (c.f., Ref. [Gri12]) and the interaction alters the momentum stored in the field while producing the final-state photon.

The conversion rate is enhanced under the condition that the photon's frequency corresponds to the cavity's resonant frequency. This resonant enhancement also tells us the frequency of the axion field that was converted in the cavity. So, a signal in this case will be proportional to the volume of the cavity (more chances to interact in the field in the cavity), the magnetic field (a more intense field will increase again the probability of conversion), and the cavity's Q-factor. The Q-factor, or *quality factor*, is the dimensionless ratio of the resonant frequency to its bandwidth, $\omega_r/\Delta\omega_r$. The bandwidth is the full width of the resonance at half of the maximum (ω_r, in this case). Large Q factors mean narrow widths relative to the central frequency, making the resonance easier to detect on top of a noise background.

As an example, consider a $m_a = 1~\mu\text{eV}/c^2$ axion. Its energy will be $E \approx m_a c^2 + \frac{1}{2}m_a\beta^2 c^2$, since $\beta \ll c$. The target energy to which a cavity experiment must be sensitive is $1~\mu\text{eV}$ (the mass), but with a correction at the level of 10^{-6} due to the kinetic contribution to the axion energy ($\beta \approx 10^{-3}$ for galactic dark matter relative to the motion of Earth). Since energy is proportional to frequency, the mass would exactly define the frequency to which any experiment would have to be sensitive. The velocity contributes to dispersion around this frequency. Thus, $Q = \omega/\Delta\omega \approx \beta^2/(\Delta\beta)^2$, which, for the Milky Way galaxy, will be about 10^6. A cavity experiment would have to target this Q factor.

(a) The ADMX Experiment

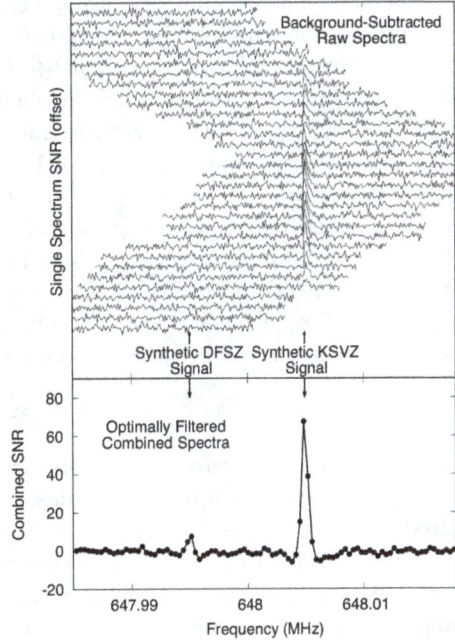

(b) An example axion signal

Figure 6.6: (left) The ADMX cavity being extracted from the magnet bore. Photo Credit: Rakshya Khatiwada, University of Washington. (right) An example of a simulated axion signal in the ADMX experiment. The spectrum on the vertical axis allows for the noise in the frequency domain to be determined and subtracted, leaving behind any residual that might be evidence of a resonant axion signal. Figure from Ref. [Du+18].

Noise in the cavity will be proportional to a couple of factors. One is the blackbody radiation present in the cavity, owing to the absorption by the oscillators (e.g., atomic electrons) in the walls of the cavity, radiation which is then re-radiated in a power spectrum given by the blackbody equation (Eqn. 2.70). In addition, since amplification will be required to detect signals in the cavity, the amplifier electronics will inevitably introduce noise into the signal read out. The ADMX experiment itself is shown in Fig. 6.6a and an example of a simulated axion signal on a noise background is shown in Fig. 6.6b.

In the cavity regime, the whole of Eqn. 6.26 is in effect. The peak power in the axion frequency [OG17; Sik83; Sik85; Sik87] will then be

$$P = \hbar^2 c^5 \varepsilon_0 g_{a\gamma\gamma}^2 V B_0^2 C_{n\ell m} \frac{\rho_{DM}}{m_a} Q \tag{6.29}$$

where $C_{n\ell m}$ is the form factor of the cavity for the $TM_{n\ell m}$ cavity mode (which is related to the overlap between the static magnetic field and the oscillating electric field of the specified mode); and f_a is the energy scale at which the PQ symmetry breaks.

Let's do a quick back-of-the-envelope calculation. For ADMX, the transverse magnetic mode TM_{001} dominates with $C_{001} = 0.69$ [Asz+10b]. We can take the local dark matter density to be the typical 0.3 GeV/cm^3 (c.f., Section 4.2). Let's assume an axion mass of 1 μeV $= 10^{-15}$ GeV. Common "off-the-shelf" high-field magnets run at the level of 5 T in strength, though there are obviously much stronger magnets than this. Let's assume a volume of $V = 1$ m^3 and a Q-factor of 10^6. We have only to guess at the coupling of the axion and photon. The coupling constant $g_{a\gamma\gamma}$ equals $(\alpha_{EM}/\pi)g_\gamma/f_a$. The factor g_γ is axion-model-dependent but is typically of order 1, so $g_{a\gamma\gamma} \approx 1/f_a$. We will further assume a PQ symmetry breaking scale of $f_a = 10^{14}$ GeV, below the GUT scale. It is helpful in Eqn. 6.29 to write $\hbar$ in units of GeV$\cdot$s, in which case its value is $\hbar = 6.582 \times 10^{-25}$ GeV$\cdot$s. This allows all appearances of GeV to neatly cancel (in $\hbar$, $g_{a\gamma\gamma}$, ρ_{DM}, and m_a). In this case, the power we could observe in the resonant frequency would be

$$P_{est} \quad \approx \quad 10^{-24} \text{ W.} \tag{6.30}$$

As the axion mass increases, the power decreases. Experiments will be more sensitive to lower axion masses than to higher ones, but such cavity experiments will be physically limited by what magnetic field they can plausibly fit into what maximum volume, and with what Q-factor they can operate the cavity. This is an exceptionally small amount of power expected to be injected by the axion field into the experiment, placing the following demands on any such project: low-temperate operations to reduce noise, high-quality and low-noise electronics, and highly symmetric cavity design.

These experiments have a narrow axion mass bandwidth (limited by the magnetic field and the physical size of any cavity), but have nonetheless set some impressive constraints on ALPs (see Fig. 6.10a at the end of this chapter). Next-generation axion experiments are expected to push through the QCD axion space, albeit in a narrow range of possible masses.

Searching for Axions in the Quasistatic Regime

We will use the ABRACADABRA experiment as an example of how searches in the quasistatic regime can work. In this regime, the idea is to look for a new component of a magnetic field that should not be there, induced by an addition to Maxwell's Equations from the long-wavelength axion oscillatory component. The goal is to establish a very-well-characterized magnetic field in such a way that, experimentally, one is very certain of the field strength in the measurement region. In our specific example, the experiment is constructed so that (under the traditional Maxwell's Equations) the field strength there is expected to be $B = 0$. One then looks for evidence of $B \neq 0$ in that region. Such an observation, beyond any uncertainty expected from experimental design, would indicate additional contributions to Maxwell's Equations.

ABRACADABRA begins with a toroidal magnetic field generated in a toroid-shaped instrument of inner radius R_1, outer radius R_2, and height h. The magnetic field strength is fixed at some value, B_0. The J_{eff} induced by the axion field will circulate parallel to the circular magnetic field confined within the toroid. The induced current oscillates with time, as indicated in Eqn. 6.25. This oscillating axion current will establish a new dipole

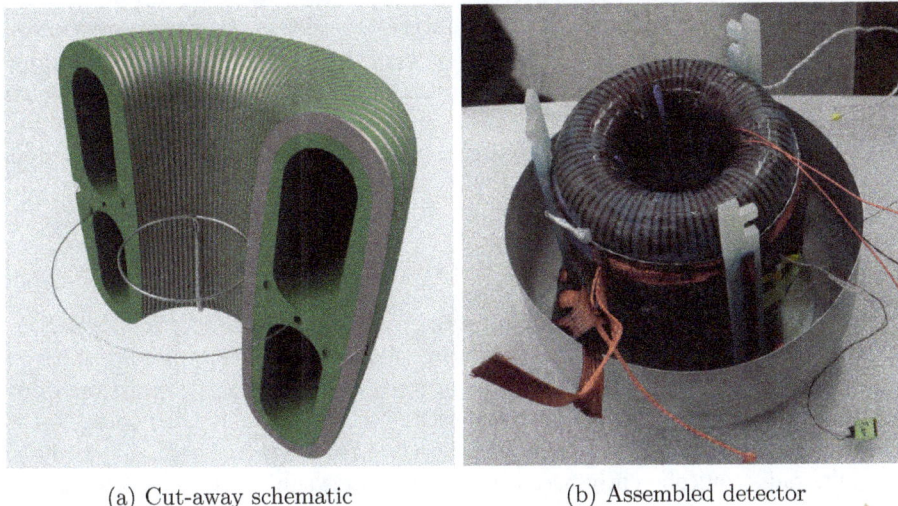

(a) Cut-away schematic (b) Assembled detector

Figure 6.7: The design of the ABRACADABRA detector is shown on the left and the assembled detector is shown on the right. Figures from Ref. [Oue+19].

magnetic field perpendicular to the plane of the toroid (e.g. as if emanating from a point in the center of the toroid hole and perpendicular to that plane). That field also oscillates, changing sign as the induced current changes sign. A schematic of ABRACADABRA and a photo of a real prototype are shown in Fig. 6.7.

The detection region is now the center of that toroid. There should be no dipole field there absent an axion coupling to the electromagnetic field. However, should this coupling exist, a pickup coil (of radius r) placed in the center of the toroid will detect this small, oscillating magnetic dipole field. The signal is expected to be very small in amplitude, so this pickup coil needs to be amplified and read out by sensitive signal electronics. SQUIDs (Superconducting Quantum Interference Devices) are a common choice. The signal is expected to be described by the equation

$$\Phi(t) = g_{a\gamma\gamma} B_{max} \sqrt{2\rho_{DM}} \cos(m_a t) \mathcal{G}_V V, \qquad (6.31)$$

where $\mathcal{G}_V$ is a geometric factor that describes the scaling factor between the effective current generated by the axion and the amount of actual, contributed flux through the pickup loop. For ABRACADABRA, this has been reported at the level of 10^{-2}. Experiments like ABRACADABRA and DM Radio Pathfinder operate on this principle. As a program, this instrumentation is still in its early days. The two demonstration experiments have recently merged collaborations into one larger DM Radio program. Results from ABRACADABRA are shown at the end of this chapter in Fig. 6.10b and indicate that these experiments can provide sensitivity in regions complementary to the cavity experiments.

Unlike the pure cavity-based experiments, instruments like ABRACADABRA are capable of a broader search in axion frequency space. Tuneable cavity experiments have a limited range of resonance frequencies determined by the size of the cavity and some of the internal hardware, so scanning over a range of frequencies requires careful planning and effort. ABRACADABRA, on the other hand, can cover a much wider range of frequencies all at once. The next-phase DM Radio program is expected to further expand this search window. This complementarity will be essential to confirmation of any claimed signal in this search portfolio.

Searching for Axions in the Radiation Regime

In an external field, the axion field (a) sources an axion-induced electric field, (E). This field is inversely proportional to the dielectric constant, and has a discontinuity at the boundary of the dielectric. This is compensated for by EM wave emission from the boundary. This effect leads to an emitted power from the boundary. To boost this effect, experiments employ many dielectric disks in a strong magnetic field. Each dielectric surface emits radiation coherently because each of them is subject to the axion field. When the axion's wavelength is shorter than the experimental apparatus it coherently oscillates inside the instrument. The contributions from the disks then interfere constructively, which provides the boost in signal amplitude.

An example of an experiment using this design is the MAgnetized Disc and Mirror AXion (MADMAX) instrument [Bru+19b] (Fig. 6.8). The MADMAX concept is a closely

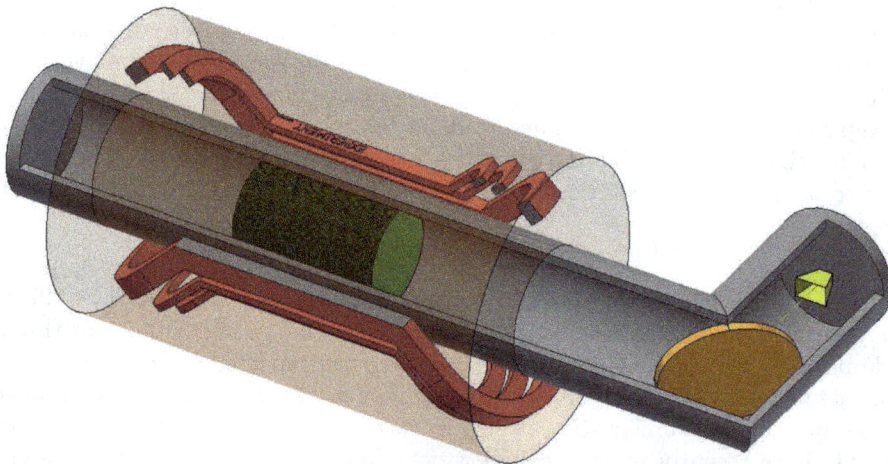

Figure 6.8: A baseline design for the MADMAX Experiment. The major components are the magnet (curving about the tube), a copper mirror (single large disk at corner), and 80 dielectric plates (closely spaced disks). Figure from Ref. [Bru+19b].

spaced series of dielectric discs (each about 1 m^2 in area) immersed in an external magnetic field. An axion field in the system induces a small, time-varying electric field perpendicular to the surfaces of the disks. The large number of disks, with the appropriate spacing, can generate constructive interference in the electromagnetic waves emitted in this process, boosting the signal. A mirror reflects the waves in such a way that they are all directed to a receiver. This system favors short-wavelength axions, and, thus, higher masses (since $m_a = \omega_a = 2\pi/\lambda_a$ in natural units).

This program, still in its early stages, will cover another complementary region of axion mass at large masses (above the cavity experiments). The collaboration has recently reported progress in their research and development and updated their sensitivity projections with this approach [Maj23]. This is illustrated at the end of this chapter in Fig. 6.10b with expected axion mass sensitivities in the range of 10^{-5}—10^{-4} eV.

Alternative Axion Detection Approaches

The axion-photon coupling term in the Lagrangian is not the only way to attempt to detect the axion field. The other terms, such as the neutron EDM term, the nuclear and electron dipole moment terms, offer alternative avenues for this detection. One example of an approach is a nuclear magnetic resonance (NMR) approach. It attempts to detect the effect of the gradient of the axion field, $\vec{\nabla}a$, on this physical process. Nuclei that are immersed in the axion field will develop an oscillating electric dipole moment that is described by the equation

$$d_n = g_d \frac{\sqrt{2\rho_{DM}}}{m_a} \cos(m_a t) \tag{6.32}$$

and/or a spin-dependent force that is described, instead, by

$$H_N \supset f_{aNN} \sqrt{2\rho_{DM}} \cos(m_a t)\vec{v} \cdot \vec{\sigma}_N. \tag{6.33}$$

The experimental approach is to polarize some of the spins of the nuclei and then observe their precession. An external magnetic field $\vec{B}_{ext}$ polarizes the nuclei. An external electric field can then be applied and one can look for evidence of an oscillating EDM. Alternatively, the axion field, with its velocity $\vec{v}$, can couple to the spin (magnetic moment) of the nuclei and induce a precession by acting as a force at an angle to the polarizing magnetic field. A resonance in the transverse magnetization of the nuclei will occur when $2\mu B_{ext} = m_a$.

An example of an experimental suite attempting these approaches is the Cosmic Axion Spin Precession Experiment (CASPEr) [Ayb+21; Bud+14; Gar+17; Gar+19; Jac+20; Wu+19]. This experiment is conducted in two major variations: CASPEr-Electric (to look for the oscillating EDM) and two other projects, CASPEr-Wind and CASPER-ZULF (to look for the induced precession from the axion velocity). In this case, the constraints obtained by such experiments will be on g_{aNN} vs m_a, and no longer involved with the axion-photon coupling. This approach tends to be most sensitive to axion dark matter with mass below the current cavity experiment region (c.f. Fig. 6.10b).

Further classes of experiment can probe the axion hypothesis independent of axion contributions to dark matter. For example, there are experiments designed to probe the

Figure 6.9: The CERN Axion Solar Telescope (CAST) experiment. Figure from Ref. [Wik24b].

axion field by searching for axions produced in the sun via solar x-rays (generated in the solar core) that are first converted (in the sun) to axions. This approach does not rely on a relic axion population passing through the earth. Instead, the axion source is through active production by the sun. This solar axion flux can, of course, pass through the earth and will appear to be directed along the line of sight from the sun to the earth. A sufficiently strong terrestrial magnetic field can convert the axion back into an x-ray, which can then be detected by conventional means.

Such an experiment is known as a *solar helioscope*. A strong magnet bore aimed at the sun and tracking its motion should promote these conversions. One notable experiment that has tried this approach is the CERN Axion Solar Telescope (CAST) shown in Fig. 6.9. The next-generation program is the International AXion Observatory (IAXO) [Arm+19].

A similar approach is to use an intense laser beam and aim it at a wall. Preceding the wall is a strong magnetic field. On the other side of the wall is another strong magnetic field, beyond which lies an optical detector. Axions can be generated in the first magnetic field, traverse the wall, and then convert back to a photon in the second magnetic field. The current generation of such "light shining through a wall" experiments is the Optical Search for QED Vacuum Bifringence, Axions and Photon Regeneration (OSQAR) [Pug+08] and the next generation is Any Light Particle Search (ALPS) [Dia+22].

As mentioned above, stars like our sun can produce axions. This provides an alternative subatomic physics pathway for energy to flow in nuclear reactions, potentially altering the energy evolution of a star or population of stars. For example, the solar luminosity (total energy emitted in photons, per unit area) is very well-known. In the standard solar model, solar photons are produced through well-defined nuclear reactions. However, if axions are also produced through reactions in the sun, that would deduct from the nominal rate of photon production. To compensate, a model would demand a higher rate of nuclear reactions in order to maintain the solar luminosity while also allowing for axion production [And+07]. This, in turn, would produce more neutrinos. Neutrino and optical astronomy can be used to constrain such possibilities. If observations fail to detect significant deviations from the standard solar model, then those can be used to constrain axion parameters such as $g_{a\gamma\gamma}$ and axion-electron couplings. Recent constraints would imply $g_{a\gamma\gamma} \leq 10^{-10}$ [Vin+15].

Constraints on the axion couplings are obtained by studying various populations of stars, such as those in the process of moving off the main branch of the Herzsprung-Russel diagram, as well as red giants and white dwarves. The energy loss in these populations is expected to be affected by axion production, and different populations will produce axions at different rates. This permits the use of relative measurements to constrain axion couplings. Axion-photon couplings tend to be excluded down to the level of 10^{-11} at the current best sensitivities, while axion-electron couplings have been contrained at least down to the level of 10^{-13} [Par22].

6.5 Perspectives on Non-Particle Dark Matter Methodologies

The detection of a long-wavelength phenomenon is inherently challenging, especially if the exact interaction mechanism is not known. Axions, or axion-like particles, offer an interesting class of dark matter candidate. They are motivated originally by the need to explain an observed feature of the neutron (its extremely small electric dipole moment). However, the PQ or QCD axion is not the only framework in which this kind of particle solution can be formulated. Regardless of the exact theoretical details, the nature of the axion allows a coupling to SM particles such as the photon. This provides multiple avenues for detection through modifications to Maxwell's equations.

The recent state of constraints on axion properties are summarized by the illustrations in Fig. 6.10. Axion searches by axion-photon coupling, or alternative means, makes for a very vibrant and open field of investigation having many opportunities for new ideas and approaches in the coming decades. The broader context of axion search efforts is part of the next and final chapter.

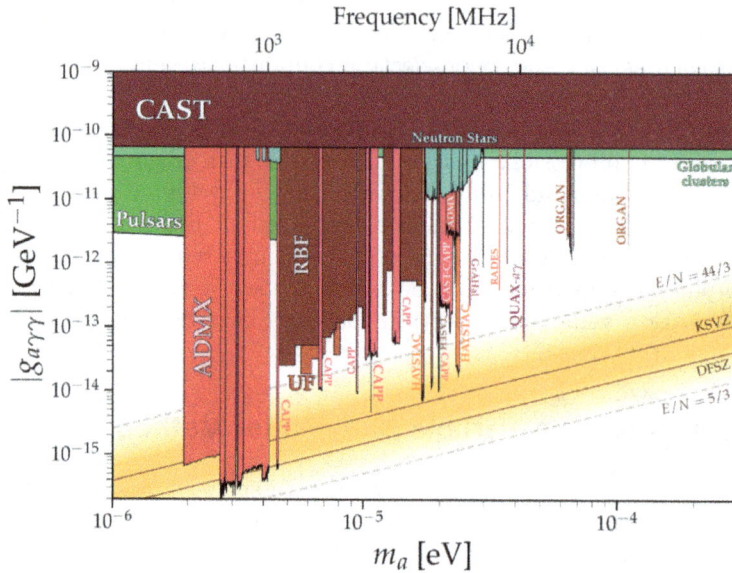

(a) Cavity-based searches

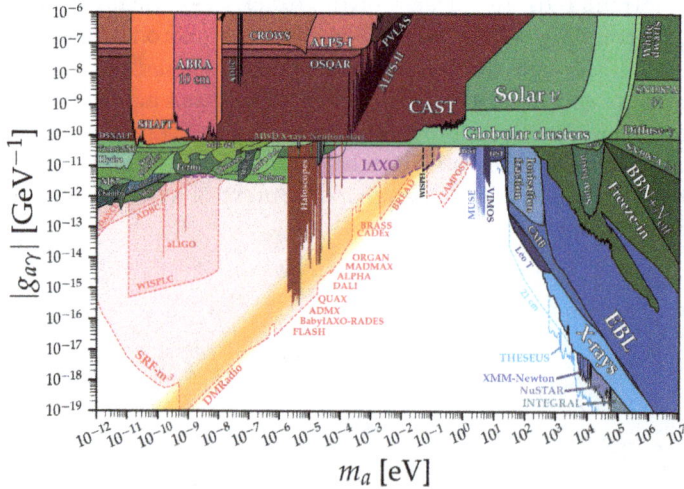

(b) Various axion search techniques

Figure 6.10: Results from a series of cavity-based axion searches (above) in a narrow region m_a and from (below) a broader range of ALP searches across a very wide m_a. The KSVZ and DFSZ diagonal shaded bands indicate regions permitted by the PQ axion (QCD axion) framework. No experiment has observed a definitive signal. Parameter exclusion regions are shown in the shaded regions. Projections from proposed or forthcoming experiments are shown using dashed lines. Figures from Ref. [OHa20; Wor+22].

PART III

UNANSWERED QUESTIONS AND OPEN PROBLEMS

CHAPTER 7

Future Directions

There have been no conclusive detections of dark matter. Given the space available to explore (Fig. 4.1), it is perhaps no surprise that the field of dark matter research has rapidly expanded with new ideas to widen the search and new technology to enable the greater chance of discovery. Until recently, the "nightmare scenario" has been dark matter that interacts with normal matter only via a gravitational interaction. However, new ideas and technology seem to be emerging that make even this possibility a fruitful avenue for you to investigate in laboratory conditions.

In this final chapter we take stock of a few directions of basic needs in several aspects of the search for dark matter's constituents. We catalog a few research areas where there are new directions that could lead you to breakthroughs. These are useful examples, where intrigued dark matter hunters might direct their efforts not only for the purpose of their training and their future contributions to research, but also to accelerate their pace of discovery.

There are a few questions that drive innovation in dark matter search strategies. These include, but are not limited to:

1. What if dark matter has *only* gravitational interactions with the standard model?

2. What is the smallest energy scale we can exploit in a scattering experiment in order to drop the threshold as low as achievable for low-mass dark matter?

3. How low can we make the background environment (through shielding, material selection, and material development) in order to maximize sensitivity to very-weakly-interacting dark matter?

4. What are the next observational or experimental opportunities afforded by theories of axion interactions?

5. How complex might the dark sector be? Is it one particle or many? Are there zero dark interactions, or one, or many?

6. What if dark matter is composed of neither a singular particle nor wave-like phenomenon, but rather of compact-but-large objects that evade existing constraints on such entities?

7.1 Improving the Observation of Galactic and Exogalactic Gamma Rays

We have already noted, in Section 5.4, the excitement about an observed excess of gamma rays around the Milky Way's center, one that is not explained by currently understood and verified astrophysical sources. Understanding this excess is a key target area for the investigation of dark matter. There are many promising avenues that must be explored to improve our galactic center gamma ray observations, as well as those of dwarf satellite galaxies orbiting the Milky Way, and about exogalaxies. There are opportunities, of course, for you to take these measurements further.

One such opportunity will be MeV gamma ray astronomy. There presently exists a gap in instrument sensitivity in that energy region (Fig. 7.1). The science targets for that region span a range of topics.

New measurements using dedicated instruments may produce more precise measurements of the galactic center energy spectrum. Of particular interest is the 511 keV gamma ray spectral line, indicative of electron-positron annihilation and sensitive to

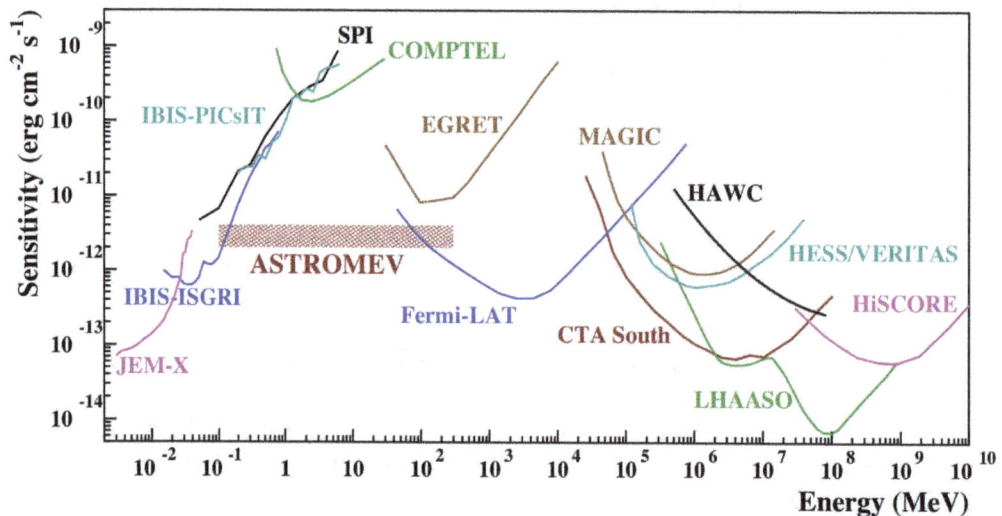

Figure 7.1: The single-photon sensitivity vs. photon energy for a range of existing or future instruments. A noticeable sensitivity "gap" exists in the MeV range. The hatched region in this gap indicates the target sensitivity for new instrumentation, assuming one-year exposure time. Such instrumentation could be provided by projects like COSI, AMEGO, and others mentioned in the text. Figure from Ref. [De +21].

excess sources of astrophysical positrons. A competing hypothesis for the source of such positrons includes the astrodynamics of pulsars; the deviations between the actual spectra of pulsars and the expected spectra from dark matter annihilation are anticipated to be their largest in that energy region.

Additional targets for future observations are nuclear gamma ray spectra that can probe nucleosynthesis of the light elements (such nuclear transitions are at the MeV scale). We emphasized the importance of BBN (Big Bang Nucleosynthesis) measurements on our knowledge of the density parameter of baryons in Section 3.4. Additional detailed information on BBN is critical to completing the picture of early cosmic history. Sensitive observatories for MeV gamma rays will also enhance multi-messenger astronomy, a field that requires the synthesis of traditional optical astronomy with full-spectrum astronomical observations, astroparticle observatories (cosmic ray and neutrino facilities), and gravitational wave observatories. Multi-messenger approaches to observing the same astrophysical phenomena benefit from a more uniform sensitivity across the electromagnetic spectrum.

The reason for the gap has been the physical constraints on new instruments meant to probe such photons at sensitivities comparable to detectors in other parts of the astronomical photon spectrum. In order to perform precise astronomy with MeV gamma rays, one needs an observation platform that provides reliable pointing information (where in the sky did the photon originate?) with a sufficiently high interaction cross-section to induce a gamma ray–instrument interaction.

The main physics channel for this detection is Compton scattering, where an MeV gamma ray ionizes an electron in an atom in the target material. However, Compton scattering comes at a price: an MeV photon will scatter multiple times inside the instrument, its direction changing with each scatter, until it is finally absorbed by the material. It is crucial to understand the pattern of this energy deposition in three dimensions in order to obtain reliable pointing information about the progenitor gamma ray. To achieve this, one wants a highly sensitive medium (one with sufficient granularity to yield 3-D position information) with greater density that is sufficient to induce the interaction. This necessitates heavy or bulky instruments that have thus far been ill-suited to space-launch constraints.

There are a wide range of instruments, however, that are proposed to exactly achieve the goal of filling in the sensitivity hole in Fig. 7.1. (The general region of target sensitivity in that figure is labeled "ASTROMEV" without reference to a specific project.) However, there are many proposed projects including AdEPT [Hun+10], AMEGO/AMEGO-X [De +21; Kie20; Rei+19; Tav+18], COSI [Tom+23], GECCO [Orl+22], GRAMS [Ara+20a; Ara+20b], MAST [DP19], and PANGU [Wu+14; Wu16]. The approximate timelines for some of these concepts was illustrated in Fig. 5.5.

Let us select a couple of examples to highlight the targets and timelines for specific concepts. The Compton Spectrometer and Imager (COSI) is approved by NASA for a target launch in 2027. Its design will yield sensitivity in the range of $0.5 - 2.0$MeV using cryogenically cooled germanium detectors. COSI is designed to provide excellent energy resolution (with uncertainy on the energy satisfying $\sigma_E/E \lesssim 1\%$) and can perform a

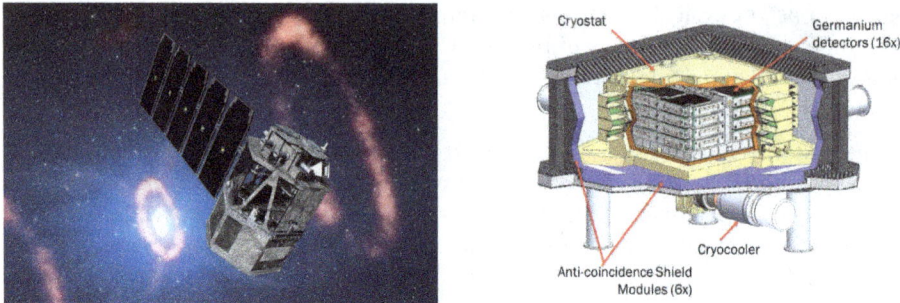

Figure 7.2: An artist's representation of the entire COSI satellite (left); and (right), a cutaway schematic view of the COSI gamma ray telescope instrument showing the 16 germanium detectors and key components such as the cryostat. Images from Ref. [NAS23]

whole-sky observation once per day. An artist's impression of the COSI instrument, along with a schematic showing the interior components of COSI, are shown in Fig. 7.2.

The All-sky Medium Energy Gamma-Ray Observatory eXplorer (AMEGO-X) is aimed for a future NASA mission portfolio. It would provide sensitivity in the range of 100 keV to 1 GeV including the ability to narrow down short MeV gamma-ray bursts or transients to within two degrees in the sky. Both COSI and AMEGO-X expect to cover a wide range of science pursuits in this area of astronomy.

Verification of a 511 keV signal from another gamma ray source (i.e., dwarf satellite galaxies) would be a major breakthrough confirming that this effect is not unique to the Milky Way. Whether confirmed in other galactic or exogalactic bodies, interpretation of the excess will be complicated by many of the same issues that make the Milky Way challenging to interpret—confounding the effects from stellar astrophysics and a potential dark matter signature. It is essential to push current instrumentation, or to develop new instrumentation that increases the sensitivity to these sources, especially the dwarf galaxies.

We want to stress that new observations alone will not be enough to make progress in this area. A key limiting factor we discussed earlier (c.f. section 2.7) is the dark matter halo model for a galaxy. This affects nearly every aspect of direct and indirect (annihilation-based) dark matter detection. For example, if a signal is observed in a terrestrial direct-detection experiment, the interpretation of that signal (constraints on dark matter interaction cross-sections) necessitates stronger constraints on the halo model (which determine local density in the galaxy). We gave an example of the importance of the halo model on interpretation of scattering in experiments in Section 4.7.

Another example is the interpretation of gamma ray excesses as evidence of dark matter annihilation. A clear and absolute excess above established astrophysical causes has been identified through gamma ray astronomy. Constraints on the dark matter interaction cross-section (i.e., with itself) will require improved knowledge of the density profile of dark matter throughout the region where the excess is observed.

We cannot neglect the need to improve the input to dark matter astrophysical models, as these play essential roles in disentangling contributions to astronomical observations and are essential for building confidence about excesses or unanticipated components that might arise from a true dark matter signature.

Better galactic diffusion models and a better understanding of cosmic ray propagation would help to reduce uncertainties in the measurement and interpretation of the data. We foresee that a combined approach using weak gravitational lensing of galactic neighbors (e.g., dwarf satellite galaxies and those in our local group) along with other astronomical observational capabilities such as massive stellar survey programs like Gaia [Pru+16] and current- and next-generation cosmic ray and neutrino telescopes will be essential to disentangling galactic structure and the properties of dark matter from a pattern of confirmed excesses across multiple instruments.

7.2 Ultra-Light Atomic Targets

The lowest-mass-particle dark matter candidates are most efficiently scattered by nuclei whose mass is comparable to that of the dark matter. This necessitates development of a new generation of low-mass dark matter targets. There are two major approaches in this category: the lightness of the nuclei itself and the tendency for the emergence of collective quantum states in ultra-cold, light-element volumes. Two examples of this approach are the use of gases such as hydrogen, helium, or neon (with atomic masses of 1, 2, and 10, respectively) or the use of light materials like helium that are known to enter superfluid states.

An example of gaseous detectors is the *New Experiments with Spheres-Gas* (NEWS-G) program. NEWS-G is a functioning concept that is planning a large-scale experiment. The program is a collaboration between Canada and France based on the idea of a spherical proportional counter (SPC). The physics is illustrated in Fig. 7.3. A sphere is engineered in which the inner volume is filled with a gas at high pressure (to increase the target density) subjected to an electric field. The field is not sufficient to ionize the gas on its own. However, when an interaction occurs in the gas, with a sufficiently high energy to ionize one of the gas atoms, the electron(s) liberated in the interaction are accelerated by the field, causing a cascade of ionization. The field draws the resulting electrons to an electrode on which the current can be read. The gas target is selected for its low mass, as well as the linearity of its ionization response, such that the current at the electrode is proportional to the original energy deposit.

The choice of gas, gas pressure, voltage, and geometry (i.e., diameter) of the sphere results in a threshold for detection in the experiment. The NEWS-G program began in 2015 at the Modane Underground Laboratory in France. The first instrument was named SEDINE and was a prototype 60 cm SPC made from ultra-pure electroformed copper. The gas target in the prototype was a mixture of neon and methane (CH_4), with the methane providing a less-flammable target than hydrogen gas alone. The mixture provides a mass range coverage for GeV-scale dark matter particle candidates (low-mass WIMPs) spanning the 1-10 GeV range between hydrogen and neon.

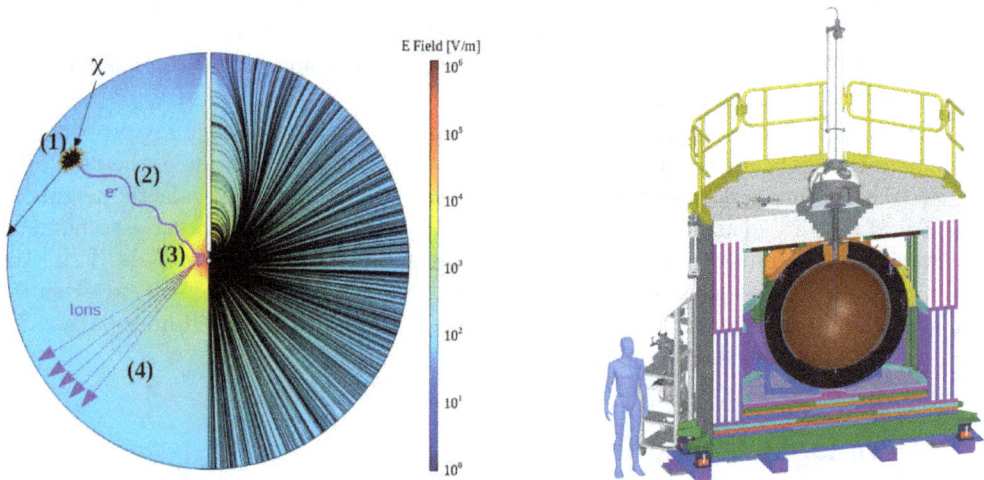

Figure 7.3: (left) The concept of a spherical proportional counter used for dark matter nuclear interaction detection. An interaction (1) results in the liberation of an electron (2), which is then drifted through the gas by an electric potential difference (mapped on the right-hand side of the image) that ultimately results in more ionization that amplifies the original signal (3). Electrons are collected in the central electrode as ions drift to the walls (4). (right) A schematic of the NEWS-G experiment at SNOLAB. In the left figure, the stages of the detection process are labeled 1–4.

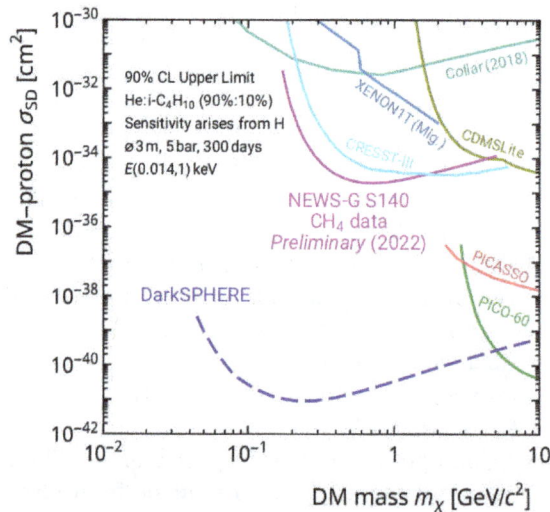

Figure 7.4: The projected sensitivity of DarkSPHERE to spin-dependent dark matter coupling, assuming scattering off the proton in hydrogen. A comparison with NEWS-G at SNOLAB, and other experiments, is also shown. Figure from Ref. [Bal+23].

Since the SEDINE prototype, the 1.4 m NEWS-G experiment has been constructed and operates at SNOLAB. The program is ongoing. The next-generation plan for the NEWS-G concept is a much larger electroformed copper sphere built entirely underground. This will reduce the presence of ^{60}Co that is produced when copper nuclei are exposed to cosmic rays, thereby reducing the risk of spontaneous beta and gamma rays when unstable cobalt nuclei transition to the stable ^{60}Ni state. These electromagnetic radiations will ionize the gas very close to the inner copper surface, leading to a background that originates from the sphere itself. It is essentially impossible to shield the gas from this effect. The half-life of ^{60}Co is over five years, so it's not possible, on project timescales, to wait for this background to decay. Underground copper electroforming promises to greatly reduce this problem.

A proposed 3 m-diameter sphere would be the basis of a project called Dark-SPHERE [Bal+23]. It is planned for construction and operation at the Boulby Underground Laboratory near Whitby in the United Kingdom. Projections of the performance of DarkSPHERE, as compared to NEWS-G and other experiments, in one of the considered scenarios[1], is shown in Fig. 7.4.

The NEWS-G program is, in many ways, a traditional dark matter search scaled to a gas rather than to a solid or a superheated liquid. This scaling required overcoming the conceptual challenge of a low-density target that is not easily kept in liquid or solid form. An additional frontier in the search for low-mass dark matter is being opened by a superfluid program. The benefit of a superfluid is that these states contain bound states, the binding energy for which is extremely small; even a small energy deposit can break the superfluid state. If this can be detected, then one has the makings of an ultrasensitive dark matter experiment. Examples of a demonstration technology that is aimed for a full-scale dark matter experiment are the SPICE and HeRALD concepts [Ant+23]. These are part of a much larger and longer-term program called Transition Edge Sensors with Sub-EV Resolution And Cryogenic Targets (TESSERACT) [Bil+24; TES22].

The Helium Roton Apparatus for Light Dark matter (HeRALD) takes advantage of several key properties of helium-4 in its superfluid state. First, helium can be made extremely radiopure at low temperature. Second, the first atomic-excited state of helium is so high in energy (about 19 eV) that a key Compton scattering background is essentially impossible in this target. Third, helium in its superfluid state forms collective quantum states including phonons and *rotons* [Gly95], whose meV-scale energies provide an opportunity for low-mass dark matter detection [KLZ17; SZ16]. When these quasiparticle states propagate to the helium superfluid surface, they can induce evaporation. A calorimeter design is used to capture the energy liberated by this process. The proposed calorimeter design is familiar: transition-edge sensor (TES) technology, an advanced and greatly tested approach to detecting, with precision, small energy deposits (c.f. Section 4.7). The breaking of quasiparticle states is accompanied in the helium target by scintillation.

---

[1]Thanks to the gas mixture used in DarkSPHERE, there is the ability to study spin-independent scenarios as well as spin-dependent dark matter scattering off of protons (in hydrogen) and neutrons (in carbon-13 in the methane component). The project also plans to study dark matter scattering off of electrons in the gas.

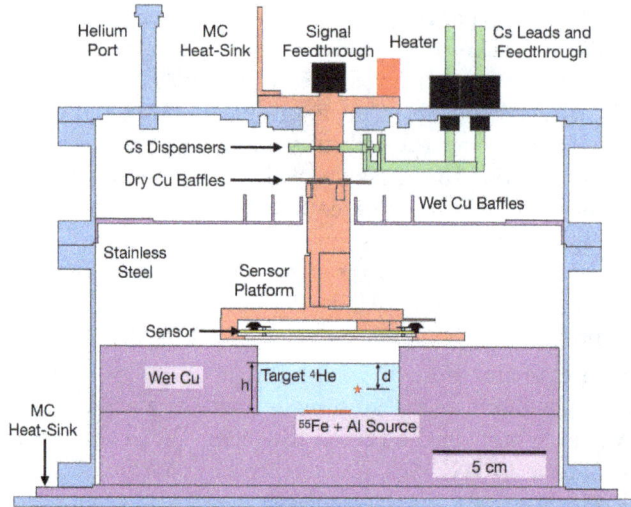

Figure 7.5: Annotated cross-sectional schematic of the HeRALD v0.1 detector. The color coding indicates regions that will be wetted by helium-4, the cesium dispenser, and regions that will remain dry of helium thanks to the cesium barrier. Figure from Ref. [Ant+23].

This should sound familiar as a similar pair of signatures is utilized by the CRESST Collaboration for higher-mass WIMP searches in solid-state detectors (Section 4.7).

There are, of course, several challenges in working with superfluid helium-4. First is its cryogenic stability and the control needed to maintain the target liquid in the superfluid state, which occurs below a temperature of 2.17 K at saturated vapor pressure. This can be achieved with a modern dilution refrigerator. Superfluid helium has no viscosity, a property that leads it to effortlessly climb the walls of its container and escape. Worse than its escape, if left unchecked, the superfluid helium-4 would wet every inner surface of the experiment, including the TESs. This would result in the complete malfunction of the calorimeter. The film layer adds heat capacity to the system, provides additional modes for energy to leave the system, and disrupts a key physical mechanism that enhances the signal. At the calorimeter surface, a van der Waals potential between the helium-4 and the calorimeter allows for adsorption of the helium; this is disrupted when a film is present.

Maintaining the helium target reservoir and preventing the wetting of the TESs becomes the main challenge. HeRALD solves this problem by vapor-depositing cesium (its stable isotope) onto a region above the reservoir and around the surfaces that lead to the calorimeter. Cesium has the property of being resistant to helium wetting. This acts as a seal to prevent superfluid helium from escaping the target volume and from wetting the calorimeter. A schematic of the HeRALD v0.1 prototype is shown in Fig. 7.5.

At the time of this writing, the superfluid helium-4 detector concept is still in its early stages. Sensitivity estimates have been determined for TESSERACT, which, in

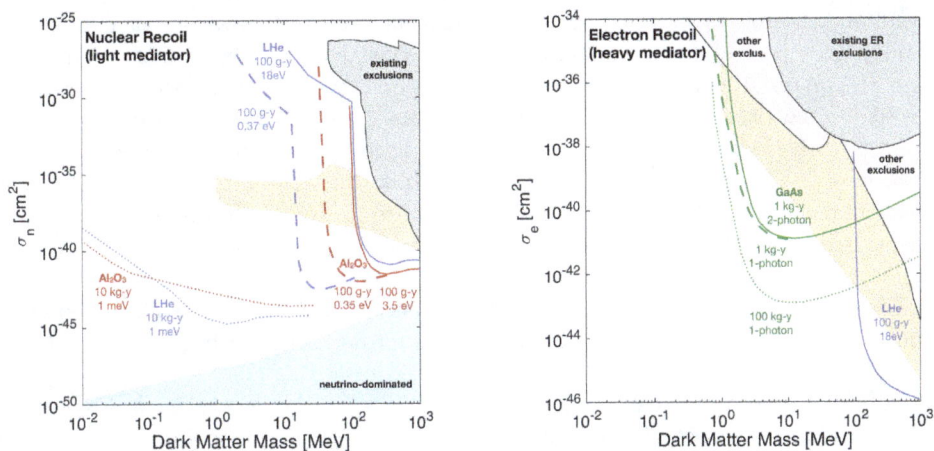

Figure 7.6: The projected sensitivity of the TESSERACT concept for 100 g and 10 kg liquid helium designs and assuming one year of live time. Projections are made for (left) nuclear recoils assuming a lower-mass dark matter mediator and (right) for electronic recoils assuming a heavier-mass mediator. Polar crystal detector projections are also shown for sapphire (Al_2O_3). Figure from Ref. [TES22].

part, imagines scaling the HeRALD prototype up to 100 g and 1 kg liquid helium targets. This target mass scale (compared to much larger superheated, liquid noble, or solid-state detector masses) illustrates the advantage of targets that are in collective quantum states that allow for a smaller mass to nevertheless deliver a very high sensitivity to an interaction. The TESSERACT concept aims to achieve sensitivity to light WIMPs in the mass range of 10 keV–1 GeV (Fig. 7.6).

A detector whose active material is dominated by its quantum behavior creates a system host to extremely weak bonds that, when disrupted, result in a change of the state of the material. If this change is detectable, it becomes the basis for an ultra-sensitive dark matter detector. This theme recurs in the next few sections in which are discussed various scales of plausibility and methodology.

7.3 New Technologies for Low-Threshold Detection

There are many ideas that have recently gained traction as plausible future directions for dark matter detection [Ess+22; KL22]. These especially focus on targets where only a small energy deposit (typically < 1 eV) is sufficient to generate a detectable signal. Such technology is needed to extend the reach of low-mass dark matter detectors. We've explored some of these, including the use of the Migdal effect in Section 4.7. We note that recent efforts have explored a *molecular Migdal effect* applied to systems of atoms and not just single atoms, an approach that might allow even lower energy thresholds for scattering experiments [Bla+22].

Another example is *nanowires*. These ultra-narrow configurations of material can be made superconducting at low temperature. When struck by a particle such as dark matter, the material will transition to be normal-conducting in that specific region. This technology allows the detection mechanism to the direct breaking of Cooper pairs, as compared to the indirect use we explored when discussing thermometry or calorimetry in Section 4.7. This nanowire-based approach shows promise for a future dark matter experimental program [Hoc+19]. Nanowires find use in single-photon detection systems and have been explored as means to detect dark matter or dark photons [Bli+24].

Another direction for new technologies could come from anisotropic organic materials [Bla+21]. These have very low excitation energies, at the eV level, and they can scintillate. This provides at least two mechanisms for depositing energy and thus offers a chance for discriminating signal and background processes.

All of these ideas—and many others—show promise for the development of small-scale (gram or kilogram) detectors that could provide proof-of-principle demonstrations and possibly even new, independent dark matter constraints. We encourage further work in all of these areas, and exploration of new ones. These will require the interaction of atomic, molecular, condensed matter, and particle physicists, as well as organic and inorganic chemists. No single discipline will be able to assess materials, develop their potential, and design a functioning experiment. The goal of these programs, in their first stage, needs to be the assessment of backgrounds and confounding effects from within the technology, as well as any problems that might affect the scaling of the experimental design.

7.4 Directional Detection of Dark Matter Interactions

We mentioned the potential importance of dark matter detection with directional capability in Section 4.3. Inasmuch as it is a more future-oriented technology, we describe it in more detail here.

Detecting both the position and movement of a dark matter particle is referred to as *tracking*. The annual modulation effect described in Section 4.3 was put forth as an overall time-dependent effect on the rate of scattering of dark matter by a detector. This was induced by the change in relative velocity, annually, between Earth and the Milky Way's dark matter halo. However, it will ultimately be necessary to exclude a background effect that better explains the modulation than does the dark matter hypothesis. A key way to do this is to demonstrate that the interactions are dominant along the line of motion of Earth through the dark matter halo. Data should show a lack of a significant modulation for interactions perpendicular to that line. Directional interaction detection would play a key role here.

There have been demonstrations of directional detection capability in the past (c.f. Refs. [Bat+17; SC09; Yak+20] and a summary in Ref. [SM09]) and there are current proposals for such detectors to be built in the coming decades (c.f. Refs [Bar+20a; Vah+20]).

These often rely on a gas medium that can be observed to look for things like the *head-to-tail* effect [Yak+20], where a recoiling ion will tend to lose most of its energy late in its flight path after being struck by a dark matter particle. The low amount of early

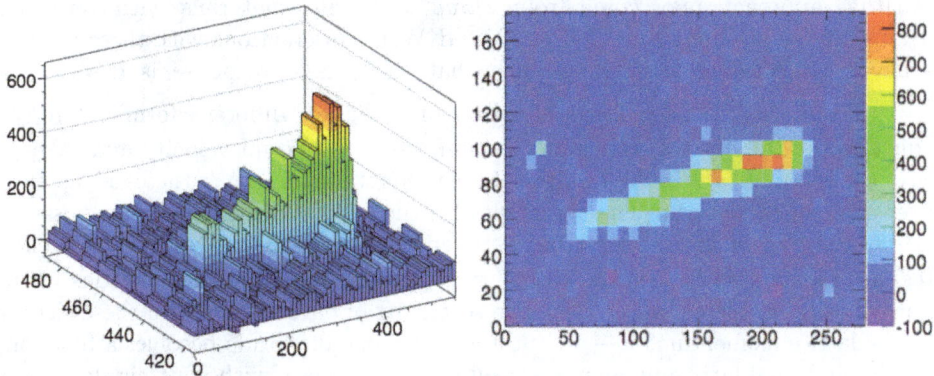

Figure 7.7: The head-to-tail effect observed by the DMTPC collaboration. Neutrons entered the gas medium from the left, and the recoil of a struck nucleus (and subsequent energy loss in the gaseous medium) is clearly visible. The path of the recoiling ion is not perfectly aligned with the incident beam, but indicates the correct flight direction. Figure from Ref. [SC09].

ionization, and the high amount of late ionization, provides an indicator of the velocity direction of the original dark matter particle (Fig. 7.7).

Gaseous directional detectors have the same issue that solid-state, superheated liquid, and liquid cryogenic detectors had before—scaling. In order to increase sensitivity to weaker and weaker dark matter interaction cross-sections, the strategy has been to make the target larger in volume. Since the currency of dark matter sensitivity is *exposure*, or active mass (M) times live time (T), one has either to run experiments longer, or put more mass in the path of dark matter. As experiments already require years to run, it has been more practical to increase the active volume of detectors. The challenge then becomes the cost of scaling any technology, including identifying (or creating) the space to house the much larger instrument. Below, we explore avenues that might facilitate tracking (or, at least, directional sensitivity) in alternative ways.

Gaseous detectors, as described above, rely on an interaction pattern that reveals the trajectory of the initiating particle. Alternatively, one can try to identify materials whose response to an interaction is enhanced in one direction over others. One such interesting approach involves *polar materials* [Cos+22]. A material like gallium arsenide or sapphire has a crystal structure that allows a large fraction of the dark matter scattering energy to be put into optical phonons (those related to out-of-phase oscillations of atoms in the crystal, as opposed to the in-phase acoustic phonons). In addition, the response of the material to scattering is maximized when the crystal's primary axis is oriented at 90° to the dark matter wind's direction (and, thus, would suppress scattering if oriented otherwise, allowing for in situ tests of a modulation of the scattering effect) [Gri+18]. Polar crystals are one of the technologies included in the TESSARACT proposal, discussed earlier.

Another approach may come from *Dirac materials*, materials with very small bandgaps (at the level of 10 meV) [Cos+21; GKW20]. Interactions with these materials also have a directional anisotropy, such that the energy response is described by $E_k^{\pm} = \pm\sqrt{v_F^2\vec{k}\cdot\vec{k} + \Delta^2}$, the Dirac equation in a slightly different form. Here, $\vec{k}$ is the direction of the crystal lattice momentum, v_F is the Fermi velocity and takes the place of the speed of light in the traditional Dirac equation, and 2Δ is the bandgap energy and replaces the mass term. The Fermi speed is the speed at which electrons possessing the Fermi energy (the energy difference between the highest and lowest occupied energy states) are moving, while $\vec{k}$ is the momentum (wave number) of the electrons in their stationary states in the crystal lattice. The scattering of dark matter with these electrons alters the lattice momentum, $\vec{k} \to \vec{k}'$, and the scattering amplitude becomes a function of the initial and final lattice momenta as well as the bandgap, with a strong dependence on the direction of the lattice vector and its angle to the momentum of the dark matter. This facilitates even the measurement of a *daily modulation* of dark matter interactions as the material changes its relationship to the dark matter wind direction over each rotation of the earth.

The future of directional detection may lie in such strong anisotropic responses of materials to the direction of the initial interacting particle. These avenues offer a great deal of promise especially in low-mass dark matter detection with directional enhancement.

7.5   Weak Chemical Bonds as Dark Matter Detectors

Chemical and molecular bonds offer another avenue for exploration in the "weak bonds equal sensitive detection" concept [ADV18; Ess+17; Ess+19]. There are many systems one can consider, but one of interest about a decade ago, recently resurfaced, is the use of deoxyribonucleic acid, or DNA, as a target medium for dark matter detection [Dru+12; OHa+22].

Earlier in this chapter we explored technologies with promise for low-mass dark matter detection, including directional detection. Each of them has an obvious set of physical features that, if exploited through a program of research and development, is likely to provide a new means to search for small energy deposits. At this time, few of those technologies has entered into a prototype detector stage (polar crystals being an exception, since they have been used in the SPICE project and are part of TESSARACT). Each technology area is a distinct place where a new generation of dark matter seekers can forge a career.

In this section, our goal is to highlight a less-mainstream concept that is radically interdisciplinary, combining biotechnology approaches with chemistry and physics to conceive of a novel device. As with all ideas, the use of DNA as a detection medium has the potential to fail. We expand on it here because it is so interesting ... and unconventional.

DNA is the encoding mechanism for genetic inheritance in living organisms on Earth. It consists of a double-helix framework, a kind of "twisted ladder" of repeating molecules. The rungs of the ladder are formed by *base pairs*—bonded pairs made from four known

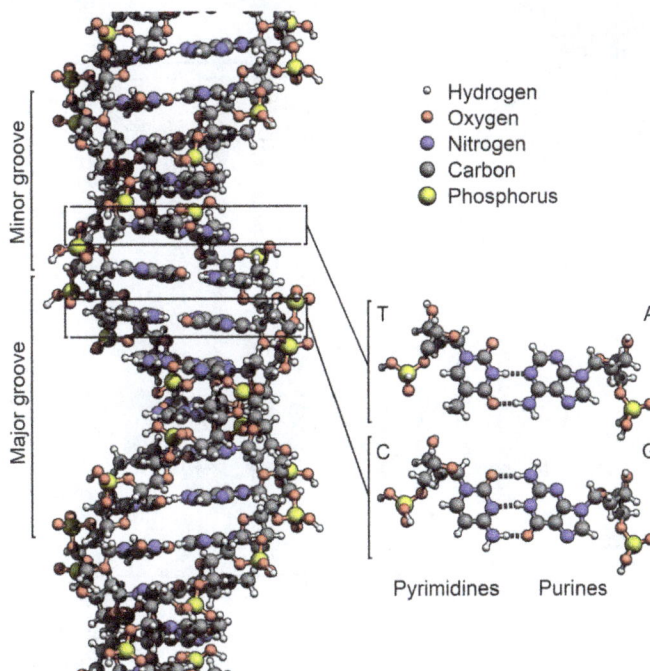

Figure 7.8: A graphic representing a small part of a DNA double-helix structure. The base pair components are indicated by the letters A, C, T, G, and U. Figure from Ref. [Wik23c].

molecules. These are the nucleic acids adenine (A), cytosine (C), guanine (G), and thiamine (T)[2]. They connect a backbone-like chemical structure that forms the long axis of the double-helix. DNA is illustrated in Fig. 7.8.

The weak hydrogen bonds between the base pairs form an attractive target medium for dark matter detection. The other attractive feature of this medium is that it does not require cryogenic operation to obtain sensitivity to feeble interactions that might break the nucleic acid bonds. In the original proposal, [Dru+12], the concept was advanced as a detector medium with sub-keV recoil energy thresholds and tracking capability. Tracking is provided by the sequence of broken base pairs, allowing (in principle) a nanometer-scale directional interaction capability.

This technology would benefit from the decades of breakthroughs in DNA sequencing that can not only establish a desired order of base pairs at a specific location in a DNA strand, but also detect individual broken base pairs along a DNA strand. This is a case where human medical interests and biological curiosity created a precise technology that can be repurposed into physics.

---

[2]Ribonucleic acid is formed during transcription, the process in which DNA is utilized in the formation of amino acids. During this process, an additional chemical, *uracil* (U), is substituted for thiamine. We mention this only because detector designs using either DNA or RNA have been conceived, and uracil's properties would play a role in the bonds of an instrument based in RNA.

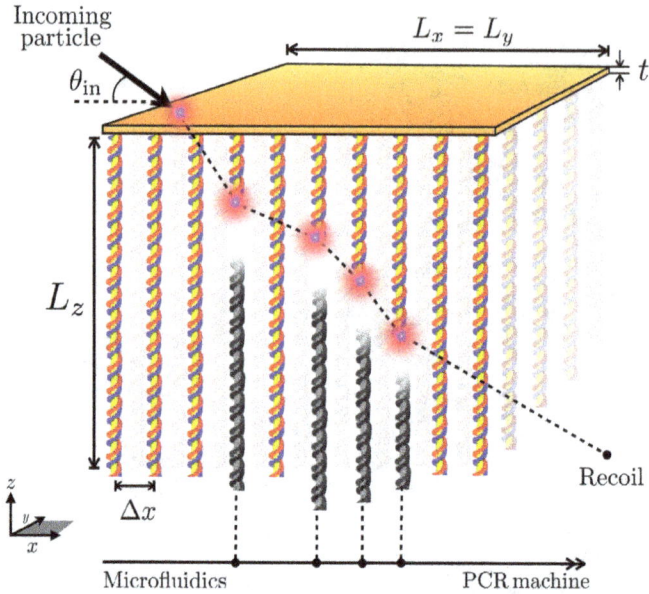

Figure 7.9: A baseline nucleic acid tracking detector concept. A gold sheet provides the target medium, but is generally passive. A gold nucleus struck by a dark matter particle then breaks hydrogen bonds in nucleic acid pairs. These pairs are suspended in strands from the gold sheet and form one of the tracking detector axes. Since hydrogen bonds are easy to break, a single nucleus will scatter with small deflections from multiple strands. This is the basis of the tracking mechanism. Figure from Ref. [OHa+22].

The baseline concepts of a *nucleic acid* (NA) tracking chamber center around the use of gold as the interaction medium. The detector is divided into small units. Each unit consists mainly of a gold sheet ($\mathcal{O}[10]$nm thick) with strands of nucleic acid pairs suspended below the sheet in a regular arrangement. The atomic number of gold is 79, larger than that of other more typical target atoms (xenon is 54, argon is 18, and germanium is 32). The gold is merely a means to liberate a nucleus which can then enter the active region of the detector, but the large mass would yield highest sensitivity to heavier dark matter candidates (albeit with potentially very small interaction cross-sections). The recoil of a gold nucleus struck by a 100 GeV WIMP will yield about 10 keV of energy (Eqn. 4.49). Estimates suggest that the NA strands can be broken by putting in as little as 10 eV of energy. This means that a single gold nucleus will break a large number of strands, approximately proportional to the ratio of kinetic energy to bond energy. This concept is illustrated in Fig. 7.9.

There are clearly a number of challenges in this design. The NA strands need to be chemically bonded to the gold and the bonding points need to be regularly spaced with control. For a nanometer-scale tracking detector, the spacing needs to be $\mathcal{O}(10)$ nm. Microfabrication of such regular chemical anchor points is straight-forward using modern techniques, so long as the chemistry of the bond method is well established. The NA strands are only tens or hundreds of nanometers long. Groups of strands, or even individual strands, need to be labeled by a unique sequence of base pairs at their base so that the location of a broken strand can be determined by processing the fragments using polymerase chain reaction (PCR) methods that are standard in modern biology. The chamber likely cannot be processed continuously, but periodically enough to maintain the precision of the tracking reconstruction. It is essential that all of this be done in an environment as free from ambient radiation as possible, so shielding will be crucial to the success of such an instrument (after the fundamental technical challenges are first solved).

We note that the original detector concept was developed in the era where high-mass WIMP searches were of most interest, however the field has since expanded to attempt to extend to lower and lower WIMP masses. Substituting a lower-atomic-mass sheet that suspends the NA strands could provide a target with more sensitivity to low-mass WIMPs while preserving the general concept of a low-threshold tracking system. One might even conceive of filling the voids between the NA strands not with only a chemically inert gas (like nitrogen), but with a chemically inert gas that supplies low-mass nuclear targets to enhance sensitivity to low-mass dark matter (akin to the strategy of the NEWS-G program). This may require a revised conception of the tracking assumptions (since the initial interaction could occur anywhere inside the 3-D volume; not just on the 2-D gold sheet). However, choosing the right materials (e.g., those based on ease of handling, fabrication, radiopurity, chemistry, etc.) will be crucial for such a detector design.

More recent work on the subject has focused on simulation of the aspects of this detector concept. While there is not currently a demonstration prototype available, the basic concept is surely worth exploration. It is part of a general approach to enhancing sensitivity to very weak interaction cross-sections at both low and high WIMP masses: utilizing systems that contain weak molecular (or other) bonds where it is possible to read out the system in a way to determine energy deposition, interaction position, or both. Such an instrument as that described in this section is highly interdisciplinary and will benefit from a diversity of teams from engineering, physics, biology, and chemistry.

7.6 Quantum Computers as Dark Matter Detectors

The Cooper pair is an example of a very weakly bound system that has been put to significant use in the search for dark matter (Section 4.7). This state of matter, possible only because of quantum mechanics, is the foundation of the calorimeter design based on TESs. Weakly bound systems that are the direct result of the quantum mechanical properties of nature are finding increasing use for the development of ultra-sensitive detection systems (*quantum sensing*). The rotons that form in superfluid helium-3 are, similarly, a collective quantum state that, when broken by an interaction, leads to a detectable consequence, boiling superfluid helium.

One popular bound system is becoming a new tool for exploring dark matter detection capability. That system is one created by the coherent quantum entanglement of elements within the system, a state capable of storing information and being utilized for computational applications: the *qubit*.

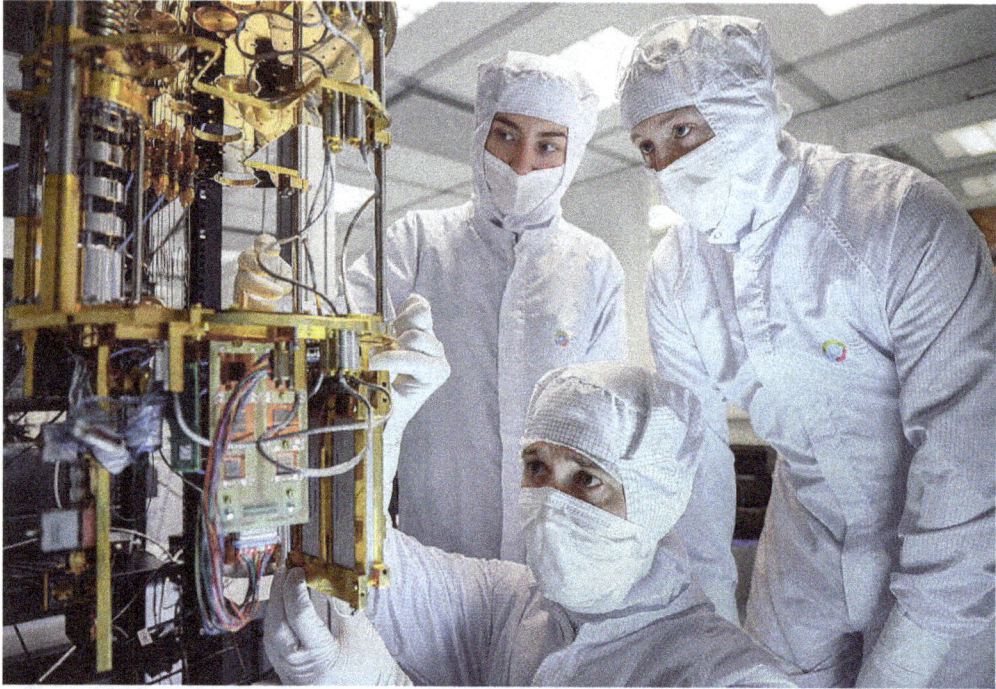

Figure 7.10: A quantum computer in 2018. The main elements here are common to modern qubit systems, especially the need to use a helium dilution refrigerator to cool the system so that it can achieve superconducting conditions (and reduce thermal noise that would otherwise upset quantum coherence). Image from Ref. [Wik23d].

Qubits are the core element of *quantum computers*. They are, quite simply, any physical system or element that is capable of storing the number 1, the number 0, or a *superposition* of these. A classical bit can only store a 1 or a 0. Qubits are subject to disruption due to the fragility of their composition. While this disruptibility makes for an unreliable quantum computer, it has the potential for a powerful dark matter search experiment.

A key performance feature of any qubit system is its *coherence time*, how long it can be practicably maintained in a coherent quantum state. Anything that upsets coherence ends the quantum state, erasing the qubit in the process. Any calculation that is in progress at the time of upset must then, in principle, be restarted from the last known stored state of the computation. At present, quantum coherence is a leading challenge to the creation of reliable and scalable quantum computers. A modern quantum computer is shown in Fig. 7.10.

As in a classical computer, it is possible to detect the loss of coherence of the quantum state. One example of a popular modern physical system for creating qubits is a *transmon* [Koc+07]. This is short for *transmission line shunted plasma oscillation*, referring to a physical system in which two superconductors are placed in close proximity to each other, separated by an insulating barrier (a *Josephson junction*). The superconductors are maintained using low temperatures. Their superconductivity arises by virtue of the formation of Cooper pairs (weakly bound electron pairs whose wavelengths are comparable under these conditions to the size of the superconductor volume). The arrangement leads to a splitting of energy levels for the Cooper pairs in the system. The lowest energy level (ground state) is identified as representing the $|0\rangle$ state, while the first excited state is used to represent $|1\rangle$. Qubit systems like this require cooling at the same level as is required by cryogenic dark matter search experiments. This means that the tools of this industry are nearly identical to the tools long used by the dark matter community.

In fact, recent work [Car+23; Har+24; McE+22; POW23] suggests that a key cause of transmon qubit upset may be phonons generated in the materials around the qubit system. These interact with the qubits and break the Cooper pairs. The tools used to explore this, including TESs, are a staple of the cryogenic solid-state dark matter search community. Identifying mechanisms like this can lead to solutions for the quantum computing community. As in dark matter searches, sources of these phonons are cosmic ray interactions (that generate a bath of phonons that propagate across the multi-qubit system, decohering them in non-random ways that are extremely challenging to correct) or even chemical or radioisotope effects in surrounding materials. Yet again, shielding and material selection appear to be key ingredients in improving quantum computers.

There is another area of exploration that has arisen, in part, out of the observations of excesses of events at the lowest energies accessible to some dark matter detectors. These excesses of events will ultimately limit the low-mass sensitivity of dark matter searches if their origin is not understood—and they have so far been inconsistent, arising solely from a hypothetical dark matter particle.

Material stress appears to be at least part of the explanation. The stress is released in the form of emissions such as phonons. These can, in turn, pollute the detection capabilities of the experiments. There is no shielding against this. Radiopurity is not the solution. Instead, this is about material design and engineering, requiring an unprecedented attention to detail in the structure, arrangement, and interlinking of materials in detector design (c.f. Ref [Ant+22]).

There are at least two kinds of particles against which no amount of shielding will protect qubits: neutrinos and weakly interaction dark matter candidates. Ultimately, what will limit quantum computers' ability to perform long and complex calculations will be their level of sensitivity to both neutrino and dark matter interactions. If they prove highly sensitive to even these weak effects, we will have identified the irreducible limiting factor that ultimately constrains the performance of any quantum computer [Hal23].

The use of qubits, and their decoherence, as an active approach to dark matter detection is an area of rapid and active development (c.f., Refs. [Che+23; DKL24; Ito+24; Mor+24]). This is an excellent area in which to make contributions. It will require the collaboration of atomic and condensed matter physicists, chemists, engineers (including electrical, mechanical, and computer science experts), particle physicists, and astrophysicists.

7.7   The Age of Axions

Having earlier covered the major types of experiment that hunt for the axionic dark matter candidate, we want to use this section to emphasize that the age of axion experiments is now upon us. In earlier decades, there were only a handful of instruments purposely designed to hunt for axion signatures. In the first decade of the 2000s, the key approaches were helioscopes and cavities. However, there is recent development into the quasistatic field and radiation regimes (Sections 6.4 6.4), as well as the use of sensitive instrumentation (e.g. SQUIDs) to detect small signals in the NMR regime (Section 6.4).

This terrestrial explosion of instrumentation is shown in the map in Fig. 7.11. The number of axion experiments now rivals, or even outnumbers, the dedicated WIMP-style search experiments. Indeed, this era feels like it is the golden age of axion search programs.

We also will use this section to indicate and describe a number of novel ideas for alternative methods of detecting the presence of axions in the cosmos. One of these, in the context of astronomy and astrophysics, is known as *axion gegenschein*.

The reflection of light by axions as a means to astronomically detect them has been advanced by several authors [AS19; BFS22; GSM20; Sun+22; Sun+24]. The basic principle is rooted in a theoretical observation. A large body of axions, such as an over-density of them in the galactic halo, can serve as a perfect back-scattering mechanism for ordinary light. The fundamental theoretical idea is stimulated emission of light from an axion. This process is allowed in any basic theory of the axion and is a higher-order effect in the theory.

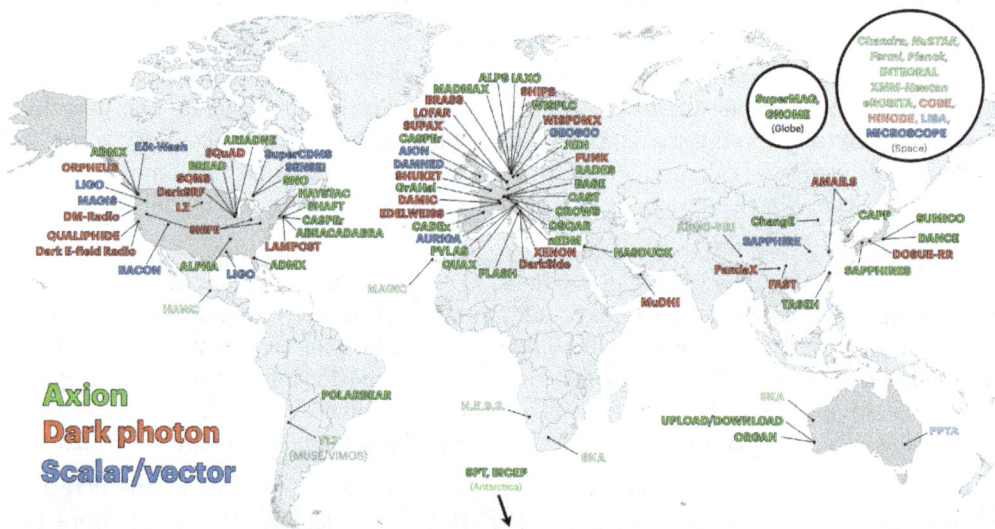

Figure 7.11: A map of dark matter and axion search experiments. Axion experiments, like ABRACADABRA or ADMX, are indicated by fainter text. Figure from Ref. [OHa20].

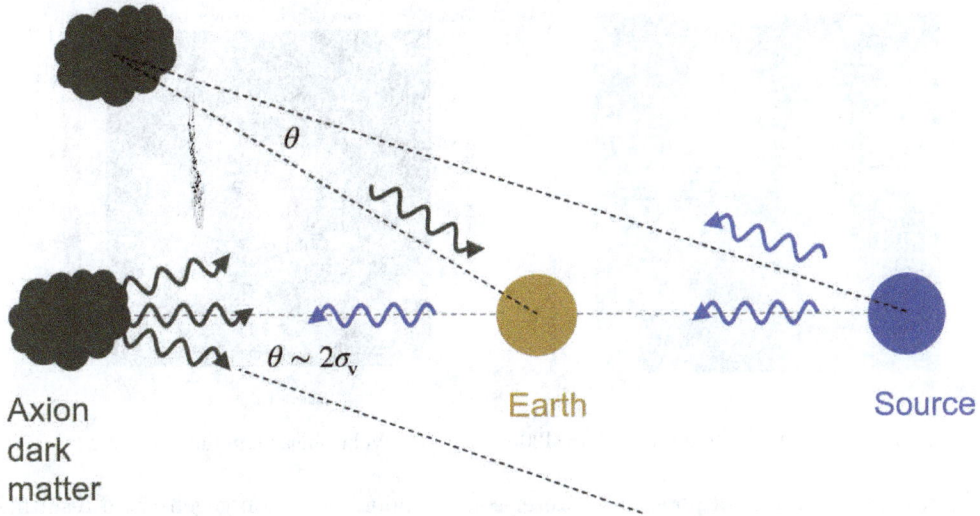

Figure 7.12: An illustration showing some of the basic ideas behind *axion gegenschein*. Figure adapted from Ref. [Sch22].

The basic principle, albeit stated overly simplistically, is represented in Fig. 7.12. Consider an axion, a_χ, in an over-dense region of the Milky Way. It will be nearly at rest compared to the speed of light, owing to the non-relativistic nature of dark matter ($\beta \ll c$). Therefore, in any interaction with an external particle of light, we can approximate $\beta_{a_\chi} \approx 0$ in the reaction. If we consider a photon incident on the axion, γ_0, then conservation of momentum dictates that the total momentum transverse to the flight direction of the photon must be zero. The only net momentum must lie along the original flight direction of γ_0.

Under the condition that the incident photon field $\vec{A}_0$ encounters an axion field, a_χ, such that the photon frequency satisfies $\omega_0 = m_a/2$ in the rest frame of the axion, this can stimulate the decay of the axion to two photons (e.g., to an outgoing photon field $\vec{A}_1$). Since the axion and photon can be described classically in this picture via Maxwell's Equations, the relationship of the incident and outgoing electromagnetic field can be written

$$(\partial_t^2 - \nabla^2)\vec{A}_1 = -g_{a\gamma\gamma}(\vec{\nabla} \times \vec{A}_0)\partial_t a_\chi. \tag{7.1}$$

The incident photon continues on its original trajectory and the axion is stimulated to decay to a photon pair, with one of the two photons aimed directly opposite the γ_0 flight direction.

This makes axions in over-dense regions (where the occupancy number of available states is very high) an "imperfect mirror" [Sch22], one that reflects back a lower intensity

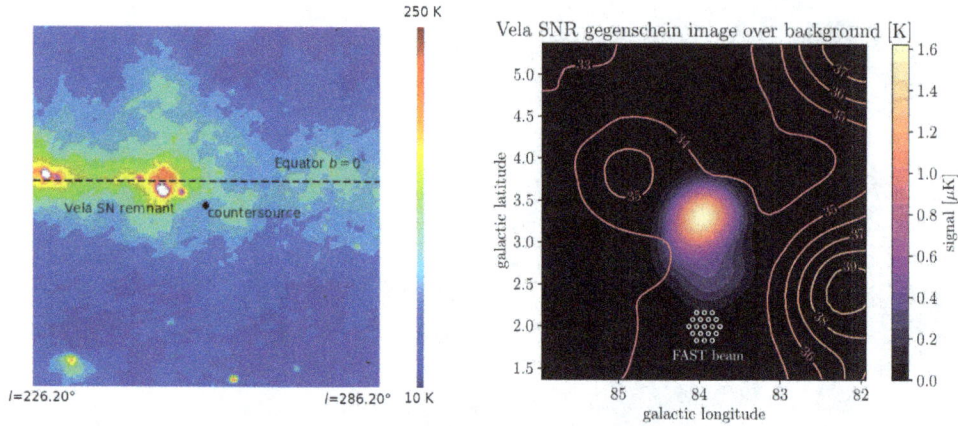

(a) Cygnus A counter-source region [GSM20]   (b) Vela pulsar simulation [Sch22]

Figure 7.13: Examples of potential counter-source regions for axion gegenschein resulting from different kinds of astronomical sources. In the right-hand figure, the scanning resolution of the FAST radio telescope is indicated on a simulation of a Vela pulsar counter-source image.

of the light than is shone upon it. The assumption that the axions are nearly at rest is, obviously, flawed, as they not only possess a net velocity relative to Earth but a velocity dispersion (σ_v). The former effect alters the sensitivity to particular wavelengths of incident light (since the stimulated emission peaks when $\omega_0 \approx m_a/2$ in the rest frame of the axion). The latter effect serves to smear the reflection with an angular spread of $\theta \approx 2\sigma_v$. However, this problem also provides an opportunity, as the reflected light from locations away from the line-of-sight between the source, the observer, and the axions will also back-scatter to the observer from locations that should not have been visible.

This effect of a weak mirroring of light by axions is called *axion gegenschein* in reference to zodiacal gegenschein, the effect of the sun's light being back-scattered toward Earth by interplanetary dust that lies between Earth and interstellar space opposite the sun. The term *gegenschein* is German and translates roughly as "counter-shine" or "counter-source." The location of a Cygnus A counter-source, compared with phenomena in the nearby region of the sky, is shown in Fig. 7.13a. A simulation of what a Vela pulsar signal might look like is shown in Fig. 7.13b.

The proponents of the use of gegenschein as a discovery tool for axions suggest looking for the reflection of intense sources of radiation along a line that is antipodal to the source relative to Earth. One must keep in mind that light's travel time is not infinitesimal on human timescales when confronted with something as immense as the Milky Way. One cannot look for recent sources of light, but rather at sources that (a) have been emitting for a long time or (b) emitted a lot of light a long time ago that has characteristic spectra.

Two examples that have been proposed to look for are radio signatures from a quasar (an active supermassive black hole at the center of another galaxy) [GSM20] or for electromagnetic reflections of supernova remnants in the Milky Way [Sun+22]. In both cases, the gegenschein will appear on the side of the Earth opposite the original source position.

Cygnus A is one such quasar and serves as a high-intensity radio source. It is the result of the jets of energy spouting from a massive black hole at the center of a galaxy that is 232 Mpc from Earth. It is the brightest radio source in the sky. Earth is presently moving around the Milky Way with its velocity vector aimed roughly in the direction of Cygnus A (this would be the direction from which the dark matter "wind" would appear to originate due to this relative motion). Cygnus A would appear as a radio point source on one side of the Earth, and one would then look for its spectrum in radio waves coming from the opposite direction. This requires powerful radio telescopes to distinguish a potential counter-source signal from the surrounding radio emissions in that region of the sky.

The second example is the use of a supernova remnant's radio emissions to look for a gegenschein signal at the antipodal point relative to Earth. The Vela pulsar (a supernova remnant roughly 300 pc from Earth) provides one potential origin of a counter-source. Observations intended to constrain the possibility of a signal in this region of the sky have been proposed using both the Five-hundred meter Aperture Spherical Telescope (FAST) and the Canadian Hydrogen Intensity Mapping Experiment (CHIME) radio telescope.

Axion gegenschein is an example of the new opportunities that can arise in partnerships between theoretical and experimental physicists. We emphasize that our examples are by no means exhaustive and we encourage the reader to pursue axion theory and instrumentation to wherever these ideas may lead. A significant benchmark ahead of axion programs is to push down to at least the parameter space (in axion mass and coupling) consistent with the QCD axion or other baseline axion scenarios. However, the axion remains a viable dark matter candidate even if it does not solve the strong CP problem. Searches are encouraged to the limits that technology will impose.

7.8 The Dark Elements

We have spent a lot of time considering very simplistic dark matter scenarios. These typically include a single dark matter candidate and either an aspect of the existing standard model interactions or a new interaction described using a simple symmetry group (e.g., U(1)). However, there is no guarantee that the universe is minimalist as to this approach. One can make a reasonable philosophical argument that, since the universe is, to our knowledge, not describable using a single particle and a single interaction, there is no reason to believe that the dark sector is any less complex. Consider, for example, that the standard model requires a minimum of six quarks and six leptons to accommodate observed phenomena such as the mixing of quark and lepton flavors, and a minimum of three quantum interactions to describe electromagnetic and nuclear forces. If the dark sector was birthed around the same time as that which we now call the standard model, why should it be expected to be more simplistic?

We recognize here the long effort expended for establishing a single unified field theory that successfully unites the electromagnetic, weak, and strong interactions. To date, no such single theory has been achieved, nor have there been clear experimental observations that definitively lead to a particular example of such a theory. In addition, we recognize the even longer effort given to unify gravity with quantum theories. This, too, has not been completed.

So, while it may ultimately be that all quantum theories of nature might unite into a single, simple framework, we have no evidence that this is the case. Regardless of the possibility of ultimate unification, we also recognize that our universe is the result of broken fundamental symmetries (Section 1.3), where a more symmetric higher-energy theory has broken at low energy into a complex series of interacting sectors. We would argue that this, also, should lead to the possibility that the dark sector might itself be broken in similar ways, and thus highly complex. We also note that a challenge to theoretical physics, ultimately, will be to recognize that there is at least one more problem in the quest to unify all field theories: the ultimate unification of gravity, the standard model, and the dark sector. This is, obviously, a known problem and we are excited by the efforts in the field to creatively address it.

The challenge, of course, is that if current experimental constraints continue to poorly limit the mass, spin, and couplings of dark matter as a single species, then those same constraints are even less restrictive on more complex sectors. However, there have been (and continue to be) serious efforts to embrace this potential complexity to see where it might lead. We highlight one recent example of this.

One can imagine that an entire dark periodic table of the elements is possible in the dark sector, just as it was possible in the standard model. In fact, there is now a body of work that leads in this direction: dark chemistry [Gur+22b; Rya+22a; Rya+22b]. Such approaches begin by expanding the complexity of a dark sector by just one step: two fermionic dark matter components, one lighter (mass m) and one heavier (mass M). They can interact via a force described by a simple gauge interaction, $U(1)$. This might look familiar: it describes a universe with an electromagnetic interaction and two particles that can form bound states. This is a theory capable of forming *dark hydrogen*.

This may seem a trivial expansion of the complexity of such a theory, but it opens a number of doors to interesting phenomena. First, we now have the possibility that dark matter consists of bound states that can be excited into higher-energy states and can decay down to its ground state. We have the possibility of molecular formation and molecular interaction. After all, H_2 is the most common molecule in the atomic universe (see Section 3.4) and it was the dynamics, primarily, of clouds of hydrogen that led to the first stars, the first galaxies, and ultimately to the richness of the cosmos. Molecular hydrogen clouds are a significant part of the "dust lanes" of galaxies and are key targets of spectroscopy in order to better understand the structure of the atomic cosmos.

In particular, dark hydrogen would be expected to exhibit one of the key properties of molecular hydrogen: cooling by dissipation of energy. Just as molecular hydrogen clouds, although hot after recombination, could clump via gravity and cool by energy dissipation through electromagnetism, so, too, could dark hydrogen similarly clump and dissipate its energy. Dark molecular hydrogen would have lost energy in the early universe through collisions, and a dark chemical theory would permit *ab initio* methods of predicting the shape(s) of dark matter halos. However, once cooled and clumped, there is a distinct change in the behavior of molecular hydrogen and dark hydrogen: there would be no nuclear fusion in dark hydrogen. Whereas standard model hydrogen, when clumped to sufficient density, will then fuse as the Coulomb repulsion barrier is overcome and the strong interaction takes over; the dark sector lacks the $SU(3)$ gauge group and no fusion would occur.

Dark hydrogen, then, would skip the step of star formation. Star formation delays the step of collapsing into black holes or other similar compact objects and does so by millions or billions of years. Dark hydrogen, however, could have cooled and collapsed into black holes very early after recombination. This theoretical possibility also leads to a key distinction between the black holes formed by standard model hydrogen and those formed by dark hydrogen: their mass spectrum [Gur+22a].

Standard model black holes result when a star about eight times more massive than the sun reaches the stage of iron fusion and can no longer sustain outward pressure from fusion to overcome gravitational inward collapse. The runaway collapse leads to the formation of a neutron star (if the stellar core is between about 1.5 and 3 solar masses) or a black hole (if the core is 3 solar masses or larger). This means, therefore, that there is a minimum mass expected of all standard model black holes: the Tolman-Oppenheimer-Volkoff (TOV) limit [OV39; Tol39]. This is the maximum size a neutron star can achieve before it collapses into a black hole. Recent gravitational wave observations of colliding neutron stars suggests, experimentally, that this limit is in the range of 2.01-2.35 solar masses [RMW18]. We will refer to this TOV limit involving SM matter as the SM TOV limit.

There will, of course, be a corresponding dark TOV limit which is analogous to the SM TOV limit: the threshold beyond which the Fermi exclusion principle (the idea that two fermions cannot occupy the same quantum state) is overcome by the force acting on the two fermions. This limit will ultimately depend on the mass of the heavier dark fermion component, M. However, there is the possibility of a lower-mass population of dark matter black holes.

In a SM-only universe where black holes can only form in the ways known from applicatiion of general relativity and nuclear physics (star birth and, eventually, death) to SM matter, there should be no black holes with masses below the SM TOV limit. The observation of such a population would suggest a mechanism in the universe that allows for black hole formation without the intermediate step of star formation. Dark hydrogen could be an explanation, should a positive observation occur. Dark chemistry work remains to fully map out the molecular states and the potential mechanisms and timescales for energy dissipation in dark hydrogen clouds. However, the other side of this problem—low-mass black hole observations—is realizable in the near future.

The currently operating gravitational wave observatories form the LIGO Scientific Collaboration and comprise the Laser Interferometer Gravitational wave Observatory (LIGO, United States), the Kamioka Gravitational Wave Detector (KAGRA, Japan), and the Virgo gravitational wave interferometer (Europe). Presently, these instruments are sensitive enough to observe mergers of objects with masses down to a few to several solar

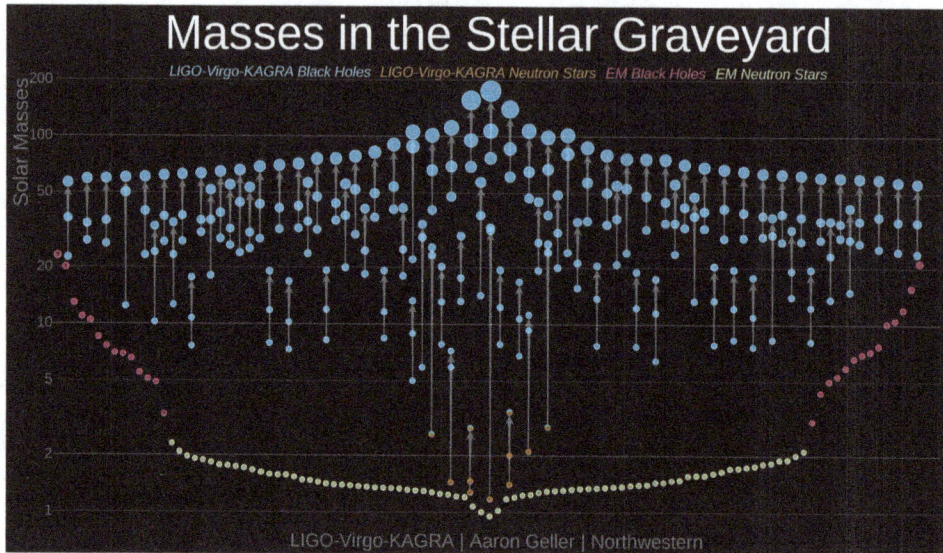

Figure 7.14: Combined information about the gravitational-wave detection and electromagnetic detection of black holes and neutron stars. The vertical triplets of points, joined by arrows, indicate black hole pair mergers; the lower pair of dots indicates the original masses and the top dot indicates the merged mass, all determined from gravitational waves. The singular dots indicate black holes and neutron stars identified by electromagnetic means. Image credit: LIGO-Virgo/Aaron Geller/Northwestern [LIG21].

masses. To date, while searches for sub-solar mass binary mergers have occurred with this instrumentation, no detections have been made. However, upgrades have been completed to existing facilities that are expected to greatly enhance the sensitivity to such events. The LIGO-observing run, resumed in 2023, opens an exciting new potential in sub-solar mass merger detection. The combined catalogue from current-generation gravitational wave observations of binary object mergers is shown in Fig. 7.14.

We pause here to give a brief overview of the principle behind the instruments in the LIGO Scientific Collaboration. All of these experiments utilize lasers and the interference of two identically prepared laser pulses in order to look for the slightest mechanical variations in the system. These variations can be induced by the stretching and squashing of spacetime by gravitational waves, and have sizes determined by the *strain* induced in spacetime by the original distortion, h_{ij}. This strain is given, in general relativity, as

$$h_{ij} \approx \frac{4GM}{c^4} \frac{f^2}{r}, \tag{7.2}$$

where M is the reduced mass of the binary system, f is the frequency of the oscillation that leads to the merger, and r is the separation of the objects. As the frequency increases and the separation decreases, the strain increases. For objects with masses around 1-100 solar masses, these strains are at the level of 10^{-21}. This places a key requirement on the interferometers: they have to be capable of detecting changes in mechanical separation that are less than the size of an atomic nucleus. Modern laser interferometry provides this capability and has been extremely well-demonstrated since the first confirmed black hole binary merger detection in 2015.

This is another example of a area of study where people can become highly active and push the limits of knowledge rather quickly. There is still much work to be done in the dark chemistry theory; there has only been a minor effort so far, led by a small group of proponents. While the LIGO Scientific Collaboration is a much larger global organization, and while detection of sub-solar mergers is part of its search portfolio, there is no standard model motivation for advancing this kind of search. Instead, dark hydrogen and its potential formation of dark matter black holes provides a strong reason to push sensitivity for such exploration. The detection of even one such binary merger could transform our understanding of both the standard model and the dark sector. We note that the observation of a population of sub-solar-mass black hole mergers would be a purely gravitational detection of dark matter, as the merger would only be detectable by the spacetime ripples caused by the merger. The primary signal will be gravitational-only; a corresponding lack of direct observation of an electromagnetic signal would corroborate that expectation.

7.9 Dark Matter via Gravity-Only Interactions

We close this outlook, and this book, with a glance forward toward a most difficult question: what if dark matter is detectable only by its gravitational interactions? We teased this in the previous section, where we noted that a dark black hole merger would be observable through gravitational waves. However, this would mark the detection of a massive compact object. If the dark $U(1)$ does not, in fact, mix with the standard model $U(1)$ field, is there any hope of ever detecting a dark hydrogen molecule on its own?

The *Windchime* proposal advances just such a concept [Att+22; Car+20], combining a number of features we proffer in this concluding chapter: laser interferometry for detecting very small mechanical distortions and quantum sensors for enhancing sensitivity to ultra-weak signals.

The principle is sound: the only interaction in which we have a guaranteed connection between dark matter and standard model matter is gravity. The force relationship on the scale of a detector, even with small components, would be Newton's law of gravitation. Consider a detector element with mass m_D and a dark matter particle with mass m_χ. The force exerted by any one detector element, labeled by the index i, on the dark matter particle will be

$$\vec{F}_{\chi,i} = -G \frac{m_D m_\chi}{r_{\chi,i}^2} \hat{r}_{\chi,i}, \tag{7.3}$$

where $\vec{r}_{i,\chi} \equiv \vec{r}_i - \vec{r}_\chi$ is the vector pointing from the dark matter particle to the detector element. The force is attractive, and if we know the positions of the detector elements with precision, we know that a dark matter signal will be distinguished from random noise in the elements by virtue of a *coherent* response by multiple elements, simultaneously,

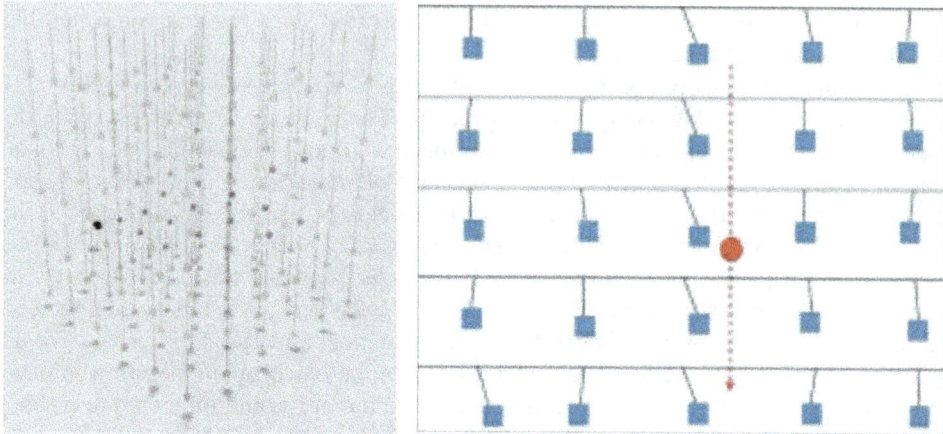

Figure 7.15: An illustration of an array of mechanical oscillators, each represented as a tiny pendulum. A particle that interacts only by gravity would cause a coherent disturbance across multiple elements, with a pattern determined by the inverse-square force law. Illustrations from Ref. [Car+20].

to the presence, and movement, of the dark matter particle. This coherence is the key distinguishing feature separating a signal from a noise background. A conceptual array of mechanical oscillators is shown in Fig. 7.15.

What about standard background particles like neutrinos, gamma rays, neutrons, alpha particles, and so forth? The single-site interaction of a particle such as a gamma ray or a neutrino with a detector element will not lead to a coherent response across multiple elements, and so, can also be distinguished from a signal.

Neutrinos would likely be the ultimate concern for such a detector, if all other backgrounds are rejected or eliminated. However, while neutrinos have mass, it is expected to be small enough that, in the particle-like domain of dark matter ($m_\chi \geq 1$ keV), the two classes should be distinguishable. Neutrons, alpha particles, and other backgrounds that can interact with multiple elements through nuclear, rather than gravitational, forces would be distinguished again by a series of single-site interactions that are far more intense in amplitude than the weak coherent gravitational signal of a dark matter particle. If one could build such a detector, the device would offer interesting possibilities both for gravity-only detection and background rejection, when combined with shielding and radiopure construction techniques to limit the number of background particles entering the active region.

The Windchime proposal is a first step toward developing the research and development needs for such a detector. Constructing an array of small mechanical oscillators is one part of the effort. Another key part is how to read out the small motions of these mechanical sensors.

The goal is to identify a sensor technology (several are proposed) that is optomechanical in nature, so that light can be used to measure the oscillator positions. The readout concept is similar to that used to detect gravitational waves: interferometry. This has often been proven to give superior sensitivity to small displacements. Current estimates suggest a goal of building gram-scale detector elements, each containing an array of oscillators, with spacing between the individual units (to provide isolation, element-to-element). A total coverage of 1 m^2 is the proposed scale of an initial functioning detector.

This concept is still in its early stages. Proponents point out that increasing the number of oscillators, each assumed to have independent and random noise, leads to a greater ability to distinguish the coherent dark matter signal that is sitting on top of the noise motion of the sensors. For N sensors, this sensitivity scales as $\sqrt{N}$. The fact that a dark matter particle will travel a straight path through the detector means that detection of this path, and the coherent way the oscillators respond along the path, should lead to superior background rejection. It is also noted that the sensitivity of an individual mechanical oscillator goes as $\sqrt{m_D}$.

Another important factor will be quantum shot noise in the laser pulses used to read out the oscillators. There are multiple modern techniques to accomplish this, for example those employed by LIGO that have pushed the limits of this noise to extremely low levels. The proponents of this technique also note that the system will be required to be very cold (achievable with large, modern dilution refrigerators) to suppress thermal noise and that the vacuum around the detector will need to be quite good to avoid buffeting the detector with molecules. Radiopurity and shielding will then play the final key role in keeping unwanted backgrounds out of the active detector elements.

In many ways, the Windchime concept combines all the "greatest hits" of the last half-century of effort to study ultra-rare interactions: highly sensitive sensor technology, interferometry, cryogenics, radiopurity, interaction and track detection, high-vacuum technology, and more. We highlight *this* idea, at the close of this book, to further underscore that there are exciting and novel paths to walk in the search for dark matter.

7.10 Closing Perspectives

Each of the subjects we have discussed in this chapter (and, indeed, the entire book) will provide new avenues for physicists, chemists, engineers, and other research professionals to provide breakthroughs. There is no doubt that any of these technologies will also lead to unexpected outcomes, either through the discovery of additional features of the natural world or through applications of technology beyond its original purpose. We hope that the search for dark matter will conclude successfully, and we look forward to the continuing need to update and add to this guidebook.

We close with a thought related to one raised by the authors of the paper, "A phenomenological profile of the Higgs boson," published in 1976 by physicists John Ellis, Mary Gaillard, and Dmitri Nanopoulos [EGN91]. After laying out an extensive picture of the possible range of properties and interactions of the Higgs particle, the authors closed their masterful work with the following commentary. "We apologize to experimentalists for having no idea what is the mass of the Higgs boson... and for not being sure of its couplings to other particles, except that they are probably all very small. For these reasons we do not want to encourage big experimental searches for the Higgs boson ...." Of course, their work would go on to influence decades of additional theoretical and experimental effort that culminated in the discovery of the Higgs particle at the LHC in 2012.

Reflecting on the body of work presented in this book, we would paraphrase that sentiment for dark matter and the modern era of science. We also apologize to both theorists and experimentalists for having no idea what is the mass of the constituents of dark matter. We, too, apologize for not being sure of its couplings to other particles, except that they might be non-existent, or very small, or even, consistent with the data, at the scale of strong nuclear interactions. For these reasons, we strongly encourage big experimental and theoretical searches for the constituents of dark matter. It is by scientific work that we will eliminate our confusion.

We invite you to take up the challenge of this hunt, and we hope that the guidance we have provided will lead you down interesting roads ... or into the wild where you might forge your own road, one that leads to the discovery of dark matter's true nature.

Appendices

APPENDIX A

Solutions to Exercises

Answer to Exercise 1

We can begin by figuring out how many centimeters are in a mile, as we have the conversion methods for centimeters (cm) to inches, inches to feet, and feet to miles. Let us step through these conversions and build those dimensions up, one by one.

First, let us convert feet into centimeters. 12 inches in a foot means that, to get to the number of centimeters:

$$2.54 \text{ cm/in} \times 12 \text{ in/foot} = C_1 \tag{A.1.1}$$

This yields $C_1 = 30.48$ cm/foot.

Note that when you multiply 2.54 cm/in times 12 in/foot, inches appears in the denominator of the first number and in the numerator of the second. Just as when multiplying fractions in that same situation, they cancel each other out.

Then, let us determine the number of cm in a mile, given that there are 5280 feet per mile:

$$5280 \text{ ft/mile} \times 30.48 \text{ cm/foot} = C_2 \tag{A.1.2}$$

This yields $C_2 = 160934.4$ cm/mile. Last, we need to convert cm to km. There are 100 cm in a meter, and 1000 meters in a kilometer (km). So:

$$C_2 \times \frac{1 \text{ m}}{100 \text{ cm}} \times \frac{1 \text{ km}}{1000 \text{ m}} = C_3 \tag{A.1.3}$$

This yields $C_3 = 1.609344$ km/mile, resulting in the conversion we sought.

A comment on significant figures: We were sloppy in making it clear how many significant figures a number like "1 inch" has. We recommend to make it clear whether a number like "1" is a 1-significant figure number or, instead, should be read as "1.00000..." with infinite significant figures (e.g., perfectly precise).

Answer to Exercise 2

We are told the radius of the earth in miles. We want to convert miles to kilometers. We know from Exercise 1 that there are about 1.609 km per mile. Based on that, we can multiply:

$$3958.8 \text{ miles} \times 1.609 \ \frac{\text{km}}{\text{mile}} \tag{A.2.1}$$

which yields 6371 km.

Answer to Exercise 3

When first encountered, this question seems odd. Its phrasing is challenging—as though the question-author is attempting to trick you—that you should simply write $\hbar c = 1$ and walk away. Rather, the author seeks here to prompt you to relate two systems of units. It turns out that doing so provides a useful relationship between the original (MKS) units and the new natural units framework. As you will see, relating $\hbar c$ back to its MKS equivalent, and then setting $\hbar c = 1$, will provide a fundamental relationship between *energy* and *length*.

Let us remind ourselves of the values of $\hbar$ and c. In MKS units, $\hbar = 1.054 \times 10^{-34}$ J·s and $c = 2.99 \times 10^8$ m/s. We have been asked to obtain a relationship described in meters and electron-volts (eV). The eV is an alternative unit of energy related to the joule that is convenient for use in atomic-scale systems. It is the energy gained by an electron of fundamental charge $e = 1.602 \times 10^{-19}$ C that experiences a 1 V electric potential difference. A volt, in turn, is equal to 1 J/C. Thus $1 \ eV = (1.602 \times 10^{-19}$ C $\cdot 1$ V $= 1.602 \times 10^{-19}$ J. The eV is closely related to the amount of energy stored in chemical or molecular bonds.

We can begin by re-framing Planck's reduced constant, showing it in eV, yielding $\hbar = 1.054 \times 10^{-34} (6.242 \times 10^{18} \text{ eV/J}) \cdot \text{s} = 6.586 \times 10^{-16}$ eV·s. We are now ready to take the final mathematical step by multiplying this with the speed of light,

$$\hbar c = (6.586 \times 10^{-16} \text{ eV} \cdot \text{s})(2.998 \times 10^8 \text{ m/s}) = 197.3 \times 10^{-9} \text{ eV} \cdot \text{m}. \tag{A.3.1}$$

If one is inclined to round such numbers to hold them in memory for quick estimates ("back-of-the-envelope calculations"), this rounds to a neat $\hbar c \approx 200 \times 10^{-9}$ eV·m.

We can then go the final step and relate natural units to this number. In natural units, $\hbar c = 1$, so we find:

$$\begin{aligned} 1 &= 197.3 \times 10^{-9} \text{ eV} \cdot \text{m} \\ 1 \text{ eV} &= 5.067 \times 10^6 \text{ m}^{-1}. \end{aligned}$$

Energy is now related to length and identified as the inverse of length. High energies correspond to short distances; and vice versa.

We have been provided some hint of this from Louis de Broglie's postulates regarding matter waves. For example, in $p = h/\lambda$, momentum (which classically relates to energy as $E = p^2/(2m)$, with m as the mass of the matter particle) is inversely related to wavelength. A higher momentum particle has a shorter wavelength, while a lower momentum particle has a longer wavelength. Together, these relationships help us to understand why higher energy particles that collide with a target probe smaller-scale features of that target.

It is useful to do one more re-framing in order to consider units that are more typical of nuclear processes. The typical energy involved in a nuclear reaction is at the scale of millions of eV (MeV), while the distances typical of a nuclear process are expressed in femtometers (fm), 10^{-15} m. With those prefixes in place, we find

$$\hbar c = 1 = 197.3 \text{ MeV} \cdot \text{fm} \approx 200 \text{ MeV} \cdot \text{fm}$$

which is rather memorable, indeed. Natural units are very convenient in nuclear-scale physics, yielding relationships with reasonable powers of ten in the resulting numerical values. From this product of fundamental constants we discover something about the nuclear scale—that an energy of just a few hundred MeV is sufficient to probe a distance scale at about the level of the proton or neutron.

Answer to Exercise 4

This is a problem that requires setting up two algebraic equations, each with two unknowns (the same two). One then substitutes one equation into the other, and solves for the unknowns.

One equation relates the total number of atoms in the universe ($N = 10^{80}$) to the number of hydrogen atoms (N_H) and the number of helium atoms (N_{He}): $N = N_H + N_{He}$. The second equation computes the total mass of hydrogen (m_H), adds to it the total mass of helium (m_{He}), and obtains from that the total mass of all of the atoms in the universe: $M = N_H m_H + N_{He} m_{He}$.

We want to find M, but we don't have the numbers of hydrogen or helium atoms. However, we do have the *fraction* of the total number of atoms represented by each population:

$$76\% = \frac{N_H}{N}$$
$$24\% = \frac{N_{He}}{N}.$$

We can solve for the number of hydrogen and helium atoms. For example, $N_H = 0.76 \times N = 7.60 \times 10^{79}$, and repeating that, we find that $N_{He} = 2.40 \times 10^{79}$.

We now have everything we need to solve for the total mass in atoms, $M = 2.87 \times 10^{53}$ kg. For comparison, the mass of the sun is only about 2×10^{30} kg. The universe is a big place.

Many observations of the atomic content of the universe have been made. The atomic matter in the universe is about 74% Hydrogen, 24% Helium, and 2% of all other kinds of atoms.

Answer to Exercise 5

We can model a neutron star as a sphere. Spheres have volumes given by $(4/3)\pi R^3$, where R is the radius of the sphere. We know the radius of our neutron star, so we can compute its volume, $V_{star} = 8.18 \times 10^{12}$ m^3. The density is the ratio of mass and volume, so:

$$\rho_{star} = \frac{M_{star}}{V_{star}} = 3.67 \times 10^{17} \text{ kg/m}^3. \tag{A.5.1}$$

In comparison, lead (symbol: Pb) is a very dense metal (a small volume is very heavy). A lead brick of a type typically used in classroom demonstrations has a mass of 25kg, but lead has a density of only 1.13×10^4 kg/m^3. This pales in comparison to the density of neutron star material.

Neutron stars are about the diameter of large city centers, but are so dense that a teaspoon of material from that star has a mass of about 200 billion kilograms and would weigh about 3 million tons on earth.

Answer to Exercise 6

This is a unit-conversion problem. We know that 3.26 light-years is the distance traveled by light in 3.26 years. All that we need do is to convert from one strange unit to another, as well as understand what the prefix *kilo* means. First, let's convert. We know that 1 pc = 3.26 ly. We know that the unknown distance to the galactic center in parsecs is X (we want to solve for this), and that this distance equals 26,673 ly. Thus:

$$1 \ pc = 3.26 \ ly \tag{A.6.1}$$
$$X = 26,673 \ ly \tag{A.6.2}$$

We can take the ratio of the top and bottom equations:

$$\frac{1 \ pc}{X} = \frac{3.26 \ ly}{26,673 \ ly}. \tag{A.6.3}$$

As it appears in both the numerator and the denominator, the unit ly cancels from the right-hand side. We then have $\frac{1 \ pc}{X} = 1.22221 \times 10^{-4}$, which we can rearrange and solve for X: $X = (1 \ pc)/(1.22221 \times 10^{-4})$. This yields $X = 8181.9$ pc.

We want the answer to be in kiloparsecs, or *thousands of parsecs*. A parsec is *smaller* than a kiloparsec by 1000, so we convert: 1 kpc = 1000 pc $\rightarrow X = 8.18$ kpc.

We can see why astronomers find the *kiloparsec* useful to talk about galactic-scale distances; those distances are typically in the range of a few to tens of kiloparsecs for a galaxy like the Milky Way. As noted in Chapter 1, distances between galaxies are typically in *Megaparsecs* and the size of the visible universe is at the scale of *Gigaparsecs*.

Answer to Exercise 7

We begin by tackling the first question regarding a radio signal sent from Earth to Proxima b. We know the speed of light, c. We can use the distance (D) and this speed to determine the time, in years, needed to make the 1-way journey.

$$t_{seconds} = D/c = 1.34 \times 10^8 \text{ s.} \tag{A.7.1}$$

We know from the comment on p. 1 that there are about $\pi \times 10^7$ seconds per year, which is also a large number of seconds. How many *years* is this journey?

Dividing the number of seconds to reach Proxima by the seconds in one year, it turns to to be 4.2 years, which makes Proxima feel close... until you consider the second question.

The second solution is identical to the first one with only the speed changed. No longer are we moving at the speed of light. In fact, compared to light, the Voyager probes are downright slow. They move at only about $(5.67 \times 10^{-3})\%$ the speed of light.

We can use the distance (D) and this speed to determine the time, in years, needed to make the one-way journey.

$$t_{seconds} = D/v = 2.36 \times 10^{12} \text{ s.} \tag{A.7.2}$$

The journey for heavier and more-human-scale objects takes far longer. How many years? It turns out to be about 75,000 years, vastly exceeding the lifespan of any single human and even transcends the amount of time modern humans have spent on Earth during the period called "civilization."

In order to send colonists to a planet like Proxima Centauri b, we must find a way to get them to reach much higher speeds than, say, the Voyager probes. Even the concept of a "generation ship" fails at speeds like this. A colony of humans, with succeeding generations living their lives out on the ship such that the final space-faring generation would land on Proxima B, would have to sustain itself longer than the time between now and the oldest known civilization on Earth.

Answer to Exercise 8

We begin by diagramming the problem. The statement of the problem gives us a few key pieces of information. There is an observer of a phenomenon. We can place ourselves in the perspective of that observer. There is a phenomenon being observed—that of a galaxy in motion. We are given a keyword for the motion—"recession." The galaxy is receding from us, which means it is moving from a position closer to us to a position farther away.

It is sensible, given these clues, to place all the elements of the problem onto a single line, given that this is a problem about motion along the radial line of sight of the observer. Let us define its axis as x, so that u refers now to velocity along this axis. Let us also place ourselves (the observer) at the origin ($x = 0$), and let us place the galaxy at an unknown position x whose value is increasing with time, such that we recognize $u > 0$ (positive velocity thus describes the "recession" effect).

We are now ready to apply the core idea of the problem: the radial Doppler shift (Eqn. 2.1). This is given for wavelength by

$$\lambda_{obs} = \lambda_{H-\alpha} \times \gamma(1 + u/c). \tag{A.8.1}$$

The velocity was given in the problem, $u = 1000$ km/s $= 10^3 \times 10^3$ m/s. The speed of light is $c = 2.998 \times 10^8$ m/s to a reasonable approximation, given the precision of the galaxy's velocity. The wavelength of the $H - \alpha$ Balmer series line is 656.279 nm $= 6.56279 \times 10^{-9}$ m.

We need to compute the *Lorentz gamma factor*, γ, for this motion of the galaxy relative to the observer. This is given in special relativity by

$$\gamma \equiv \frac{1}{\sqrt{1 - u^2/c^2}}. \tag{A.8.2}$$

If we insert our velocity and the speed of light, we obtain:

$$\gamma = \frac{1}{\sqrt{1 - (1 \times 10^6 \text{ m/s})^2/(2.998 \times 10^8 \text{ m/s})^2}} \tag{A.8.3}$$

$$= \frac{1}{\sqrt{1 - (1 \times 10^{12})/(8.988 \times 10^{16})}} \tag{A.8.4}$$

$$= \frac{1}{\sqrt{1 - (1.11259 \times 10^{-5})}} \tag{A.8.5}$$

$$= \frac{1}{\sqrt{0.999988874059}} \tag{A.8.6}$$

$$= \frac{1}{0.999994437014} \tag{A.8.7}$$

$$= 1.00000556302. \tag{A.8.8}$$

We intentionally kept an excessive amount of precision in the last few steps because calculations of gamma factors with such "low" velocities, compared to the speed of light, can often cause problems when calculating by hand or using a computer. In fact, unless the reader is careful about the precision with which numbers are reported from each step of the computation, it is likely that this problem would lead to a nonsensical answer like $\gamma = 1$ (in other words, $u = 0$ m/s) because of truncation in the precision of the numbers.

Calculating gamma factors under these conditions is often better handled by the use of mathematical techniques to *approximate* the exact expression to reasonable numerical precision. One way of doing this for the gamma factor is the use of a *binomial expansion* of this expression into a series of terms, each smaller than the next, but whose sum (in the limit of keeping infinite terms) converges to the exact value. In practice, it is not necessary to keep more than a few terms (at most). Let us proceed with this reasonable approximate method and confirm the above result.

The gamma factor has the form

$$f(x) = (1 + x)^\alpha \tag{A.8.9}$$

where x is a number such that $|x| < 1$. In our case, $x = -(u/c)^2$ (meeting that criterion). Also for our case, $\alpha = -1/2$ (one over the square-root of the argument). The binomial expansion of this function yields

$$f(x) = \sum_{k=0}^{\infty} \binom{\alpha}{k} x^k. \tag{A.8.10}$$

where the term in parenthesis is known as the *binomial coefficient*, defined as

$$\binom{\alpha}{k} \equiv \frac{\alpha!}{(\alpha - k)!k!}. \tag{A.8.11}$$

We need to next deal with the fact that we will have to compute $\alpha!$, which, in our case, involves a non-integer number, $\alpha = -1/2$. In calculus, we can define

$$n! \equiv \int_0^\infty x^n e^{-x} dx. \tag{A.8.12}$$

In our specific case, we have

$$\left(-\frac{1}{2}\right)! = \int_0^\infty x^{-1/2} e^{-x} dx. \tag{A.8.13}$$

We perform this integral using integration by parts with $u = e^{-x}$ and $dv = x^{-1/2} dx$. We find that

$$\left(-\frac{1}{2}\right)! = \int_0^\infty x^{-1/2} e^{-x} dx \tag{A.8.14}$$

$$= uv\Big|_0^\infty - \int_0^\infty v \, du \tag{A.8.15}$$

$$= (e^{-x})(2x^{1/2})\Big|_0^\infty - \int_0^\infty (2x^{1/2})(-e^{-x}) dx. \tag{A.8.16}$$

The first term in the integration by parts yields exactly 0, leaving just:

$$\left(-\frac{1}{2}\right)! = \int_0^\infty 2x^{1/2} e^{-x} dx \tag{A.8.17}$$

$$= 2 \int_0^\infty x^{1/2} e^{-x} dx. \tag{A.8.18}$$

Staring at this integral for a moment, we realize this is just $(1/2)!$. At this point, we can repeat the integration-by-parts trick. For brevity, we will skip this here, but note that this yields $(1/2)! = \sqrt{\pi}/2$. Thus we find $(-1/2)! = \sqrt{\pi}$. We could have simply looked this up in a mathematics reference, but we feel it is appropriate to remind the reader of some of the basic tricks provided by calculus.

We can finally write the few terms in our series expansion,

$$\gamma = \frac{-\frac{1}{2}!}{(-\frac{1}{2} - 0)!0!}\left(-\frac{u^2}{c^2}\right)^0 + \frac{-\frac{1}{2}!}{(-\frac{1}{2} - 1)!1!}\left(-\frac{u^2}{c^2}\right)^1 + \frac{-\frac{1}{2}!}{(-\frac{1}{2} - 2)!2!}\left(-\frac{u^2}{c^2}\right)^2 + \dots \tag{A.8.19}$$

We observe that we are going to need other fractional factorials, such as $(-3/2)!$, $(-5/2)!$, etc. These are best obtained from a reference, and we find $(-3/2)! = -2\sqrt{\pi}$ and

$(-5/2)! = 4\sqrt{\pi}/3$. Thus

$$\gamma = \frac{(-\frac{1}{2})!}{(-\frac{1}{2}-0)!0!}\left(\frac{u^2}{c^2}\right)^0 + \frac{(-\frac{1}{2})!}{(-3/2)!1!}\left(\frac{u^2}{c^2}\right)^1 + \frac{(-\frac{1}{2})!}{(-5/2)!2!}\left(\frac{u^2}{c^2}\right)^2 + \dots \quad \text{(A.8.20)}$$

$$= 1 + \frac{\sqrt{\pi}}{-2\sqrt{\pi}}\frac{u^2}{c^2} + \frac{\sqrt{\pi}}{-4\sqrt{\pi}/3 \cdot 2}\frac{u^4}{c^4} + \dots \quad \text{(A.8.21)}$$

$$= 1 + \frac{1}{2}\beta^2 + \frac{3}{8}\beta^4 + \dots \quad \text{(A.8.22)}$$

where $\beta \equiv u/c$.

We begin to get a sense of how this series builds. Each term increases the power of β by 2, and, since $\beta < 1$, this results in an increasingly smaller number multiplying the binomial coefficient each time. We also see the coefficient is decreasing in size as well, from $1 \to \frac{1}{2} \to \frac{3}{8}$. Thus, each subsequent term is only important as $\beta \to 1$. For velocities close to the speed of light, more and more terms will matter; for velocities much smaller than that of light, only the leading terms will matter.

Let us begin by assuming we can approximate this as

$$\gamma \approx 1 + \frac{1}{2}\beta^2 \quad \text{(A.8.23)}$$

and throw out all other terms. In this case, $\beta = u/c = 0.003336$, so $\frac{1}{2}\beta^2 = 5.5630 \times 10^{-6}$. This yields

$$\gamma \approx 1 + 5.5630 \times 10^{-6} = 1.0000055630. \quad \text{(A.8.24)}$$

Within the precision we have retained for this approximation, we recover the exact result. This confirms that we kept sufficient numerical precision in the exact method, assessed using this reasonable, but approximate, method.

We are ready to perform our calculation of the *observed* wavelength,

$$\lambda_{obs} = (6.56279 \times 10^{-9} \text{ m}) \times (1.0000055630) \times (1 + 0.003336) = 6.58472 \times 10^{-9} \text{ m}. \quad \text{(A.8.25)}$$

In terms of the fractional change in wavelength relative to the original wavelength,

$$\frac{\Delta\lambda}{\lambda_{H-\alpha}} = -\frac{\lambda_{obs} - \lambda_{H-\alpha}}{\lambda_{H-\alpha}} = 0.0033416 \quad \text{(A.8.26)}$$

or a 0.33% increase in the wavelength. Notice, by the way, that $\frac{\Delta\lambda}{\lambda_{H-\alpha}} \approx u/c = \beta$! For the condition that $\beta \ll 1$, and using the binomial series expansion of γ, you can show that $\Delta\lambda/\lambda \approx \beta$. You are encouraged to check this. Given the small speed of our galaxy relative to that of light, this approximation would have saved us a lot of work. However, we argue that the effort put into solving for the observed wavelength teaches us the value and limitations of the approximation, which for low speeds is a useful shortcut.

This raises a fair question: can astronomers (during the days of Fritz Zwicky compared to now) measure such a small shift in the wavelength of the Balmer series α line? Based on Zwicky's own work [Zwi33; Zwi37], his precision on inferred recessional velocities (at the level of 2%) implies a measurement precision on wavelength far better than this degree of percent shift. Using error propagation, it is possible to estimate that the original uncertainty on the observed wavelengths was at the level of one part in 100,000, which is better precision than the degree of difference between the observed and emitted wavelengths. Modern instruments would be superior to what Zwicky possessed.

Answer to Exercise 9

The virial theorem assumes a system in equilibrium. We will make this assumption when treating A2142. In that case, we can immediately apply the results of the virial theorem in the context of a number of gravitating objects co-orbiting a common center of gravity. The gravitational mass is related to the observed radius of the cluster (R), the velocity dispersion of its galaxies (σ_r), and constants or numbers via Eqn. 2.37,

$$M = \frac{5R\sigma_r^2}{3G}. \tag{A.9.1}$$

We can insert our knowns into this equation, and we find that (up to resolving any issues with units) we are ready to solve for the gravitational mass.

$$M = \frac{5}{3G} (1371 \text{ kpc}) (1062 \text{ km/s})^2 \tag{A.9.2}$$

$$= \frac{5}{3G} (1371 \times 10^3 \text{ pc}) (1062 \text{ km/s})^2 \tag{A.9.3}$$

$$= \frac{5}{3G} (1371 \times 10^3 \times (3.09 \times 10^{16} \text{ m})) (1062 \times 10^3 \text{ m/s})^2 \tag{A.9.4}$$

$$= \frac{5}{3} \frac{(1371 \times 10^3 \times (3.09 \times 10^{16} \text{ m})) (1062 \times 10^3 \text{ m/s})^2}{6.67 \times 10^{-11} \text{ m}^3/(\text{kg} \cdot \text{s}^2)} \tag{A.9.5}$$

$$= \frac{5}{3} \frac{(1371 \times 3.09 \times 10^{19} \text{ m}) (1127844 \times 10^6 \text{ m}^2/\text{s}^2)}{6.67 \times 10^{-11} \text{ m}^3/(\text{kg} \cdot \text{s}^2)} \tag{A.9.6}$$

$$= \frac{5}{3} \frac{(4240 \times 10^{19} \text{ m}) (1.128 \times 10^{12} \text{ m}^2/\text{s}^2)}{6.67 \times 10^{-11} \text{ m}^3/(\text{kg} \cdot \text{s}^2)} \tag{A.9.7}$$

$$= \frac{5}{3} \frac{(4.240 \times 10^{22} \text{ m}) (1.128 \times 10^{12} \text{ m}^2/\text{s}^2)}{6.67 \times 10^{-11} \text{ m}^3/(\text{kg} \cdot \text{s}^2)} \tag{A.9.8}$$

$$= \frac{5}{3} \frac{(4.240 \text{ m}) (1.128 \text{ m}^2/\text{s}^2)}{6.67 \text{ m}^3/(\text{kg} \cdot \text{s}^2)} \times 10^{22} \times 10^{12} \times 10^{11} \tag{A.9.9}$$

$$= \frac{5}{3} \times 0.717 \text{ kg} \times 10^{45} \tag{A.9.10}$$

$$= 1.195 \times 10^{45} \text{ kg} \tag{A.9.11}$$

Let us now evaluate this in solar masses, where the mass of our sun (1 solar mass) is $1 \text{ M}_\odot = 1.98847 \times 10^{30}$ kg. Thus $M = 6.010 \times 10^{14} \text{ M}_\odot$. This is the gravitational mass, so, finally, let us compare this to the visible ("baryonic") mass,

$$\log(M_{bar}) = 14.142 \tag{A.9.12}$$

$$M_{bar} = 10^{14.142} \ M_{\odot}. \tag{A.9.13}$$

This allows us to compare the mass ratio of gravitational-to-visible mass, $M/M_{bar} = 6.010 \times 10^{14}/10^{14.142} = 4.334$.

A fair question at this point, given that the gravitational mass "only" exceeds the visible mass by a factor of about four, is whether or not this is actually significant. Let's do some basic error propagation to answer this question.

For a quantity that is the result of multiplying or dividing a group of numbers, and assuming that the uncertainties (Δ) on those numbers are independent (we'll return to this point shortly), the error propagation formula for the gravitational mass (inferred from the virial theorem) is

$$\frac{\Delta_M}{M} \equiv \sqrt{\left(\frac{\Delta_G}{G}\right)^2 + \left(2\frac{\Delta_{\sigma_r}}{\sigma_r}\right)^2 + \left(\frac{\Delta_R}{R}\right)^2} \tag{A.9.14}$$

We identify the ratios inside the square root as the *relative uncertainties* on each of the quantities (the uncertainty divided by the value). We were not given the uncertainty on the radius of the cluster; this can be assumed to be so small on the scale of this number as to be negligible. Similarly, the uncertainty on Newton's gravitational constant is typically well below the sub-percent level. Let's compare that to the relative uncertainty on the velocity dispersion, $70/1062 = 0.066$. This would clearly dominate in the calculation (note it is also the term that has the factor of 2, which increases its contribution again). Since it is the only term that would effectively survive, we can estimate the relative uncertainty on the gravitational mass to, similarly, be about 9.3%.

The uncertainty on the baryonic mass would be given by varying $log(M_{bar})$ and taking the difference between the upward (downward) variation and the nominal value to determine an upper (lower) bound on the mass. This yields an upper (lower) baryonic mass of $M_{bar}^{upper} = 10^{14.145} = 1.396 \times 10^{14} \ M_{\odot} \ (M_{bar}^{upper} = 10^{14.139} = 1.377 \times 10^{14} \ M_{\odot})$. The difference between these and the nominal value yields $M_{bar} = 1.387^{+0.009}_{-0.010} \times 10^{14} \ M_{\odot}$. This translates into a relative uncertainty of about 0.7%.

The gravitational and baryonic masses differ by 400%, and we should compare that to the relative uncertainties on each which are at the 10% level, at most. This difference cannot be explained only by uncertainties on the observed or inferred quantities.

There is one last point to make. We assumed that the uncertainties on the reported quantities (Newton's gravitational constant, cluster radius, and velocity dispersion) were all independent. Even if they were fully correlated in the worst possible ways *and* all the uncertainties were about the same relative size, this would, at most, inflate the total uncertainty by something like factors of 2–3. Even that would be insufficient to cover the 400% difference between the gravitational and baryonic (visible) masses.

Answer to Exercise 10

The first task is to estimate the gravitational mass enclosed by the orbit of the sun about the Milky Way's center. We are provided with the distance between that point and our sun, a distance estimated through various measurements to be 24,000–28,000 ly. The conversion provided in Eqn. 1.2, combined with the relationship 100 cm = 1 m, allows us to convert this range to $(2.271—2.649) \times 10^{20}$ m.

Assuming that the galaxy is free from external forces, and is in a state of overall equilibrium, we can apply the virial theorem first to the minimum and then to the maximum possible orbital radius:

$$M(r)_{min} = \frac{v^2 r_{min}}{G} \tag{A.10.1}$$

$$= \frac{(240 \times 10^3 \text{ m/s})^2 \times (2.271 \times 10^{20} \text{ m})}{6.674 \times 10^{-1} \text{N} \cdot \text{m}^2/\text{kg}^2} \tag{A.10.2}$$

$$= \frac{(5.76 \times 10^{10} \text{ m}^2/\text{s}^2) \times (2.271 \times 10^{20} \text{ m})}{6.674 \times 10^{-1} \text{m}^3/(\text{kg} \cdot \text{s}^2)} \tag{A.10.3}$$

$$= \frac{1.308 \times 10^{31} \text{ m}^3/\text{s}^2}{6.674 \times 10^{-1} \text{m}^3/(\text{kg} \cdot \text{s})^2} \tag{A.10.4}$$

$$= 1.960 \times 10^{41} \text{ kg.} \tag{A.10.5}$$

We should convert this into more astronomer-friendly units, which typically involves using *solar masses*. The mass of the sun is about $M_\odot = 1.989 \times 10^{30}$ kg, so we find the enclosed gravitational mass to be $M_{min} = 9.853 \times 10^{10}$ $M_\odot$, or roughly 99 billion solar masses. Repeating the calculation with the maximum radius, we arrive at the maximum value of the enclosed mass $M_{max} = 11.50 \times 10^{10}$ kg, or about 115 billion solar masses. We can then state, from this calculation, that, since the radius of the orbit is known to about 10%, we can similarly bound the enclosed gravitational mass to the same level of uncertainty.

Let's take stock of this estimate for a moment. Astronomical estimates of the number of stars in the Milky Way suggests there are about 200 billion stars. There is also a supermassive black hole at the galactic center with a mass estimated from gravitational approaches to be about 4 million solar masses, representing a small fraction of the mass of our galaxy alone. Gas and dust have been surveyed via techniques such as radio astronomy and found to comprise about 10–15% of the mass of the Milky Way, most of that in the form of gas (hydrogen and some helium).

Most of the stars, gas, and dust in the Milky Way lie inside our orbit; we are located in the Sagittarius spiral arm of the galaxy, approximately two-thirds of the way from the center to the most apparent outermost stars. If we are generous, and assume that, on average, the 200 billion stars of our galaxy each have the mass of our sun, then we would expect that something like half the total mass of the galaxy is enclosed by our orbit. If we go twice as far out from the center of the galaxy, we can then assume that something more like 100% of the stars, gas, and dust will now be enclosed by an imagined stellar orbit. Stars orbiting the center of the Milky Way have been found at distances farther than that [Boc+14]. For example, star ULAS J001535.72+015549.6 has been estimated to be 274 ± 74 kpc from earth, putting it at a similar distance from the galactic center (Earth is only about $8 - 9$ kpc from the galactic center). This is a star 32 times the orbital radius of our sun.

Using this star, we can then estimate the speed with which a star at that distance would be orbiting. We again employ the virial theorem, this time estimating that $M_{2r_{sun}} = 2M(r_{sun})$. Let $r = r_{sun}$. You can show that it doesn't matter which assumed solar orbital radius and enclosed gravitational mass we assume in this calculation since the two linearly track one another and those assumptions cancel out in what now follows. We then solve for the speed of an orbiting star at this new distance,

$$v_{2r} = \sqrt{\frac{GM(2r)}{32r}} \tag{A.10.6}$$

$$= \sqrt{\frac{(6.674 \times 10^{11} \text{ kg} \cdot \text{m}^3/\text{s}^2) \times 2(11.50 \times 10^{10} \text{ kg})}{2 \times 7.947 \times 10^{20} \text{ m}}} \tag{A.10.7}$$

$$= \sqrt{3.840 \times 10^{10} \text{ m}^2/\text{s}^2} \tag{A.10.8}$$

$$= 196 \text{ km/s} \tag{A.10.9}$$

The above answer would lead us to the conclusion that a star at that distance should orbit about 20% slower than does the sun, yet observations of stars at distances comparable to this reveal that they orbit as fast as the sun. This, in turn, implies that as the orbital radii increase, so does the enclosed mass.

We provide a final observation on the subject of using kinematics alone to determine the mass enclosed in an orbit about the center of a galaxy. There are other methods that can be used to determine the mass profile of a galaxy such as the Milky Way. A recent measurement of the galactic mass enclosed by the sun's orbit was performed using 26 binary spinning neutron stars (binary pulsars) [Don+24]. The observed orbital periods of pairs of pulsars are Doppler-shifted relative to their true values due to the acceleration experienced by the pairs in their region of the Milky Way. Combining observations from a large number of pulsars allows for an independent mapping of the acceleration field inside our own galaxy. This approach has suggested that:

If an NFW model is applied to the interpretation of the data, the mass enclosed by the orbit of the sun about the galactic center is estimated to be 2.3×10^{11} $M_\odot$ (230 billion solar masses), about a factor of two higher than the maximum value we estimated above.

We note that our estimated maximum, calculated above by kinematic means, is consistent with previous professional kinematic estimates of the enclosed mass. The more-recent binary pulsar result is larger than those accepted values and based on a non-kinematic method complementary to past approaches. The authors of the pulsar work did not propagate uncertainties into their determined value, so it is difficult to assess the consistency of past results with this new one. Nevertheless, while future efforts using such approaches will likely mature in both their technical detail and the available population of observed binary pulsars, we note that such approaches have much promise for improved mapping of the Milky Way's mass distribution.

Answer to Exercise 11

Wien's displacement law is about asking and answering the following mathematics question: at what frequency (wavelength) does Planck's spectral radiance function take its maximal value? This stems from a fundamental question of calculus: given a function $f(x)$ dependent on a variable x, at what value of x does $f(x)$ take its maximal values?

In calculus, this question is answered using the first and second derivatives. The first derivative tells you the slope of the function at any point, x, while the second derivative tells you the kind of curvature (positive, negative, or zero) at x. We are looking for places in the function where the slope is *zero*, meaning it is a flat line at that point. Such features occur at maxima and minima, where the function "turns over," meaning it changes the direction of its trajectory in x. A positive (negative) second derivative tells us that the point is a minimum (maximum), as on either side of x the function increases (decreases) in value. A zero second derivative means there is no change in slope at this point, which can mean it is completely flat or a *saddle point*: a point with positive curvature on one side and negative curvature on the other.

We can graph Planck's spectral radiance function for a fixed temperature (say, 6000 K, the temperature of the sun's photosphere) and look at $B_f(E)$ as a function of f.

Let us now employ the desire to find when $B_f(E)$ is *maximal*. For this to be true, we must find the f for a fixed T, such that

$$\left. \frac{dB_f(E)}{dE} \right|_{f=f_{max}} = 0. \tag{A.11.1}$$

The spectral radiance is not a linear function in f, making it challenging to handle, but we can simplify its format by recognizing that all other numbers in the formula are either constant or are held constant (T), allowing us to absorb them into singular constants for the purpose of this exercise. Let us write it

$$B_f(E) = \frac{2f^2}{c^2} \frac{hf}{e^{hf/k_BT} - 1} = \alpha \frac{f^3}{e^{\beta f} - 1}. \tag{A.11.2}$$

We can then immediately see that this function, up to the constants $\alpha = 2h/c^2$ and $\beta = h/(k_BT)$ is of the form

$$f(x) = \frac{x^3}{e^{ax} - 1}. \tag{A.11.3}$$

We need solve only for x_0 such that $df/dx = 0$ when $x = x_0$. Let us begin by taking the derivative of this function. There are several ways to do this, including from first principles, but those are typically the domain of dedicated mathematics courses. Instead, we will employ computational tools that perform the integration symbolically and return the result in equation form. A good example of this is sympy, a symbolic mathematics library freely available in Python, an excellent open-source programming language. Many computational physics courses now use Python as the basis of their teaching.

A simple piece of code for executing symbolic first derivatives on functions of this type is as follows:

```
import sympy

spectral_radiance_template="x**3 / (exp(a*x) - 1)"
first_derivative = sympy.diff(spectral_radiance_template, "x")
sympy.print_latex(first_derivative)
```

This prints the LaTeX code for the mathematical formula resulting from the symbolic differentiation, yielding

$$-\frac{ax^3 e^{ax}}{(e^{ax} - 1)^2} + \frac{3x^2}{e^{ax} - 1} \tag{A.11.4}$$

The next step is to set the above equal to zero and rearrange the equation to solve for the value of x, x_0, that achieves this result.

$$0 = -\frac{ax^3 e^{ax}}{(e^{ax} - 1)^2} + \frac{3x^2}{e^{ax} - 1} \tag{A.11.5}$$

$$\frac{ax^3 e^{ax}}{(e^{ax} - 1)^2} = \frac{3x^2}{e^{ax} - 1} \tag{A.11.6}$$

$$\frac{axe^{ax}}{(e^{ax} - 1)^2} = \frac{3}{e^{ax} - 1} \tag{A.11.7}$$

$$\frac{axe^{ax}}{(e^{ax} - 1)} = 3 \tag{A.11.8}$$

$$axe^{ax} = 3(e^{ax} - 1) \tag{A.11.9}$$

$$ax = 3(1 - e^{-ax}) \tag{A.11.10}$$

$$ax + 3e^{-ax} = 3 \tag{A.11.11}$$

$$0 = 3 - ax - 3e^{-ax} \tag{A.11.12}$$

This is a nasty polynomial whose roots we now have to determine. Here we jump to computational techniques once more,

```
polynomial = "3 - a*x - 3*exp(-a*x)"
roots = sympy.solvers.solvers.solve(polynomial, "x")
print(roots)
sympy.print_latex(roots)
```

and find the roots to be of the form

$$\left[\frac{W\left(-\frac{3}{e^3}\right) + 3}{a}, \frac{W_{-1}\left(-\frac{3}{e^3}\right) + 3}{a}\right] \tag{A.11.13}$$

where W is the Lambert W function or *product logarithm*. This is a complicated subject, but we can also tackle this computationally,

```
from scipy.special import lambertw
from math import exp
root1 = lambertw(-3*exp(-3))+3
```

```
    root2 = lambertw(-3*exp(-3),-1)+3
    print(root1)
    print(root2)
```

which yields the first root, $2.8214393721220787/a$, and the second root, 0. We then determine that, since $a = \beta = h/(k_B T)$, for frequency, the maximum likely occurs at

$$f_0 \approx 2.8214 k_B T / h. \tag{A.11.14}$$

We have just solved for the location of the maximum ("peak") of the spectral radiance as a function of frequency. But that is not what is typically referred to as *Wien's law*, which instead relates temperature and wavelength. We must be careful here. Finding the wavelength-equivalent form of the above is not as simple as substituting $c = \lambda f$. In fact, you will get wrong results. Why is this so? It is because the wavelength is the reciprocal of the frequency, so the spectral density function for wavelength has *different units* than that for frequency. To convert between them, one must use calculus to transform one function to the other.

The spectral density in terms of wavelength and temperature looks, instead, like this:

$$B_\lambda(E) = \frac{2hc^2}{\lambda^5} \frac{1}{e^{hc/\lambda k_B T} - 1} = \alpha' \frac{1}{\lambda^5 \left(e^{\beta'/\lambda} - 1\right)}. \tag{A.11.15}$$

where $\alpha' = 2hc^2$ and $\beta' = hc/(k_B T)$. This is of the form

$$f(x) = \frac{1}{x^5 \left(e^{a/x} - 1\right)}. \tag{A.11.16}$$

We can immediately reuse our program and change the function we are differentiating,

```
import sympy

spectral_radiance_template="1/(x**5 * (exp(a/x) - 1))"
first_derivative = sympy.diff(spectral_radiance_template, "x")
sympy.print_latex(first_derivative)
```

yielding

$$\frac{ae^{\frac{a}{x}}}{x^7 \left(e^{\frac{a}{x}} - 1\right)^2} - \frac{5}{x^6 \left(e^{\frac{a}{x}} - 1\right)}. \tag{A.11.17}$$

We leave it to the reader to simplify this to an equation whose roots can then be determined, but we assert that this, the resulting polynomial, has the form

$$u - 5 + 5e^{-u} = 0 \tag{A.11.18}$$

where $u \equiv a/x$. We can update our root-finding program again,

```
polynomial = "u - 5 + 5*exp(-u)"
roots = sympy.solvers.solvers.solve(polynomial, "u")
print(roots)
sympy.print_latex(roots)
```

which yields

$$\left[0,\ W\left(-\frac{5}{e^5}\right)+5\right] \tag{A.11.19}$$

for the roots. We again compute the results using the Lambert W function,

```
from scipy.special import lambertw
from math import exp
root2 = lambertw(-5*exp(-5))+5
print(root2)
```

from which we find $u_0 = 4.965114231744276$. Since $u = a/x$, $x = \lambda$ in this part of the derivation, and $a = \beta' = hc/(k_BT)$, we arrive at

$$\lambda_0 = \frac{a}{u_0} = \frac{hc}{k_B u_0 T} \approx \frac{2.8996 \times 10^{-3}\ \mathrm{m \cdot K}}{T} \tag{A.11.20}$$

where the number 2.8996×10^{-3} m $\cdot$ K is often denoted by the letter b such that the peak wavelength is $\lambda_{peak} = b/T$. This is traditionally what is known as *Wien's displacement law*.

Answer to Exercise 12

We will approximate a solution to this problem using the equation for the Einstein ring angular size. We recognize this is not quite correct but it provides a useful baseline approximation to the correct answer.

We have been provided in the text with the distance to the lens ($D_d = 1400$ Mpc) and the distance to the original quasar that lies just about along the line of sight ($D_{ds} = 4400$ Mpc). From this, we can infer the distance from the observer (us) to the quasar, $D_s = 9800$ Mpc.

The Einstein ring angular deflection equation (Eqn. 2.96) can then be rearranged to solve for the mass of the lens in terms of the other quantities in the problem,

$$\theta_{Einstein} = \sqrt{\frac{D_{ds}}{D_s D_d}\frac{4GM(\xi)}{c^2}} \tag{A.12.1}$$

$$\theta_{Einstein}^2 = \frac{D_{ds}}{D_s D_d}\frac{4GM(\xi)}{c^2} \tag{A.12.2}$$

$$\frac{\theta_{Einstein}^2 c^2 D_s D_d}{4G D_{ds}} = M(\xi). \tag{A.12.3}$$

We have all the numbers we need to solve the problem. This yields a lens mass of about 7.6×10^{42} kg, or about 4 trillion solar masses. A galaxy like ours or the Andromeda galaxy contains something like a trillion solar masses, so this lens is a large galaxy. However, the inferred lens mass—even using this Einstein ring approximations—is not at all inconsistent with galactic masses.

Answer to Exercise 13

We learned from the solution to Exercise 11 that there is a direct relationship between the temperature of a system and the most prominent frequency of radiation emitted by it, when the system can be modeled as a blackbody. Evidence from microwave and radio astronomy suggests that the ambient radiation field of the universe can be almost entirely explained by assuming it was emitted by a near-perfect blackbody. As the universe expanded, the frequency distribution has shifted from its original shape but maintains the near-perfect blackbody spectrum. Using the frequency spectral radiance function, we can relate the most prominent emitted frequency to the temperature of the blackbody by

$$f_0 \approx 2.8214 k_B T / h. \tag{A.13.1}$$

We can invert this equation to instead solve for the temperature of that system knowing the most prominent frequency in the emitted electromagnetic radiation,

$$T = \frac{h f_0}{2.8214 k_B}. \tag{A.13.2}$$

We answer the question by utilizing the relationship, *Wien's displacement law*. The other terms in this equation are constants of nature: Planck's constant (h), and the Boltzmann constant (k_B). We find that the current temperature of the universe must therefore be around 2.7 K. The currently accepted measurement of 2.72548 ± 0.00057 K [Fix09] is based on a combination of data from the COBE and WMAP satellites.

This question is "simple" in that it doesn't require difficult mathematics (once one understands the blackbody model) to answer. We include it here to demonstrate that a little basic knowledge goes a long way in physics. Relating microscopic properties (a specific and most prominent emitted frequency) to the temperature of a much larger and more complex system is a powerful skill. Not all systems will allow its description by this model, but many common ones do—the photospheres of stars, even the universe itself.

Answer to Exercise 14

This problem is meant to prompt a few actions from the reader. First, we want you to build a habit of looking up measurements from comprehensive resources. Second, we want you to learn to incorporate the usage of probability into your calculations.

In the book's text, we have not provided sufficient information for you to answer some initial questions:

- How often does the Higgs particle decay to a pair of W bosons?

- How often does the W boson decay to a final state that includes at least one of the lightest charged leptons, the electron or muon?

We recommend that you keep handy (either electronically or in paper form) a copy of *The Review of Particle Physics* [Wor+22]. This resource is curated by the Particle Data Group and is formally updated biennially (including publication in a peer-reviewed journal) with incremental electronic updates in the intermediate years. It contains a comprehensive list of the state of knowledge of all fundamental and composite particle states, beginning with the quarks, leptons, and fundamental forces.

We consulted the latest version of the document (electronically, on the web) and found:

$$\mathcal{B}(H^0 \to W^*W) = (25.7 \pm 2.5)\% \tag{A.14.1}$$

$$\mathcal{B}(W^+ \to e^+\nu_e) = (10.71 \pm 0.16)\% \tag{A.14.2}$$

$$\mathcal{B}(W^+ \to \mu^+\nu_\mu) = (10.63 \pm 0.15)\%. \tag{A.14.3}$$

$$\tag{A.14.4}$$

We can learn something immediately from looking at these numbers. First, the branching fraction for the Higgs is less-well-known than for the W bosons. The relative uncertainty on the Higgs branching fraction is $2.5/25.7 = 9.7\%$, while the relative uncertainty on either of the two W branching fractions is 1.4%. Why are these so different?

While it is generally true that reconstructing the Higgs decay is more challenging than reconstructing the W decay (the latter involves two final-state particles while the former involves four when one is using the WW final state with decay to light leptons), the simple fact is that it takes time to improve measurements. The W boson was discovered in the early 1980s and has been directly studied by at least a half-dozen experiments since then. The Higgs was only discovered in 2012 and has been studied by two experiments.

The other lesson is one built into the SM: *lepton universality* of the W boson interactions. This part of the SM is blind to the lepton to which the W boson couples. At the level of the fundamental weak coupling, the W boson doesn't know the difference between an electron, a muon, or a tau lepton. Kinematically, there are differences in physical W-lepton processes: the tau is much heavier than the electron, limiting the kinematic phase space allowed if the W decays to the tau. While this can act to suppress the rate at which the W decays to the tau, at the fundamental level of the interaction itself the W cannot discriminate between leptons. We can see this in the branching fractions to the electron and muon: they are effectively identical within the limits of the measurements.

Let us pause and engage some statistics. Assuming that the W boson measurements are completely independent (with no statistical or systematic uncertainties in common between them), then we can also treat the uncertainties as completely independent. This allows us to combine errors in simple quadrature when performing arithmetic operations on the two numbers.

For example, let us subtract the two branching fractions. This will provide a measure of how close to zero is this difference (once we incorporate the combined uncertainties). Ignoring correlations, the total uncertainty on either a sum or a difference of two numbers with errors is

$$\Delta = \mathcal{B}_1 - \mathcal{B}_2 \longrightarrow \sigma_\Delta = \sqrt{\sigma_{\mathcal{B}_1}^2 + \sigma_{\mathcal{B}_2}^2}. \tag{A.14.5}$$

We remind the reader that despite this being the error on the *difference* of two numbers, the uncertainty on that difference (Δ) is nonetheless the result of *summing* the squared-errors. We recommend that the reader review error analysis, which is grounded in calculus and probability theory (c.f. Ref. [BR03]). In short, the above numbers are not true values and their errors are not absolute. The branching fractions are estimates of the mean of these numbers and their errors are related to the variance of measurements around the mean. When we combine branching fractions, we have to combine the means and the variances carefully.

Error analysis gives us a framework to compute the variance on Δ based on the variances of the branching fractions,

$$\sigma_\Delta^2 \approx \sqrt{\left(\frac{\partial \Delta}{\partial \mathcal{B}_1}\right)^2 \sigma_{\mathcal{B}_1}^2 + \left(\frac{\partial \Delta}{\partial \mathcal{B}_2}\right)^2 \sigma_{\mathcal{B}_2}^2}. \tag{A.14.6}$$

The partial derivative of each branching fraction is computed by fixing the other branching fractions at their mean values (those determined from measurement). Thus $\partial \Delta / \partial \mathcal{B}_1 = 1$ and $\partial \Delta / \partial \mathcal{B}_2 = -1$. However, each of these is squared in the computation of the variance of Δ, so we obtain Eqn. A.14.5.

When we apply Eqn. A.14.5 we find that $\sigma_\Delta = 0.22\%$. Thus $\Delta = (0.08 \pm 0.22)\%$. This result tells us that the observed difference between these two branching fractions is highly consistent with zero. The measurement errors of each number would need to be reduced below 0.056% to begin determining whether the 0.08% difference is significant.

We return to the original goal of this question: to estimate how often $H \to WW$ can fake $H \to$ invisible. In order for this to happen, a particle detector would have to fail to observe the electrons and/or muons from the W decays. We begin by counting the number of possible unique final states of the $H \to WW$ decay process, accounting only for the lighter lepton decays,

$$H \to WW \to e\nu_e e\nu_e \tag{A.14.7}$$
$$\to e\nu_e \mu\nu_\mu \tag{A.14.8}$$
$$\to \mu\nu_\mu e\nu_e \tag{A.14.9}$$
$$\to \mu\nu_\mu \mu\nu_\mu. \tag{A.14.10}$$
$$\tag{A.14.11}$$

We are saving space with a loose shorthand here. We know that one of the W bosons must decay to a positively charged lepton and the other to a negatively charged lepton in order to ensure charge conservation at every stage of the decay. We have dropped the charge (and particle/antiparticle) distinctions to save time, but please keep in mind that charge conservation and lepton number conservation (the conservation of flavor) is implied throughout this calculation.

We note that there are four overall final states. We will never detect the neutrinos (they produce real missing energy), so we need only focus on the charged leptons and whether or not we lose them. To simplify the rules of this game, if we identify either of the two charged leptons, we would reject this process from consideration as $H \to$ invisible. Therefore, *both* charged leptons have to be lost for this process to become a background.

The probability of the W boson to decay to the electron (muon) final state can be denoted $\mathcal{B}_e$ ($\mathcal{B}_\mu$). The probability of *detecting* electrons (muons) can be denoted P_e (P_μ). Since there are only two outcomes—detect the lepton or don't detect the lepton—we can write the probability to *lose the electron (muon)* as $Q_e = 1 - P_e$ ($Q_\mu = 1 - P_\mu$). This is known as a *binomial probability* for each lepton (a two-outcome system). The probability to lose the neutrinos is 100%, so we can universally write $Q_\nu = 1$ for both kinds of neutrino in this problem.

Probability theory gives us the final tool with which to convert these ideas into calculations. In probability theory, we can consider two possible outcomes. Let us label them A and B. Let the probability of A's occurring be P_A and the probability of B's occurring be P_B. If these are uncorrelated (the probability of one does not depend on the other, and vice-versa) then the composition of probabilities will be as follows. If we want to know the probability that A occurs and then B occurs, this is given as $P_{AB} = P_A \cdot P_B$. In our case, if a W boson decays to $e\nu_e$ and we want to know the probability that the electron is missed and then that the neutrino is missed, we need only multiply those probabilities (since the loss of the electron has nothing to do with the loss of the neutrino), $Q_e \cdot Q_\nu$.

If we have two W bosons and we want to know the probability that one decays to the electron and the other to the muon, we can write $\mathcal{B}_e \cdot \mathcal{B}_\mu$ (since the decay of one W is not influenced by the decay of the other, as they are independent processes). If we want to quantify the probability of this decay occurring and that we subsequently lose the electron and the muon, we can compose this as $\mathcal{B}_e Q_e \mathcal{B}_\mu Q_\mu$.

We are now ready to replace the above decay outcomes with equations representing the probability of losing each one, thus mis-classifying the decays as invisible. For each particle written above we can replace it with the *probability to lose the particle*,

$$P(H \rightarrow WW \Longrightarrow H \rightarrow invisible) = \mathcal{B}_e Q_e Q_\nu \mathcal{B}_e Q_e Q_\nu + \tag{A.14.12}$$
$$\mathcal{B}_e Q_e Q_\nu \mathcal{B}_\mu Q_\mu Q_\nu +$$
$$\mathcal{B}_\mu Q_\mu Q_\nu \mathcal{B}_e Q_e Q_\nu +$$
$$\mathcal{B}_\mu Q_\mu Q_\nu \mathcal{B}_\mu Q_\mu Q_\nu.$$

$$\tag{A.14.13}$$

This can be further simplified to

$$P(H \rightarrow WW \Longrightarrow H \rightarrow invisible) = (\mathcal{B}_e Q_e Q_\nu)^2 + \tag{A.14.14}$$
$$2\mathcal{B}_e Q_e Q_\nu \mathcal{B}_\mu Q_\mu Q_\nu +$$
$$(\mathcal{B}_\mu Q_\mu Q_\nu)^2.$$

$$\tag{A.14.15}$$

We can now use the numbers we have been given (or looked up) to compute this probability, $P(H \rightarrow WW \Longrightarrow H \rightarrow invisible) = 0.0056$ (about a half-percent chance).

We are finally prepared to answer the original question. The Higgs particle decays to WW 25.7% of the time. When that happens there is a 0.56% chance that $H \rightarrow WW$ will result in lighter leptonic final states that can get misreconstructed as "invisible." Thus the probability (out of all Higgs decays) that this entire chain of events happens is $25.7\% \times 0.0056 = 0.14\%$.

What does this mean? It means that if we have accounted for all ways of losing electrons and muons from a process like this, then the ultimate limit of sensitivity for $H \to invisible$, based on this background alone, would be a branching fraction of 0.14%. We would have a very hard time seeing the invisible Higgs decay at a precision better than this because we would continue to misreconstruct these Higgs decays as if they were invisible.

We note that this is a simplification of the real problem with these Higgs decays. The numbers we provided for P_e and P_μ are only the typical *identification* probabilities for electrons and muons. That means that there is the chance that an electron or muon that passes through the detector systems would be identified correctly. We have neglected the *acceptance* probabilities—the possibility that the electron or muon actually misses the detector components entirely, and thus never be identified.

Acceptance depends on the typical angles, relative to the colliding beams, with which electrons and muons leave the parent process. Generally, the smaller those angles, the lower the acceptance, since particle detectors at colliders have less instrumentation closer to the beamline.

In fact, the probability of losing an electron or muon is somewhat higher than what we have used here. However, we did this because acceptance depends heavily on the parent process (the momentum of the Higgs, the orientation of the W bosons relative to the beamline when they are produced, etc.). This is a simplified view of the calculation, but it can be made more complicated as needed.

Bibliography

[Aab+16]   Aaboud, M. et al. "Measurement of the Inelastic Proton-Proton Cross Section at $\sqrt{s} = 13$ TeV with the ATLAS Detector at the LHC." In: *Phys. Rev. Lett.* vol. 117, no. 18 (2016), p. 182002. arXiv: 1606.02625 [hep-ex].

[Aab+18a]   Aaboud, M. et al. "Search for an invisibly decaying Higgs boson or dark matter candidates produced in association with a Z boson in pp collisions at $\sqrt{s} = 13$ TeV with the ATLAS detector." In: *Phys. Lett. B* vol. 776 (2018), pp. 318–337. arXiv: 1708.09624 [hep-ex].

[Aab+18b]   Aaboud, M. et al. "Performance of missing transverse momentum reconstruction with the ATLAS detector using proton-proton collisions at $\sqrt{s} = 13$ TeV." In: *Eur. Phys. J. C* vol. 78, no. 11 (2018), p. 903. arXiv: 1802.08168 [hep-ex].

[Aab+20]   Aab, A. et al. "Features of the Energy Spectrum of Cosmic Rays above 2.5×10^{18} eV Using the Pierre Auger Observatory." In: *Phys. Rev. Lett.* vol. 125, no. 12 (2020), p. 121106. arXiv: 2008.06488 [astro-ph.HE].

[Aad+22]   Aad, G. et al. "Search for invisible Higgs-boson decays in events with vector-boson fusion signatures using 139 fb^{-1} of proton-proton data recorded by the ATLAS experiment." In: *JHEP* vol. 08 (2022), p. 104. arXiv: 2202.07953 [hep-ex].

[Aad+23a]   Aad, G. et al. "Combination of searches for invisible decays of the Higgs boson using 139 fb-1 of proton-proton collision data at s=13 TeV collected with the ATLAS experiment." In: *Phys. Lett. B* vol. 842 (2023), p. 137963. arXiv: 2301.10731 [hep-ex].

[Aad+23b]   Aad, G. et al. "Improving topological cluster reconstruction using calorimeter cell timing in ATLAS." In: (Oct. 2023). arXiv: 2310.16497 [physics.ins-det].

[Aai+20]   Aaij, R. et al. "Search for $A^{'} \to \mu^+\mu^-$ Decays." In: *Phys. Rev. Lett.* vol. 124 (4 Jan. 2020), p. 041801.

[Aal+18]   Aalseth, C. E. et al. "DarkSide-20k: A 20 tonne two-phase LAr TPC for direct dark matter detection at LNGS." In: *Eur. Phys. J. Plus* vol. 133 (2018), p. 131. arXiv: 1707.08145 [physics.ins-det].

[Aal+22]   Aalbers, J. et al. *First Dark Matter Search Results from the LUX-ZEPLIN (LZ) Experiment.* 2022.

[Aar+17]   Aartsen, M. G. et al. "The IceCube Neutrino Observatory: Instrumentation and Online Systems." In: *JINST* vol. 12, no. 03 (2017), P03012. arXiv: 1612.05093 [astro-ph.IM].

[Abb+05]   Abbott, T. et al. "The Dark Energy Survey." In: (Oct. 2005). arXiv: astro-ph/0510346.

[Abb+23a]   Abbasi, R. et al. "Observation of high-energy neutrinos from the Galactic plane." In: *Science* vol. 380, no. 6652 (2023), adc9818. arXiv: 2307.04427 [astro-ph.HE].

[Abb+23b]   Abbott, R. et al. "Search for gravitational-lensing signatures in the full third observing run of the LIGO-Virgo network." In: (Apr. 2023). arXiv: 2304.08393 [gr-qc].

[Abd+18a]   Abdalla, H. et al. "The H.E.S.S. Galactic plane survey." In: *Astron. Astrophys.* vol. 612 (2018), A1. arXiv: 1804.02432 [astro-ph.HE].

[Abd+18b]   Abdallah, H. et al. "Search for γ-Ray Line Signals from Dark Matter Annihilations in the Inner Galactic Halo from 10 Years of Observations with H.E.S.S." In: *Phys. Rev. Lett.* vol. 120, no. 20 (2018), p. 201101. arXiv: 1805.05741 [astro-ph.HE].

[Abd+19]   Abdelhameed, A. H. et al. "First results from the CRESST-III low-mass dark matter program." In: *Physical Review D* vol. 100, no. 10 (Nov. 2019).

[Abe+20a]   Abel, C. et al. "Measurement of the permanent electric dipole moment of the neutron." In: *Phys. Rev. Lett.* vol. 124, no. 8 (2020), p. 081803. arXiv: 2001.11966 [hep-ex].

[Abe+20b]   Abel, C. et al. "Measurement of the Permanent Electric Dipole Moment of the Neutron." In: *Phys. Rev. Lett.* vol. 124 (8 Feb. 2020), p. 081803.

[Abe+23]   Abeysekara, A. U. et al. "The High-Altitude Water Cherenkov (HAWC) observatory in México: The primary detector." In: *Nucl. Instrum. Meth. A* vol. 1052 (2023), p. 168253. arXiv: 2304.00730 [astro-ph.HE].

[Abl+17]   Ablikim, M. et al. "Dark Photon Search in the Mass Range Between 1.5 and 3.4 GeV/c^2." In: *Phys. Lett. B* vol. 774 (2017), pp. 252–257. arXiv: 1705.04265 [hep-ex].

[Abr+04]   Abraham, J. et al. "Properties and performance of the prototype instrument for the Pierre Auger Observatory." In: *Nuclear Instruments and Methods in Physics Research Section A: Accelerators, Spectrometers, Detectors and Associated Equipment* vol. 523, no. 1 (2004), pp. 50–95.

[Abr+19]   Abramoff, O. et al. "SENSEI: Direct-Detection Constraints on Sub-GeV Dark Matter from a Shallow Underground Run Using a Prototype Skipper-CCD." In: *Phys. Rev. Lett.* vol. 122, no. 16 (2019), p. 161801. arXiv: 1901.10478 [hep-ex].

[Acc+08]   Acciari, V. A. et al. "VERITAS Observations of the gamma-Ray Binary LS I +61 303." In: *Astrophys. J.* vol. 679 (2008), p. 1427. arXiv: 0802.2363 [astro-ph].

[ACO89]   Abell, G. O., Corwin Harold G., J., and Olowin, R. P. "A Catalog of Rich Clusters of Galaxies." In: *The Astrophysical Journal Supplement Series* vol. 70 (May 1989), p. 1.

[Ada+15]   Adare, A. et al. "Search for dark photons from neutral meson decays in $p + p$ and $d +$ Au collisions at $\sqrt{s_{NN}} = 200$ GeV." In: *Phys. Rev. C* vol. 91, no. 3 (2015), p. 031901. arXiv: 1409.0851 [nucl-ex].

[Ada+23]   Adari, P. et al. "SENSEI: First Direct-Detection Results on sub-GeV Dark Matter from SENSEI at SNOLAB." In: (Dec. 2023). arXiv: 2312.13342 [astro-ph.CO].

[Adh+19]   Adhikari, G. et al. "Search for a Dark Matter-Induced Annual Modulation Signal in NaI(Tl) with the COSINE-100 Experiment." In: *Phys. Rev. Lett.* vol. 123, no. 3 (2019), p. 031302. arXiv: 1903.10098 [astro-ph.IM].

[Adh+20]   Adhikari, P. et al. "Constraints on dark matter-nucleon effective couplings in the presence of kinematically distinct halo substructures using the DEAP-3600 detector." In: *Phys. Rev. D* vol. 102 (8 Oct. 2020), p. 082001.

[Adh+21]   Adhikari, G. et al. "Strong constraints from COSINE-100 on the DAMA dark matter results using the same sodium iodide target." In: *Science Advances* vol. 7, no. 46 (2021), eabk2699. eprint: https://www.science.org/doi/pdf/10.1126/sciadv.abk2699.

[Adh+22a]   Adhikari, G. et al. "Three-year annual modulation search with COSINE-100." In: *Phys. Rev. D* vol. 106, no. 5 (2022), p. 052005. arXiv: 2111.08863 [hep-ex].

[Adh+22b]   Adhikari, P. et al. "Erratum: Constraints on dark matter-nucleon effective couplings in the presence of kinematically distinct halo substructures using the DEAP-3600 detector [Phys. Rev. D 102, 082001 (2020)]." In: *Phys. Rev. D* vol. 105 (2 Jan. 2022), p. 029901.

[ADV18]   Arvanitaki, A., Dimopoulos, S., and Van Tilburg, K. "Resonant absorption of bosonic dark matter in molecules." In: *Phys. Rev. X* vol. 8, no. 4 (2018), p. 041001. arXiv: 1709.05354 [hep-ph].

[AFH14] Anand, N., Fitzpatrick, A. L., and Haxton, W. C. "Weakly interacting massive particle-nucleus elastic scattering response." In: *Phys. Rev. C* vol. 89, no. 6 (2014), p. 065501. arXiv: 1308.6288 [hep-ph].

[Aga+14] Agakishiev, G. et al. "Searching a Dark Photon with HADES." In: *Phys. Lett. B* vol. 731 (2014), pp. 265–271. arXiv: 1311.0216 [hep-ex].

[Age+11] Ageron, M. et al. "ANTARES: the first undersea neutrino telescope." In: *Nucl. Instrum. Meth. A* vol. 656 (2011), pp. 11–38. arXiv: 1104.1607 [astro-ph.IM].

[Age+20] Ageron, M. et al. "Dependence of atmospheric muon flux on seawater depth measured with the first KM3NeT detection units: The KM3NeT Collaboration." In: *Eur. Phys. J. C* vol. 80, no. 2 (2020), p. 99. arXiv: 1906.02704 [physics.ins-det].

[Agh+20a] Aghanim, N. et al. "Planck 2018 results. I. Overview and the cosmological legacy of Planck." In: *Astron. Astrophys.* vol. 641 (2020), A1. arXiv: 1807.06205 [astro-ph.CO].

[Agh+20b] Aghanim, N. et al. "Planck 2018 results. VI. Cosmological parameters." In: *Astron. Astrophys.* vol. 641 (2020). [Erratum: Astron.Astrophys. 652, C4 (2021)], A6. arXiv: 1807.06209 [astro-ph.CO].

[Agn+13] Agnese, R. et al. "Demonstration of surface electron rejection with interleaved germanium detectors for dark matter searches." In: *Applied Physics Letters* vol. 103, no. 16 (Oct. 2013), p. 164105. eprint: https://pubs.aip.org/aip/apl/article-pdf/doi/10.1063/1.4826093/13676279/164105_1_online.pdf.

[Agn+16] Agnese, R. et al. "New Results from the Search for Low-Mass Weakly Interacting Massive Particles with the CDMS Low Ionization Threshold Experiment." In: *Phys. Rev. Lett.* vol. 116, no. 7 (2016), p. 071301. arXiv: 1509.02448 [astro-ph.CO].

[Agn+17a] Agnese, R. et al. "Projected Sensitivity of the SuperCDMS SNOLAB experiment." In: *Phys. Rev. D* vol. 95, no. 8 (2017), p. 082002. arXiv: 1610.00006 [physics.ins-det].

[Agn+17b] Agnese, R. et al. "Projected Sensitivity of the SuperCDMS SNOLAB experiment." In: *Phys. Rev. D* vol. 95, no. 8 (2017), p. 082002. arXiv: 1610.00006 [physics.ins-det].

[Agn+18a] Agnes, P. et al. "Constraints on Sub-GeV Dark-MatterElectron Scattering from the DarkSide-50 Experiment." In: *Phys. Rev. Lett.* vol. 121, no. 11 (2018), p. 111303. arXiv: 1802.06998 [astro-ph.CO].

[Agn+18b] Agnes, P. et al. "Measurement of the liquid argon energy response to nuclear and electronic recoils." In: *Phys. Rev. D* vol. 97, no. 11 (2018), p. 112005. arXiv: 1801.06653 [physics.ins-det].

[Agn+18c] Agnese, R. et al. "First Dark Matter Constraints from a SuperCDMS Single-Charge Sensitive Detector." In: *Phys. Rev. Lett.* vol. 121, no. 5 (2018). [Erratum: Phys.Rev.Lett. 122, 069901 (2019)], p. 051301. arXiv: 1804.10697 [hep-ex].

[Agn+18d] Agnese, R. et al. "Low-mass dark matter search with CDMSlite." In: *Phys. Rev. D* vol. 97 (2 Jan. 2018), p. 022002.

[Agn+20] Agnes, P. et al. "Effective field theory interactions for liquid argon target in DarkSide-50 experiment." In: *Phys. Rev. D* vol. 101 (6 Mar. 2020), p. 062002.

[Agn+23] Agnes, P. et al. "Search for Dark-Matter–Nucleon Interactions via Migdal Effect with DarkSide-50." In: *Phys. Rev. Lett.* vol. 130 (10 Mar. 2023), p. 101001.

[Ago+03] Agostinelli, S. et al. "Geant4 - a simulation toolkit." In: *Nuclear Instruments and Methods in Physics Research Section A: Accelerators, Spectrometers, Detectors and Associated Equipment* vol. 506, no. 3 (2003), pp. 250–303.

[Agu+02] Aguilar, M. et al. "The Alpha Magnetic Spectrometer (AMS) on the International Space Station. I: Results from the test flight on the space shuttle." In: *Phys. Rept.* vol. 366 (2002). [Erratum: Phys.Rept. 380, 97–98 (2003)], pp. 331–405.

[Agu+15a] Aguilar, M. et al. "Precision Measurement of the Proton Flux in Primary Cosmic Rays from Rigidity 1 GV to 1.8 TV with the Alpha Magnetic Spectrometer on the International Space Station." In: *Phys. Rev. Lett.* vol. 114 (17 Apr. 2015), p. 171103.

[Agu+15b] Aguilar-Arevalo, A. et al. "Measurement of radioactive contamination in the high-resistivity silicon CCDs of the DAMIC experiment." In: *JINST* vol. 10, no. 08 (2015), P08014. arXiv: 1506.02562 [astro-ph.IM].

[Agu+17] Aguilar-Arevalo, A. et al. "First Direct-Detection Constraints on eV-Scale Hidden-Photon Dark Matter with DAMIC at SNOLAB." In: *Phys. Rev. Lett.* vol. 118, no. 14 (2017), p. 141803. arXiv: 1611.03066 [astro-ph.CO].

[Agu+19] Aguilar-Arevalo, A. et al. "Constraints on Light Dark Matter Particles Interacting with Electrons from DAMIC at SNOLAB." In: *Phys. Rev. Lett.* vol. 123, no. 18 (2019), p. 181802. arXiv: 1907.12628 [astro-ph.CO].

[Agu+20] Aguilar-Arevalo, A. et al. *Results on low-mass weakly interacting massive particles from a 11 kg-day target exposure of DAMIC at SNOLAB*. 2020. arXiv: 2007.15622 [astro-ph.CO].

[Agu+21] Aguilar, M. et al. "The Alpha Magnetic Spectrometer (AMS) on the international space station: Part II Results from the first seven years." In: *Phys. Rept.* vol. 894 (2021), pp. 1–116.

[Agu+22] Aguilar-Arevalo, A. et al. *The Oscura Experiment*. 2022. arXiv: 2202.10518 [astro-ph.IM].

[Aja+19a] Ajaj, R. et al. "Search for dark matter with a 231-day exposure of liquid argon using DEAP-3600 at SNOLAB." In: *Phys. Rev. D* vol. 100, no. 2 (2019), p. 022004. arXiv: 1902.04048 [astro-ph.CO].

[Aja+19b] Ajaj, R. et al. "Search for dark matter with a 231-day exposure of liquid argon using DEAP-3600 at SNOLAB." In: *Phys. Rev. D* vol. 100 (2 July 2019), p. 022004.

[Ake+17] Akerib, D. S. et al. "Limits on Spin-Dependent WIMP-Nucleon Cross Section Obtained from the Complete LUX Exposure." In: *Phys. Rev. Lett.* vol. 118 (25 June 2017), p. 251302.

[Alb+17] Albert, A. et al. "Searching for Dark Matter Annihilation in Recently Discovered Milky Way Satellites with Fermi-LAT." In: *Astrophys. J.* vol. 834, no. 2 (2017), p. 110. arXiv: 1611.03184 [astro-ph.HE].

[Alb+22] Albakry, M. F. et al. "Ionization yield measurement in a germanium CDMSlite detector using photo-neutron sources." In: *Phys. Rev. D* vol. 105 (12 June 2022), p. 122002.

[Alb+23] Albakry, M. F. et al. "Search for low-mass dark matter via bremsstrahlung radiation and the Migdal effect in SuperCDMS." In: *Phys. Rev. D* vol. 107, no. 11 (2023), p. 112013. arXiv: 2302.09115 [hep-ex].

[Ale+16] Alekhin, S. et al. "A facility to Search for Hidden Particles at the CERN SPS: the SHiP physics case." In: *Rept. Prog. Phys.* vol. 79, no. 12 (2016), p. 124201. arXiv: 1504.04855 [hep-ph].

[Alk+20] Alkhatib, I. et al. "Light Dark Matter Search with a High-Resolution Athermal Phonon Detector Operated Above Ground." In: (July 2020). arXiv: 2007.14289 [hep-ex].

[All+06] Allison, J. et al. "Geant4 developments and applications." In: *IEEE Transactions on Nuclear Science* vol. 53, no. 1 (2006), pp. 270–278.

[All+16] Allison, J. et al. "Recent developments in Geant4." In: *Nuclear Instruments and Methods in Physics Research Section A: Accelerators, Spectrometers, Detectors and Associated Equipment* vol. 835 (2016), pp. 186–225.

[Ama+19a] Amaré, J. et al. "Performance of ANAIS-112 experiment after the first year of data taking." In: *Eur. Phys. J. C* vol. 79, no. 3 (2019), p. 228. arXiv: 1812.01472 [astro-ph.IM].

[Ama+19b]   Amaudruz, P.-A. et al. "Design and construction of the DEAP-3600 dark matter detector." In: *Astroparticle Physics* vol. 108 (2019), pp. 1–23.

[Ama+20]   Amaral, D. W. et al. "Constraints on low-mass, relic dark matter candidates from a surface-operated SuperCDMS single-charge sensitive detector." In: *Phys. Rev. D* vol. 102, no. 9 (2020), p. 091101. arXiv: 2005.14067 [hep-ex].

[Ama+21]   Amare, J. et al. "Annual Modulation Results from Three Years Exposure of ANAIS-112." In: (Mar. 2021). arXiv: 2103.01175 [astro-ph.IM].

[Amo+15]   Amole, C. et al. "Dark Matter Search Results from the PICO-2L C_3F_8 Bubble Chamber." In: *Phys. Rev. Lett.* vol. 114 (23 June 2015), p. 231302.

[Amo+16]   Amole, C. et al. "Improved dark matter search results from PICO-2L Run 2." In: *Phys. Rev. D* vol. 93 (6 Mar. 2016), p. 061101.

[Amo+17]   Amole, C. et al. "Dark Matter Search Results from the PICO$-$60 C_3F_8 Bubble Chamber." In: *Phys. Rev. Lett.* vol. 118 (25 June 2017), p. 251301.

[Amo+19]   Amole, C. et al. "Dark matter search results from the complete exposure of the PICO-60 C3F8 bubble chamber." In: *Physical Review D* vol. 100, no. 2 (July 2019).

[Ana+15]   Anastasi, A. et al. "Limit on the production of a low-mass vector boson in $e^+e^- \to U\gamma$, $U \to e^+e^-$ with the KLOE experiment." In: *Phys. Lett. B* vol. 750 (2015), pp. 633–637. arXiv: 1509.00740 [hep-ex].

[Ana+16]   Anastasiou, C. et al. "High precision determination of the gluon fusion Higgs boson cross-section at the LHC." In: *JHEP* vol. 05 (2016), p. 058. arXiv: 1602.00695 [hep-ph].

[Ana+18]   Anastasi, A. et al. "Combined limit on the production of a light gauge boson decaying into $\mu^+\mu^-$ and $\pi^+\pi^-$." In: *Phys. Lett. B* vol. 784 (2018), pp. 336–341. arXiv: 1807.02691 [hep-ex].

[And+07]   Andriamonje, S. et al. "An Improved limit on the axion-photon coupling from the CAST experiment." In: *JCAP* vol. 04 (2007), p. 010. arXiv: hep-ex/0702006.

[and16]   and, A. F. "Distillation column for the XENON1T experiment." In: *Journal of Physics: Conference Series* vol. 718 (May 2016), p. 042020.

[Ang+19]   Angloher, G. et al. "Limits on dark matter effective field theory parameters with CRESST-II." In: *The European Physical Journal C* vol. 79, no. 1 (Jan. 2019), p. 43.

[Ant+19]   Antonello, M. et al. "The SABRE project and the SABRE Proof-of-Principle." In: *Eur. Phys. J. C* vol. 79, no. 4 (2019), p. 363. arXiv: 1806.09340 [physics.ins-det].

[Ant+22]   Anthony-Petersen, R. et al. "A Stress Induced Source of Phonon Bursts and Quasiparticle Poisoning." In: (Aug. 2022). arXiv: 2208.02790 [physics.ins-det].

[Ant+23]   Anthony-Petersen, R. et al. "Applying Superfluid Helium to Light Dark Matter Searches: Demonstration of the HeRALD Detector Concept." In: (July 2023). arXiv: 2307.11877 [physics.ins-det].

[Apr+17]   Aprile, E. et al. "Effective field theory search for high-energy nuclear recoils using the XENON100 dark matter detector." In: *Phys. Rev. D* vol. 96 (4 Aug. 2017), p. 042004.

[Apr+19a]   Aprile, E. et al. "Search for Light Dark Matter Interactions Enhanced by the Migdal Effect or Bremsstrahlung in XENON1T." In: *Phys. Rev. Lett.* vol. 123, no. 24 (2019), p. 241803. arXiv: 1907.12771 [hep-ex].

[Apr+19b]   Aprile, E. et al. "First Results on the Scalar WIMP-Pion Coupling, Using the XENON1T Experiment." In: *Phys. Rev. Lett.* vol. 122 (7 Feb. 2019), p. 071301.

[Apr+19c]   Aprile, E. et al. "Light Dark Matter Search with Ionization Signals in XENON1T." In: *Phys. Rev. Lett.* vol. 123 (25 Dec. 2019), p. 251801.

[Apr+23]   Aprile, E. et al. "First Dark Matter Search with Nuclear Recoils from the XENONnT Experiment." In: (Mar. 2023). arXiv: 2303.14729 [hep-ex].

[Ara+20a] Aramaki, T. et al. "Snowmass 2021 Letter of Interest: The GRAMS Project: MeV Gamma-Ray Observations and Antimatter-Based Dark Matter Searches." In: (Sept. 2020). arXiv: 2009.03754 [astro-ph.HE].

[Ara+20b] Aramaki, T. et al. "Dual MeV Gamma-Ray and Dark Matter Observatory - GRAMS Project." In: *Astropart. Phys.* vol. 114 (2020), pp. 107–114. arXiv: 1901.03430 [astro-ph.HE].

[Arc+12] Archambault, S. et al. "Constraints on low-mass WIMP interactions on 19F from PICASSO." In: *Physics Letters B* vol. 711, no. 2 (2012), pp. 153–161.

[Ark+09] Arkani-Hamed, N. et al. "A theory of dark matter." In: *Phys. Rev. D* vol. 79 (1 Jan. 2009), p. 015014.

[Arm+19] Armengaud, E. et al. "Physics potential of the International Axion Observatory (IAXO)." In: *JCAP* vol. 06 (2019), p. 047. arXiv: 1904.09155 [hep-ph].

[Arm+22] Armengaud, E. et al. "Search for sub-GeV dark matter via the Migdal effect with an EDELWEISS germanium detector with NbSi transition-edge sensors." In: *Phys. Rev. D* vol. 106 (6 Sept. 2022), p. 062004.

[Arn+18a] Arnaud, Q. et al. "Optimizing EDELWEISS detectors for low-mass WIMP searches." In: *Phys. Rev. D* vol. 97, no. 2 (2018), p. 022003. arXiv: 1707.04308 [physics.ins-det].

[Arn+18b] Arnaud, Q. et al. "First results from the NEWS-G direct dark matter search experiment at the LSM." In: *Astroparticle Physics* vol. 97 (2018), pp. 54–62.

[Arn+18c] Arnaud, Q. et al. "Optimizing EDELWEISS detectors for low-mass WIMP searches." In: *Phys. Rev. D* vol. 97 (2 Jan. 2018), p. 022003.

[Arn+20] Arnaud, Q. et al. "First germanium-based constraints on sub-MeV Dark Matter with the EDELWEISS experiment." In: *Phys. Rev. Lett.* vol. 125, no. 14 (2020), p. 141301. arXiv: 2003.01046 [astro-ph.GA].

[Arn+24] Arnquist, I. et al. "The DAMIC-M Low Background Chamber." In: (July 2024). arXiv: 2407.17872 [physics.ins-det].

[AS19] Arza, A. and Sikivie, P. "Production and detection of an axion dark matter echo." In: *Phys. Rev. Lett.* vol. 123, no. 13 (2019), p. 131804. arXiv: 1902.00114 [hep-ph].

[Asz+10a] Asztalos, S. J. et al. "SQUID-Based Microwave Cavity Search for Dark-Matter Axions." In: *Phys. Rev. Lett.* vol. 104 (4 Jan. 2010), p. 041301.

[Asz+10b] Asztalos, S. J. et al. "SQUID-Based Microwave Cavity Search for Dark-Matter Axions." In: *Physical Review Letters* vol. 104, no. 4 (Jan. 2010).

[ATL24] ATLAS Collaboration. "Standard Model Summary Plots June 2024." In: *ATL-PHYS-PUB-2024-011* (2024).

[Att+22] Attanasio, A. et al. "Snowmass 2021 White Paper: The Windchime Project." In: *Snowmass 2021.* Mar. 2022. arXiv: 2203.07242 [hep-ex].

[Atw+09] Atwood, W. B. et al. "THE LARGE AREA TELESCOPE ON THE FERMI GAMMA-RAY SPACE TELESCOPE MISSION." In: *The Astrophysical Journal* vol. 697, no. 2 (May 2009), p. 1071.

[Aub+09] Aubert, B. et al. "Search for Dimuon Decays of a Light Scalar Boson in Radiative Transitions Upsilon —> gamma A0." In: *Phys. Rev. Lett.* vol. 103 (2009), p. 081803. arXiv: 0905.4539 [hep-ex].

[Ayb+21] Aybas, D. et al. "Search for Axionlike Dark Matter Using Solid-State Nuclear Magnetic Resonance." In: *Phys. Rev. Lett.* vol. 126, no. 14 (2021), p. 141802. arXiv: 2101.01241 [hep-ex].

[Bal+17] Ballesteros, G. et al. "Standard ModelaxionseesawHiggs portal inflation. Five problems of particle physics and cosmology solved in one stroke." In: *JCAP* vol. 08 (2017), p. 001. arXiv: 1610.01639 [hep-ph].

[Bal+21]   Balogh, L. et al. "Copper electroplating for background suppression in the NEWS-G experiment." In: *Nucl. Instrum. Meth. A* vol. 988 (2021), p. 164844. arXiv: 2008.03153 `[physics.ins-det]`.

[Bal+23]   Balogh, L. et al. *Exploring light dark matter with the DarkSPHERE spherical proportional counter electroformed underground at the Boulby Underground Laboratory.* 2023. arXiv: 2301.05183 `[hep-ex]`.

[Bal+99]   Balser, D. S. et al. "3He in the Milky Way Interstellar Medium: Abundance Determinations." In: *The Astrophysical Journal* vol. 510 (1999), pp. 759–783.

[Bar+20a]   Baracchini, E. et al. "CYGNO: a gaseous TPC with optical readout for dark matter directional search." In: *JINST* vol. 15, no. 07 (2020), p. C07036. arXiv: 2007.12627 `[physics.ins-det]`.

[Bar+20b]   Barak, L. et al. "SENSEI: Direct-Detection Results on sub-GeV Dark Matter from a New Skipper-CCD." In: *Phys. Rev. Lett.* vol. 125, no. 17 (2020), p. 171802. arXiv: 2004.11378 `[astro-ph.CO]`.

[bas16]   basicavisual. *Satellite.* OpenClipArt.org. [Online; accessed 4-April-2024]. 2016.

[Bat+17]   Battat, J. B. R. et al. "Low Threshold Results and Limits from the DRIFT Directional Dark Matter Detector." In: *Astropart. Phys.* vol. 91 (2017), pp. 65–74. arXiv: 1701.00171 `[astro-ph.IM]`.

[Bau+22]   Baudis, L. et al. "Snowmass 2021 Underground Facilities and Infrastructure Overview Topical Report." In: (Dec. 2022). arXiv: 2212.07037 `[hep-ex]`.

[Bax+21]   Baxter, D. et al. "Recommended conventions for reporting results from direct dark matter searches." In: *Eur. Phys. J. C* vol. 81, no. 10 (2021), p. 907. arXiv: 2105.00599 `[hep-ex]`.

[BCT07]   Barbeau, P. S., Collar, J. I., and Tench, O. "Large-mass ultralow noise germanium detectors: performance and applications in neutrino and astroparticle physics." In: *Journal of Cosmology and Astroparticle Physics* vol. 2007, no. 09 (Sept. 2007), pp. 009–009.

[Beh+12]   Behnke, E. et al. "First dark matter search results from a 4-kgCF3Ibubble chamber operated in a deep underground site." In: *Physical Review D* vol. 86, no. 5 (Sept. 2012).

[Bel+18]   Belokurov, V. et al. "Co-formation of the disc and the stellar halo." In: *Monthly Notices of the Royal Astronomical Society* vol. 478, no. 1 (June 2018), pp. 611–619. eprint: https://academic.oup.com/mnras/article-pdf/478/1/611/25005885/sty982.pdf.

[Bel04]   Bellerive, A. "Review of solar neutrino experiments." In: vol. 19. Mar. 2004, pp. 385–397.

[Ber+18]   Bernabei, R. et al. "First model independent results from DAMA/LIBRA-phase2." In: *Nucl. Phys. Atom. Energy* vol. 19, no. 4 (2018), pp. 307–325. arXiv: 1805.10486 `[hep-ex]`.

[Bez+18]   Bezrukov, L. B. et al. "New Low-Background Laboratory in the Pyhäsalmi Mine, Finland." In: *Phys. Part. Nucl.* vol. 49, no. 4 (2018), pp. 769–773.

[BFS14]   Billard, J., Figueroa-Feliciano, E., and Strigari, L. "Implication of neutrino backgrounds on the reach of next generation dark matter direct detection experiments." In: *Physical Review D* vol. 89, no. 2 (Jan. 2014).

[BFS20]   Buschmann, M., Foster, J. W., and Safdi, B. R. "Early-Universe Simulations of the Cosmological Axion." In: *Phys. Rev. Lett.* vol. 124, no. 16 (2020), p. 161103. arXiv: 1906.00967 `[astro-ph.CO]`.

[BFS22]   Buen-Abad, M. A., Fan, J., and Sun, C. "Axion echoes from the supernova graveyard." In: *Phys. Rev. D* vol. 105, no. 7 (2022), p. 075006. arXiv: 2110.13916 `[hep-ph]`.

[BH06]   Boulay, M. and Hime, A. "Technique for direct detection of weakly interacting massive particles using scintillation time discrimination in liquid argon." In: *Astroparticle Physics* vol. 25, no. 3 (2006), pp. 179–182.

[BH18]   Bertone, G. and Hooper, D. "History of dark matter." In: *Rev. Mod. Phys.* vol. 90, no. 4 (2018), p. 045002. arXiv: 1605.04909 `[astro-ph.CO]`.

[BHK08]   Bae, K. J., Huh, J.-H., and Kim, J. E. "Update of axion CDM energy." In: *JCAP* vol. 09 (2008), p. 005. arXiv: 0806.0497 [hep-ph].

[Bil+24]   Billard, J. et al. "Transition Edge Sensors with Sub-eV Resolution And Cryogenic Targets (TESSERACT) at the underground laboratory of Modane (LSM)." In: *Nucl. Phys. B* vol. 1003 (2024), p. 116465.

[Bla+21]   Blanco, C. et al. "Dark Matter Daily Modulation With Anisotropic Organic Crystals." In: *Phys. Rev. D* vol. 104 (2021), p. 036011. arXiv: 2103.08601 [hep-ph].

[Bla+22]   Blanco, C. et al. "Molecular Migdal effect." In: *Phys. Rev. D* vol. 106, no. 11 (2022), p. 115015. arXiv: 2208.09002 [hep-ph].

[Bli+24]   Blinov, N. et al. "Dark Matter Searches on a Photonic Chip." In: (Jan. 2024). arXiv: 2401.17260 [hep-ph].

[BM12]   Barker, D. and Mei, D.-M. "Germanium detector response to nuclear recoils in searching for dark matter." In: *Astroparticle Physics* vol. 38 (2012), pp. 1–6.

[Bo+24]   Bo, Z. et al. "Dark Matter Search Results from 1.54 Tonne·Year Exposure of PandaX-4T." In: (Aug. 2024). arXiv: 2408.00664 [hep-ex].

[Boc+14]   Bochanski, J. J. et al. "The Most Distant Stars in the Milky Way." In: *The Astrophysical Journal* vol. 790, no. 1 (July 2014), p. L5.

[Bor+16]   Borsanyi, S. et al. "Calculation of the axion mass based on high-temperature lattice quantum chromodynamics." In: *Nature* vol. 539, no. 7627 (2016), pp. 69–71. arXiv: 1606.07494 [hep-lat].

[Boy+14]   Boyarsky, A. et al. "Unidentified Line in X-Ray Spectra of the Andromeda Galaxy and Perseus Galaxy Cluster." In: *Phys. Rev. Lett.* vol. 113 (25 Dec. 2014), p. 251301.

[Boz+16]   Bozorgnia, N. et al. "Simulated Milky Way analogues: implications for dark matter direct searches." In: *JCAP* vol. 05 (2016), p. 024. arXiv: 1601.04707 [astro-ph.CO].

[BP78]   Bryman, D. and Picciotto, C. "Double Beta Decay." In: *Rev. Mod. Phys.* vol. 50 (1978), pp. 11–21.

[BPZ20]   Bae, K. J., Park, M., and Zhang, M. "Demystifying freeze-in dark matter at the LHC." In: *Phys. Rev. D* vol. 101, no. 11 (2020), p. 115036. arXiv: 2001.02142 [hep-ph].

[BR03]   Bevington, P. R. and Robinson, D. K. *Data Reduction and Error Analysis for the Physical Sciences.* McGraw-Hill Higher Education, 2003.

[Bru+19a] Brun, P. et al. "A new experimental approach to probe QCD axion dark matter in the mass range above 40 μeV." In: *Eur. Phys. J. C* vol. 79, no. 3 (2019), p. 186. arXiv: 1901.07401 [physics.ins-det].

[Bru+19b] Brun, P. et al. "A new experimental approach to probe QCD axion dark matter in the mass range above 40 μeV." In: *Eur. Phys. J. C* vol. 79, no. 3 (2019), p. 186. arXiv: 1901.07401 [physics.ins-det].

[BS05]   Bahcall, J. N. and Serenelli, A. M. "How do uncertainties in the surface chemical abundances of the Sun affect the predicted solar neutrino fluxes?" In: *Astrophys. J.* vol. 626 (2005), p. 530. arXiv: astro-ph/0412096.

[Bud+14]   Budker, D. et al. "Proposal for a Cosmic Axion Spin Precession Experiment (CASPEr)." In: *Phys. Rev. X* vol. 4, no. 2 (2014), p. 021030. arXiv: 1306.6089 [hep-ph].

[Bug+98]   Bugaev, E. V. et al. "Atmospheric muon flux at sea level, underground, and underwater." In: *Phys. Rev. D* vol. 58 (5 July 1998), p. 054001.

[Bul+14]   Bulbul, E. et al. "Detection of An Unidentified Emission Line in the Stacked X-ray spectrum of Galaxy Clusters." In: *Astrophys. J.* vol. 789 (2014), p. 13. arXiv: 1402.2301 [astro-ph.CO].

[Bul+16]   Bulbul, E. et al. "SEARCHING FOR THE 3.5 keV LINE IN THE STACKED SUZAKU OBSERVATIONS OF GALAXY CLUSTERS." In: *The Astrophysical Journal* vol. 831, no. 1 (Oct. 2016), p. 55.

[Bus+95]   Buskulic, D. et al. "Performance of the ALEPH detector at LEP." In: *Nuclear Instruments and Methods in Physics Research Section A: Accelerators, Spectrometers, Detectors and Associated Equipment* vol. 360, no. 3 (1995), pp. 481–506.

[CA13]   Chepel, V. and Araújo, H. "Liquid noble gas detectors for low energy particle physics." In: *Journal of Instrumentation* vol. 8, no. 04 (Apr. 2013), R04001–R04001.

[Cah+12]   Cahill-Rowley, M. W. et al. "The new look pMSSM with neutralino and gravitino LSPs." In: *The European Physical Journal C* vol. 72, no. 9 (Sept. 2012).

[Cal+21]   Calaprice, F. et al. "High sensitivity characterization of an ultrahigh purity NaI(Tl) crystal scintillator with the SABRE proof-of-principle detector." In: *Phys. Rev. D* vol. 104, no. 2 (2021), p. L021302. arXiv: 2105.09225 [physics.ins-det].

[Cao+14]   Cao, J. et al. "A light SUSY dark matter after CDMS-II, LUX and LHC Higgs data." In: *Journal of High Energy Physics* vol. 2014, no. 5 (May 2014).

[Car+20]   Carney, D. et al. "Proposal for gravitational direct detection of dark matter." In: *Phys. Rev. D* vol. 102, no. 7 (2020), p. 072003. arXiv: 1903.00492 [hep-ph].

[Car+23]   Cardani, L. et al. "Disentangling the sources of ionizing radiation in superconducting qubits." In: *Eur. Phys. J. C* vol. 83, no. 1 (2023), p. 94. arXiv: 2211.13597 [quant-ph].

[CCW15]   Calore, F., Cholis, I., and Weniger, C. "Background Model Systematics for the Fermi GeV Excess." In: *JCAP* vol. 03 (2015), p. 038. arXiv: 1409.0042 [astro-ph.CO].

[Cha+90]   Chandler, C. E. et al. "Sub-electron noise charge-coupled devices." In: *Charge-Coupled Devices and Solid State Optical Sensors*. Ed. by Blouke, M. M. Vol. 1242. International Society for Optics and Photonics. SPIE, 1990, pp. 238–251.

[Che+21]   Cheng, C. et al. "Search for Light Dark Matter–Electron Scattering in the PandaX-II Experiment." In: *Phys. Rev. Lett.* vol. 126 (21 May 2021), p. 211803.

[Che+23]   Chen, S. et al. "Quantum Enhancement in Dark Matter Detection with Quantum Computation." In: (Nov. 2023). arXiv: 2311.10413 [hep-ph].

[Cli+22]   Cline, E. et al. "Searching for New Physics with DarkLight at the ARIEL Electron-Linac." In: *J. Phys. Conf. Ser.* vol. 2391, no. 1 (2022), p. 012010. arXiv: 2208.04120 [nucl-ex].

[CMS19]   CMS Collaboration. "Pileup mitigation at CMS in 13 TeV data." In: *CMS-PAS-JME-18-001* (2019).

[Col18]   Collar, J. I. "Search for a nonrelativistic component in the spectrum of cosmic rays at Earth." In: *Phys. Rev. D* vol. 98, no. 2 (2018), p. 023005. arXiv: 1805.02646 [astro-ph.CO].

[Col23]   Collaboration, C. *Detector*. [Online; accessed 16-Nov-2023]. 2023.

[Coo+18]   Cooley, J. et al. "Input comparison of radiogenic neutron estimates for ultra-low background experiments." In: *Nuclear Instruments and Methods in Physics Research Section A: Accelerators, Spectrometers, Detectors and Associated Equipment* vol. 888 (2018), pp. 110–118.

[Cos+21]   Coskuner, A. et al. "Directional Dark Matter Detection in Anisotropic Dirac Materials." In: *Phys. Rev. D* vol. 103, no. 1 (2021), p. 016006. arXiv: 1909.09170 [hep-ph].

[Cos+22]   Coskuner, A. et al. "Directional detectability of dark matter with single phonon excitations: Target comparison." In: *Phys. Rev. D* vol. 105, no. 1 (2022), p. 015010. arXiv: 2102.09567 [hep-ph].

[CS00]   Corbelli, E. and Salucci, P. "The extended rotation curve and the dark matter halo of M33." In: *Monthly Notices of the Royal Astronomical Society* vol. 311, no. 2 (Jan. 2000), pp. 441–447.

[CW13]   Cohen, T. and Wacker, J. G. "Here be dragons: the unexplored continents of the CMSSM." In: *Journal of High Energy Physics* vol. 2013, no. 9 (Sept. 2013).

[Cyb+16]   Cyburt, R. H. et al. "Big Bang Nucleosynthesis: 2015." In: *Rev. Mod. Phys.* vol. 88 (2016), p. 015004. arXiv: 1505.01076 [astro-ph.CO].

[Dai+19]   Dainese, A. et al., eds. *Report on the Physics at the HL-LHC,and Perspectives for the HE-LHC.* Vol. 7/2019. CERN Yellow Reports: Monographs. Geneva, Switzerland: CERN, 2019.

[Dar00]   Dar, S. "The Neutron EDM in the SM: A Review." In: (Aug. 2000). eprint: hep-ph/0008248.

[Day+16]   Daylan, T. et al. "The characterization of the gamma-ray signal from the central Milky Way: A case for annihilating dark matter." In: *Phys. Dark Univ.* vol. 12 (2016), pp. 1–23. arXiv: 1402.6703 [astro-ph.HE].

[De +21]   De Angelis, A. et al. "Gamma-ray astrophysics in the MeV range: The ASTROGAM concept and beyond." In: *Experimental Astronomy* vol. 51, no. 3 (June 2021), pp. 1225–1254.

[Del14]   Del Popolo, A. "Nonbaryonic Dark Matter in Cosmology." In: *Int. J. Mod. Phys. D* vol. 23 (2014), p. 1430005. arXiv: 1305.0456 [astro-ph.CO].

[DFS86]   Drukier, A. K., Freese, K., and Spergel, D. N. "Detecting cold dark-matter candidates." In: *Phys. Rev. D* vol. 33 (12 Jan. 1986), pp. 3495–3508.

[Dia+22]   Diaz Ortiz, M. et al. "Design of the ALPS II optical system." In: *Physics of the Dark Universe* vol. 35 (2022), p. 100968.

[Dib15]   Dib, C. O. "ANDES: An Underground Laboratory in South America." In: *Physics Procedia* vol. 61 (2015). 13th International Conference on Topics in Astroparticle and Underground Physics, TAUP 2013, pp. 534–541.

[DKG07]   Duda, G., Kemper, A., and Gondolo, P. "Model Independent Form Factors for Spin Independent Neutralino-Nucleon Scattering from Elastic Electron Scattering Data." In: *JCAP* vol. 04 (2007), p. 012. arXiv: hep-ph/0608035.

[DKL24]   Das, A., Kurinsky, N., and Leane, R. K. "Dark Matter Induced Power in Quantum Devices." In: *Phys. Rev. Lett.* vol. 132, no. 12 (2024), p. 121801. arXiv: 2210.09313 [hep-ph].

[Don+24]   Donlon II, T. et al. *Galactic Structure From Binary Pulsar Accelerations: Beyond Smooth Models.* 2024. arXiv: 2401.15808 [astro-ph.GA].

[DP19]   Dzhatdoev, T. and Podlesnyi, E. "Massive Argon Space Telescope (MAST): A concept of heavy time projection chamber for γ-ray astronomy in the 100 MeV1 TeV energy range." In: *Astropart. Phys.* vol. 112 (2019), pp. 1–7. arXiv: 1902.01491 [astro-ph.HE].

[Drl+15]   Drlica-Wagner, A. et al. "Eight Ultra-faint Galaxy Candidates Discovered in Year Two of the Dark Energy Survey." In: *Astrophys. J.* vol. 813, no. 2 (2015), p. 109. arXiv: 1508.03622 [astro-ph.GA].

[DRS20]   Dessert, C., Rodd, N. L., and Safdi, B. R. "The dark matter interpretation of the 3.5-keV line is inconsistent with blank-sky observations." In: *Science* vol. 367, no. 6485 (2020), pp. 1465–1467. arXiv: 1812.06976 [astro-ph.CO].

[Dru+12]   Drukier, A. et al. "New Dark Matter Detectors using DNA or RNA for Nanometer Tracking." In: (June 2012). arXiv: 1206.6809 [astro-ph.IM].

[Du+18]   Du, N. et al. "Search for Invisible Axion Dark Matter with the Axion Dark Matter Experiment." In: *Phys. Rev. Lett.* vol. 120 (15 Apr. 2018), p. 151301.

[EGN91]   Ellis, J., Gaillard, M. K., and Nanopoulos, D. "A Phenomenological Profile of the Higgs Boson." In: *The Standard Model Higgs Boson.* Ed. by EINHORN, M. Vol. 8. Current PhysicsSources and Comments. Elsevier, 1991, pp. 24–72.

[Ein65]   Einasto, J. "Kinematics and Dynamics of Stellar Systems." In: *Trudy Inst. Astrofiz. Alma-Ata* vol. 5 (1965), p. 87.

[EK18]   Emken, T. and Kouvaris, C. "How blind are underground and surface detectors to strongly interacting dark matter?" In: *Phys. Rev. D* vol. 97 (11 June 2018), p. 115047.

[EMV12]   Essig, R., Mardon, J., and Volansky, T. "Direct detection of sub-GeV dark matter." In: *Phys. Rev. D* vol. 85 (7 Apr. 2012), p. 076007.

[EPR24]   Ema, Y., Pospelov, M., and Ray, A. "Probing earth-bound dark matter with nuclear reactors." In: *JHEP* vol. 07 (2024), p. 094. arXiv: 2402.03431 [hep-ph].

[Esc+05]   Esch, E.-I. et al. "The Cosmic ray muon flux at WIPP." In: *Nucl. Instrum. Meth. A* vol. 538 (2005), pp. 516–525. arXiv: astro-ph/0408486.

[Ess+17]   Essig, R. et al. "Detection of sub-GeV Dark Matter and Solar Neutrinos via Chemical-Bond Breaking." In: *Phys. Rev. D* vol. 95, no. 5 (2017), p. 056011. arXiv: 1608.02940 [hep-ph].

[Ess+19]   Essig, R. et al. "Direct Detection of Spin-(In)dependent Nuclear Scattering of Sub-GeV Dark Matter Using Molecular Excitations." In: *Phys. Rev. Research.* vol. 1 (2019), p. 033105. arXiv: 1907.07682 [hep-ph].

[Ess+22]   Essig, R. et al. "Snowmass2021 Cosmic Frontier: The landscape of low-threshold dark matter direct detection in the next decade." In: *Snowmass 2021.* Mar. 2022. arXiv: 2203.08297 [hep-ph].

[ESY18]   Essig, R., Sholapurkar, M., and Yu, T.-T. "Solar Neutrinos as a Signal and Background in Direct-Detection Experiments Searching for Sub-GeV Dark Matter With Electron Recoils." In: *Phys. Rev. D* vol. 97, no. 9 (2018), p. 095029. arXiv: 1801.10159 [hep-ph].

[EVY17]   Essig, R., Volansky, T., and Yu, T.-T. "New Constraints and Prospects for sub-GeV Dark Matter Scattering off Electrons in Xenon." In: *Phys. Rev. D* vol. 96, no. 4 (2017), p. 043017. arXiv: 1703.00910 [hep-ph].

[FD20]   Filippi, A. and De Napoli, M. "Searching in the dark: the hunt for the dark photon." In: *Reviews in Physics* vol. 5 (2020), p. 100042.

[Fer+18]   Fernández, V. et al. "Primordial helium abundance determination using sulphur as metallicity tracer." In: *Monthly Notices of the Royal Astronomical Society* vol. 478, no. 4 (May 2018), pp. 5301–5319.

[FGL20]   Fabbrichesi, M., Gabrielli, E., and Lanfranchi, G. "The Dark Photon." In: (May 2020). arXiv: 2005.01515 [hep-ph].

[Fie+20]   Fields, B. D. et al. "Big-Bang Nucleosynthesis after Planck." In: *JCAP* vol. 03 (2020). [Erratum: JCAP 11, E02 (2020)], p. 010. arXiv: 1912.01132 [astro-ph.CO].

[Fit+12]   Fitzpatrick, A. L. et al. "Model Independent Direct Detection Analyses." In: (Nov. 2012). arXiv: 1211.2818 [hep-ph].

[Fit+13]   Fitzpatrick, A. L. et al. "The effective field theory of dark matter direct detection." In: *Journal of Cosmology and Astroparticle Physics* vol. 2013, no. 02 (Feb. 2013), pp. 004–004.

[Fix09]   Fixsen, D. J. "THE TEMPERATURE OF THE COSMIC MICROWAVE BACKGROUND." In: *The Astrophysical Journal* vol. 707, no. 2 (Nov. 2009), pp. 916–920.

[Flo+17]   Florian, D. de et al. *Handbook of LHC Higgs Cross Sections: 4. Deciphering the Nature of the Higgs Sector.* CERN Yellow Reports: Monographs. 869 pages, 295 figures, 248 tables and 1645 citations. Working Group web page: https://twiki.cern.ch/twiki/bin/view/LHCPhysics/LHCHXSWG. Geneva: CERN, 2017.

[FLS13]   Freese, K., Lisanti, M., and Savage, C. "Colloquium: Annual modulation of dark matter." In: *Rev. Mod. Phys.* vol. 85 (2013), pp. 1561–1581. arXiv: 1209.3339 [astro-ph.CO].

[Fu+17]   Fu, C. et al. "Spin-Dependent Weakly-Interacting-Massive-Particle–Nucleon Cross Section
            Limits from First Data of PandaX-II Experiment." In: *Phys. Rev. Lett.* vol. 118 (7 Feb.
            2017), p. 071301.

[Fuk+03]   Fukuda, S. et al. "The Super-Kamiokande detector." In: *Nuclear Instruments and Methods
            in Physics Research Section A: Accelerators, Spectrometers, Detectors and Associated
            Equipment* vol. 501, no. 2 (2003), pp. 418–462.

[Gar+17]   Garcon, A. et al. "The Cosmic Axion Spin Precession Experiment (CASPEr): a dark-
            matter search with nuclear magnetic resonance." In: (July 2017). arXiv: 1707.05312
            [physics.ins-det].

[Gar+19]   Garcon, A. et al. "Constraints on bosonic dark matter from ultralow-field nuclear magnetic
            resonance." In: *Science Advances* vol. 5, no. 10 (2019), eaax4539.

[Gas+21]   Gascon, J. et al. *Low-mass Dark Matter searches with EDELWEISS.* 2021. arXiv:
            2112.05467 [physics.ins-det].

[Gel15]   Gelmini, G. B. "The Hunt for Dark Matter." In: *Theoretical Advanced Study Institute in
            Elementary Particle Physics: Journeys Through the Precision Frontier: Amplitudes for
            Colliders.* Feb. 2015. arXiv: 1502.01320 [hep-ph].

[GF23]   Grohs, E. and Fuller, G. M. "Big Bang Nucleosynthesis." In: *Handbook of Nuclear
            Physics.* Ed. by Tanihata, I., Toki, H., and Kajino, T. 2023, pp. 1–21. arXiv: 2301.12299
            [astro-ph.CO].

[GKW20]   Geilhufe, R. M., Kahlhoefer, F., and Winkler, M. W. "Dirac Materials for Sub-MeV Dark
            Matter Detection: New Targets and Improved Formalism." In: *Phys. Rev. D* vol. 101,
            no. 5 (2020), p. 055005. arXiv: 1910.02091 [hep-ph].

[Gly95]   Glyde, H. R. *Excitations in Liquid and Solid Helium.* Oxford University Press, Apr. 1995.

[Gon96]   Gondolo, P. "Phenomenological introduction to direct dark matter detection." In: *31st
            Rencontres de Moriond: Dark Matter and Cosmology, Quantum Measurements and
            Experimental Gravitation.* 1996. arXiv: hep-ph/9605290.

[GP19]   Gates, S. and Pelletier, C. *Proving Einstein Right: The Daring Expeditions that Changed
            How We Look at the Universe.* PublicAffairs, 2019.

[Gri+16]   Grilli di Cortona, G. et al. "The QCD axion, precisely." In: *JHEP* vol. 01 (2016), p. 034.
            arXiv: 1511.02867 [hep-ph].

[Gri+18]   Griffin, S. et al. "Directional Detection of Light Dark Matter with Polar Materials." In:
            Phys. Rev. D vol. 98, no. 11 (2018), p. 115034. arXiv: 1807.10291 [hep-ph].

[Gri12]   Griffiths, D. J. "Resource Letter EM-1: Electromagnetic Momentum." In: *American
            Journal of Physics* vol. 80, no. 1 (Jan. 2012), pp. 7–18. eprint: https://pubs.aip.org/
            aapt/ajp/article-pdf/80/1/7/13118704/7_1_online.pdf.

[GSM20]   Ghosh, O., Salvado, J., and Miralda-Escudé, J. "Axion Gegenschein: Probing Back-
            scattering of Astrophysical Radio Sources Induced by Dark Matter." In: (Aug. 2020).
            arXiv: 2008.02729 [astro-ph.CO].

[Guo+21]   Guo, Z. et al. "Muon flux measurement at China Jinping Underground Laboratory." In:
            Chin. Phys. C vol. 45, no. 2 (2021), p. 025001. arXiv: 2007.15925 [physics.ins-det].

[Gur+22a]   Gurian, J. et al. "A Lower Bound on the Mass of Compact Objects from Dissipative
            Dark Matter." In: *Astrophys. J. Lett.* vol. 939, no. 1 (2022). [Erratum: Astrophys.J.Lett.
            949, L44 (2023), Erratum: Astrophys.J. 949, L44 (2023)], p. L12. arXiv: 2209.00064
            [astro-ph.CO].

[Gur+22b]   Gurian, J. et al. "Molecular Chemistry for Dark Matter II: Recombination, Molecule
            Formation, and Halo Mass Function in Atomic Dark Matter." In: *Astrophys. J.* vol. 934
            (2022), p. 121. arXiv: 2110.11964 [astro-ph.CO].

[Hal+10]   Hall, L. J. et al. "Freeze-In Production of FIMP Dark Matter." In: *JHEP* vol. 03 (2010), p. 080. arXiv: 0911.1120 [hep-ph].

[Hal23]   Hall, J. "An intersection of radiation detection and superconducting electronics." In: *AIP Conf. Proc.* vol. 2908, no. 1 (2023), p. 040002.

[Har+24]   Harrington, P. M. et al. "Synchronous Detection of Cosmic Rays and Correlated Errors in Superconducting Qubit Arrays." In: (Feb. 2024). arXiv: 2402.03208 [quant-ph].

[Hei22]   Heise, J. "The Sanford Underground Research Facility." In: *Snowmass 2021*. Mar. 2022. arXiv: 2203.08293 [hep-ex].

[Hel+18]   Helmi, A. et al. "The merger that led to the formation of the Milky Way's inner stellar halo and thick disk." In: *Nature* vol. 563, no. 7729 (Nov. 2018), pp. 85–88.

[Hel56]   Helm, R. H. "Inelastic and Elastic Scattering of 187-Mev Electrons from Selected Even-Even Nuclei." In: *Phys. Rev.* vol. 104 (5 Dec. 1956), pp. 1466–1475.

[Hes15]   Hesketh, G. "How does an experiment at the Large Hadron Collider work?" In: *Phys.org* (June 2015).

[HM06]   Herr, W. and Muratori, B. "Concept of luminosity." In: (2006).

[HM84]   Halzen, F. and Martin, A. D. *Quarks & Leptons: An Introductory Course in Modern Particle Physics.* John Wiley & Sons, 1984.

[Hoc+19]   Hochberg, Y. et al. "Detecting Sub-GeV Dark Matter with Superconducting Nanowires." In: *Phys. Rev. Lett.* vol. 123, no. 15 (2019), p. 151802. arXiv: 1903.05101 [hep-ph].

[Hol86]   Holdom, B. "Two U(1)'s and ϵ charge shifts." In: *Physics Letters B* vol. 166, no. 2 (1986), pp. 196–198.

[HOS08]   Hernandez-Garcia, C., O'Shea, P. G., and Stutzman, M. L. "Electron sources for accelerators." In: *Physics Today* vol. 61, no. 2 (Feb. 2008), pp. 44–49.

[HS16]   Hartman, N. and Sekula, S. "The Solar Solution: Tracking the Sun with Low Energy Neutrinos." In: (May 2016). arXiv: 1605.01787 [hep-ph].

[Hun+10]   Hunter, S. D. et al. "Development of the Advance Energetic Pair Telescope (AdEPT) for medium-energy gamma-ray astronomy." In: *Space Telescopes and Instrumentation 2010: Ultraviolet to Gamma Ray.* Ed. by Arnaud, M., Murray, S. S., and Takahashi, T. Vol. 7732. International Society for Optics and Photonics. SPIE, 2010, p. 773221.

[Hus16]   Husdal, L. "On Effective Degrees of Freedom in the Early Universe." In: *Galaxies* vol. 4, no. 4 (2016), p. 78. arXiv: 1609.04979 [astro-ph.CO].

[HWX12]   Hooper, D., Weiner, N., and Xue, W. "Dark forces and light dark matter." In: *Phys. Rev. D* vol. 86 (5 Sept. 2012), p. 056009.

[Ibe+18]   Ibe, M. et al. "Migdal Effect in Dark Matter Direct Detection Experiments." In: *JHEP* vol. 03 (2018), p. 194. arXiv: 1707.07258 [hep-ph].

[Ito+24]   Ito, A. et al. "Quantum entanglement of ions for light dark matter detection." In: *JHEP* vol. 02 (2024), p. 124. arXiv: 2311.11632 [hep-ph].

[Ive+19]   Ivezi, . et al. "LSST: from Science Drivers to Reference Design and Anticipated Data Products." In: *Astrophys. J.* vol. 873, no. 2 (2019), p. 111. arXiv: 0805.2366 [astro-ph].

[J L63a]   J. Lindhard V. Nielsen, M. S. "Integral Equations Governing Radiation Effects." In: *Matematisk-fysiske Meddelelse rudgivet af Det Kongelige Danske Videnskabernes Selskab* vol. 33, no. 10 (1963).

[J L63b]   J. Lindhard V. Nielsen, M. S. "Range Concepts and Heavy Ion Ranges." In: *Matematisk-fysiske Meddelelse rudgivet af Det Kongelige Danske Videnskabernes Selskab* vol. 33, no. 13 (1963).

[J L68]   J. Lindhard V. Nielsen, M. S. "Approximation Method in Classical Scattering by Screened Coulomb Fields." In: *Matematisk-fysiske Meddelelse rudgivet af Det Kongelige Danske Videnskabernes Selskab* vol. 36, no. 10 (1968).

[Jac+20]   Jackson Kimball, D. F. et al. "Overview of the Cosmic Axion Spin Precession Experiment (CASPEr)." In: *Springer Proc. Phys.* vol. 245 (2020). Ed. by Carosi, G. and Rybka, G., pp. 105–121. arXiv: 1711.08999 `[physics.ins-det]`.

[Jan+90]   Janesick, J. R. et al. "New advancements in charge-coupled device technology: subelectron noise and 4096 x 4096 pixel CCDs." In: *Charge-Coupled Devices and Solid State Optical Sensors.* Ed. by Blouke, M. M. Vol. 1242. International Society for Optics and Photonics. SPIE, 1990, pp. 223–237.

[JK75]   Jones, K. W. and Kraner, H. W. "Energy lost to ionization by 254-eV ^{73}Ge atoms stopping in Ge." In: *Phys. Rev. A* vol. 11 (4 Apr. 1975), pp. 1347–1353.

[Kah+18]   Kahlhoefer, F. et al. "Model-independent comparison of annual modulation and total rate with direct detection experiments." In: *JCAP* vol. 05 (2018), p. 074. arXiv: 1802.10175 `[hep-ph]`.

[Kap22]   Kapteyn, J. C. "First Attempt at a Theory of the Arrangement and Motion of the Sidereal System." In: *Astrophys. J.* vol. 55 (1922), pp. 302–328.

[KD23]   K., G. and Desai, S. "Constraints on Self-Interacting dark matter from relaxed galaxy groups." In: *Phys. Dark Univ.* vol. 42 (2023), p. 101291. arXiv: 2307.05880 `[astro-ph.CO]`.

[Kha+17]   Khachatryan, V. et al. "Searches for invisible decays of the Higgs boson in pp collisions at $\sqrt{s} = 7$, 8, and 13 TeV." In: *JHEP* vol. 02 (2017), p. 135. arXiv: 1610.09218 `[hep-ex]`.

[Kie20]   Kierans, C. A. "AMEGO: Exploring the Extreme Multimessenger Universe." In: *Proc. SPIE Int. Soc. Opt. Eng.* vol. 11444 (2020), p. 1144431. arXiv: 2101.03105 `[astro-ph.IM]`.

[KL22]   Kahn, Y. and Lin, T. "Searches for light dark matter using condensed matter systems." In: *Rept. Prog. Phys.* vol. 85, no. 6 (2022), p. 066901. arXiv: 2108.03239 `[hep-ph]`.

[KLZ17]   Knapen, S., Lin, T., and Zurek, K. M. "Light Dark Matter in Superfluid Helium: Detection with Multi-excitation Production." In: *Phys. Rev. D* vol. 95, no. 5 (2017), p. 056019. arXiv: 1611.06228 `[hep-ph]`.

[Koc+07]   Koch, J. et al. "Charge-insensitive qubit design derived from the Cooper pair box." In: *Phys. Rev. A* vol. 76 (4 Oct. 2007), p. 042319.

[KT90]   Kolb, E. W. and Turner, M. S. *The Early Universe.* Vol. 69. 1990.

[Kud19]   Kudryavtsev, V. A. "Recent Results from LUX and Prospects for Dark Matter Searches with LZ." In: *Universe* vol. 5, no. 3 (2019).

[Kun+97]   Kundic, T. et al. "A Robust determination of the time delay in 0957+561a,b and a measurement of the global value of Hubble's constant." In: *Astrophys. J.* vol. 482 (1997), p. 75. arXiv: astro-ph/9610162.

[Lee+14]   Lees, J. P. et al. "Search for a Dark Photon in e^+e^- Collisions at BaBar." In: *Phys. Rev. Lett.* vol. 113, no. 20 (2014), p. 201801. arXiv: 1406.2980 `[hep-ex]`.

[Lee+17]   Lees, J. P. et al. "Search for Invisible Decays of a Dark Photon Produced in e^+e^- Collisions at BaBar." In: *Phys. Rev. Lett.* vol. 119, no. 13 (2017), p. 131804. arXiv: 1702.03327 `[hep-ex]`.

[Lee+24]   Lee, A. J. et al. "The Chicago-Carnegie Hubble Program: The JWST J-region Asymptotic Giant Branch (JAGB) Extragalactic Distance Scale." In: (Aug. 2024). arXiv: 2408.03474 `[astro-ph.GA]`.

[LIG21]   LIGO-Virgo-KAGRA Collaboration. *O3b Catalog.* https://www.ligo.org/detections/O3bcatalog.php. [Online; accessed 4-March-2024]. 2021.

[Lis+18]   Lisanti, M. et al. "Search for Dark Matter Annihilation in Galaxy Groups." In: *Phys. Rev. Lett.* vol. 120, no. 10 (2018), p. 101101. arXiv: 1708.09385 `[astro-ph.CO]`.

[Liu+20]   Liu, C. P. et al. "Model-independent determination of the Migdal effect via photoabsorption." In: *Phys. Rev. D* vol. 102, no. 12 (2020), p. 121303. arXiv: 2007.10965 `[hep-ph]`.

[Liu+22]   Liu, Z. Z. et al. "Studies of the Earth shielding effect to direct dark matter searches at the China Jinping Underground Laboratory." In: *Phys. Rev. D* vol. 105 (5 Mar. 2022), p. 052005.

[LS96]   Lewin, J. and Smith, P. "Review of mathematics, numerical factors, and corrections for dark matter experiments based on elastic nuclear recoil." In: *Astroparticle Physics* vol. 6, no. 1 (1996), pp. 87–112.

[LW19]   Livan, M. and Wigmans, R. "The Energy Resolution of Calorimeters." In: *Calorimetry for Collider Physics, an Introduction.* Cham: Springer International Publishing, 2019, pp. 111–140.

[LYZ12]   Lin, T., Yu, H.-B., and Zurek, K. M. "Symmetric and asymmetric light dark matter." In: *Physical Review D* vol. 85, no. 6 (Mar. 2012).

[Maj23]   Majorovits, B. "The search of Axion Dark Matter with a dielectric halo-scope: MADMAX." In: *EPJ Web Conf.* vol. 282 (2023), p. 01008.

[Man+10]   Manzur, A. et al. "Scintillation efficiency and ionization yield of liquid xenon for monoenergetic nuclear recoils down to 4 keV." In: *Phys. Rev. C* vol. 81 (2 Feb. 2010), p. 025808.

[Man+20]   Mancuso, M. et al. "Searches for Light Dark Matter with the CRESST-III Experiment." In: *J. Low Temp. Phys.* vol. 199, no. 1-2 (2020), pp. 547–555.

[Mar16]   Marsh, D. J. E. "Axion Cosmology." In: *Phys. Rept.* vol. 643 (2016), pp. 1–79. arXiv: 1510.07633 [astro-ph.CO].

[May+16]   Mayet, F. et al. "A review of the discovery reach of directional Dark Matter detection." In: *Physics Reports* vol. 627 (Apr. 2016), pp. 1–49.

[McE+22]   McEwen, M. et al. "Resolving catastrophic error bursts from cosmic rays in large arrays of superconducting qubits." In: *Nature Phys.* vol. 18, no. 1 (2022), pp. 107–111. arXiv: 2104.05219 [quant-ph].

[Meh21]   Mehlhase, S. "ATLAS detector slice (and particle visualisations)." In: (2021).

[MH06]   Mei, D.-M. and Hime, A. "Muon-induced background study for underground laboratories." In: *Phys. Rev. D* vol. 73 (5 Mar. 2006), p. 053004.

[mig12]   mightyman. *Green Map Marker.* [Online; accessed 12-October-2024]. 2012.

[Mig41]   Migdal, A. B. "Ionization of atoms accompanying α- and β-decay." In: *J. Phys. USSR* vol. 4 (1941), p. 449.

[Mon12]   Mondal, N. K. "India-Based Neutrino Observatory (INO)." In: *Eur. Phys. J. Plus* vol. 127 (2012), p. 106.

[Mor+24]   Moretti, R. et al. "Design and Simulation of a Transmon Qubit Chip for Axion Detection." In: *IEEE Trans. Appl. Supercond.* vol. 34, no. 3 (2024), p. 1700705. arXiv: 2310.05238 [quant-ph].

[Mye+18a] Myeong, G. C. et al. "The Milky Way Halo in Action Space." In: *The Astrophysical Journal Letters* vol. 856, no. 2 (Mar. 2018), p. L26.

[Mye+18b] Myeong, G. C. et al. "The Sausage Globular Clusters." In: *The Astrophysical Journal Letters* vol. 863, no. 2 (Aug. 2018), p. L28.

[NAS23]   NASA. *COSI: Compton Spectrometer and Imager.* https://science.nasa.gov/mission/cosi. [Online; accessed 17-February-2024]. 2023.

[NFW96]   Navarro, J. F., Frenk, C. S., and White, S. D. M. "The Structure of cold dark matter halos." In: *Astrophys. J.* vol. 462 (1996), pp. 563–575. arXiv: astro-ph/9508025.

[NLB19]   Necib, L., Lisanti, M., and Belokurov, V. "Inferred Evidence for Dark Matter Kinematic Substructure with SDSSGaia." In: *The Astrophysical Journal* vol. 874, no. 1 (Mar. 2019), p. 3.

[Nor+22]   Norcini, D. et al. "Precision measurement of Compton scattering in silicon with a skipper CCD for dark matter detection." In: *Phys. Rev. D* vol. 106, no. 9 (2022), p. 092001. arXiv: 2207.00809 [physics.ins-det].

[Oak+21]   Oakes, L. et al. "Combined Dark Matter searches towards dwarf spheroidal galaxies with *Fermi*-LAT, HAWC, HESS, MAGIC and VERITAS." In: *PoS* vol. ICRC2019 (2021), p. 012. arXiv: 1909.06310 [astro-ph.HE].

[OG17]   O'Hare, C. A. J. and Green, A. M. "Axion astronomy with microwave cavity experiments." In: *Phys. Rev. D* vol. 95, no. 6 (2017), p. 063017. arXiv: 1701.03118 [astro-ph.CO].

[OHa+22]   O'Hare, C. A. J. et al. "Particle detection and tracking with DNA." In: *Eur. Phys. J. C* vol. 82, no. 4 (2022), p. 306. arXiv: 2105.11949 [physics.ins-det].

[OHa20]   O'Hare, C. *cajohare/AxionLimits: AxionLimits*. https://cajohare.github.io/AxionLimits/. Version v1.0. July 2020.

[OLu13]   O'Luanaigh, C. "Heavy metal: Refilling the lead source for the LHC." In: *CERN* (Feb. 2013).

[Oor32]   Oort, J. H. "The force exerted by the stellar system in the direction perpendicular to the galactic plane and some related problems." In: *Bull. Astron. Inst. Netherlands* vol. 6 (1932), pp. 249–287.

[Oor60]   Oort, J. H. "Note on the determination of K_z and on the mass density near the Sun." In: *Bull. Astron. Inst. Netherlands* vol. 15 (Feb. 1960), p. 45.

[OP73]   Ostriker, J. P. and Peebles, P. J. E. "A Numerical Study of the Stability of Flattened Galaxies: or, can Cold Galaxies Survive?" In: *Astrophysical Journal* vol. 186 (Dec. 1973), pp. 467–480.

[Orl+22]   Orlando, E. et al. "Exploring the MeV sky with a combined coded mask and Compton telescope: the Galactic Explorer with a Coded aperture mask Compton telescope (GECCO)." In: *JCAP* vol. 07, no. 07 (2022), p. 036. arXiv: 2112.07190 [astro-ph.HE].

[Oue+19]   Ouellet, J. L. et al. "Design and implementation of the ABRACADABRA-10 cm axion dark matter search." In: *Phys. Rev. D* vol. 99, no. 5 (2019), p. 052012. arXiv: 1901.10652 [physics.ins-det].

[OV39]   Oppenheimer, J. R. and Volkoff, G. M. "On Massive Neutron Cores." In: *Phys. Rev.* vol. 55 (4 Feb. 1939), pp. 374–381.

[Ove+12]   Overman, N. R. et al. "Majorana Electroformed Copper Mechanical Analysis." In: *OSTI Technical Report* (Apr. 2012).

[Par22]   Particle Data Group. "Review of Particle Physics." In: *Progress of Theoretical and Experimental Physics* vol. 2022, no. 8 (Aug. 2022), p. 083C01.

[Pec08]   Peccei, R. D. "The Strong CP problem and axions." In: *Lect. Notes Phys.* vol. 741 (2008). Ed. by Kuster, M., Raffelt, G., and Beltran, B., pp. 3–17. arXiv: hep-ph/0607268.

[Pet+20]   Petricca, F. et al. "First results on low-mass dark matter from the CRESST-III experiment." In: *J. Phys. Conf. Ser.* vol. 1342, no. 1 (2020). Ed. by Clark, K. et al., p. 012076. arXiv: 1711.07692 [astro-ph.CO].

[PH19]   Posti, Lorenzo and Helmi, Amina. "Mass and shape of the Milky Ways dark matter halo with globular clusters from Gaia and Hubble." In: *A&A* vol. 621 (2019), A56.

[PJM22]   Porter, T. A., Johannesson, G., and Moskalenko, I. V. "The GALPROP Cosmic-ray Propagation and Nonthermal Emissions Framework: Release v57." In: *Astrophys. J. Supp.* vol. 262, no. 1 (2022), p. 30. arXiv: 2112.12745 [astro-ph.HE].

[POW23]   Ponce, F., Orrell, J. L., and Wang, Z. *Radiation-induced secondary emissions in solid-state devices as a possible contribution to quasiparticle poisoning of superconducting circuits.* 2023. arXiv: 2301.08239 [physics.app-ph].

[PQ77a]   Peccei, R. D. and Quinn, H. R. "CP Conservation in the Presence of Pseudoparticles." In: *Phys. Rev. Lett.* vol. 38 (25 June 1977), pp. 1440–1443.

[PQ77b]   Peccei, R. D. and Quinn, H. R. "Constraints imposed by CP conservation in the presence of pseudoparticles." In: *Phys. Rev. D* vol. 16 (6 Sept. 1977), pp. 1791–1797.

[Pri+21]   Prihtiadi, H. et al. "Measurement of the cosmic muon annual and diurnal flux variation with the COSINE-100 detector." In: *JCAP* vol. 02 (2021), p. 013. arXiv: 2005.13672 [physics.ins-det].

[Pru+16]   Prusti, T. et al. "TheGaiamission." In: *Astronomy & Astrophysics* vol. 595 (Nov. 2016), A1.

[PRV08]   Pospelov, M., Ritz, A., and Voloshin, M. "Secluded WIMP dark matter." In: *Physics Letters B* vol. 662, no. 1 (2008), pp. 53–61.

[Pug+08]   Pugnat, P. et al. "First results from the OSQAR photon regeneration experiment: No light shining through a wall." In: *Phys. Rev. D* vol. 78 (2008), p. 092003. arXiv: 0712.3362 [hep-ex].

[Ras+09]   Rasmussen, J. et al. "Hot gas halos around disk galaxies: Confronting cosmological simulations with observations." In: *The Astrophysical Journal* vol. 697, no. 1 (Apr. 2009), pp. 79–93.

[Rea14]   Read, J. I. "The local dark matter density." In: *Journal of Physics G: Nuclear and Particle Physics* vol. 41, no. 6 (May 2014), p. 063101.

[Rei+19]   Reitze, D. et al. "Cosmic Explorer: The U.S. Contribution to Gravitational-Wave Astronomy beyond LIGO." In: *Bull. Am. Astron. Soc.* vol. 51, no. 7 (2019), p. 035. arXiv: 1907.04833 [astro-ph.IM].

[RMW18]   Rezzolla, L., Most, E. R., and Weih, L. R. "Using Gravitational-wave Observations and Quasi-universal Relations to Constrain the Maximum Mass of Neutron Stars." In: *The Astrophysical Journal Letters* vol. 852, no. 2 (Jan. 2018), p. L25.

[Rup+14]   Ruppin, F. et al. "Complementarity of dark matter detectors in light of the neutrino background." In: *Phys. Rev. D* vol. 90 (8 Oct. 2014), p. 083510.

[Rya+22a]   Ryan, M. et al. "Molecular Chemistry for Dark Matter." In: *Astrophys. J.* vol. 934 (2022), p. 120. arXiv: 2106.13245 [astro-ph.CO].

[Rya+22b]   Ryan, M. et al. "Molecular Chemistry for Dark Matter III: DarkKROME." In: *Astrophys. J.* vol. 934 (2022), p. 122. arXiv: 2110.11971 [astro-ph.CO].

[Ryd03]   Ryden, B. *Introduction to cosmology*. 2003.

[S N+17]   S. Nisi et al. "ICP-MS measurement of natural radioactivity at LNGS." In: *International Journal of Modern Physics A* vol. Vol. 32, no. No. 30 (2017), p. 1743003.

[Saa02]   Saab, T. "Search for weakly interacting massive particles with the Cryogenic Dark Matter Search experiment." PhD thesis. Stanford U., 2002.

[Sal10]   Salam, G. P. "Towards Jetography." In: *Eur. Phys. J. C* vol. 67 (2010), pp. 637–686. arXiv: 0906.1833 [hep-ph].

[SC09]   Sciolla, G. and Collaboration, t. D. "The DMTPC project." In: *Journal of Physics: Conference Series* vol. 179 (July 2009), p. 012009.

[Sch+15a]   Schneck, K. et al. "Dark matter effective field theory scattering in direct detection experiments." In: *Phys. Rev. D* vol. 91, no. 9 (2015), p. 092004. arXiv: 1503.03379 [astro-ph.CO].

[Sch+15b]   Schneck, K. et al. "Dark matter effective field theory scattering in direct detection experiments." In: *Phys. Rev. D* vol. 91 (9 May 2015), p. 092004.

[Sch11]   Schnee, R. W. "Introduction to dark matter experiments." In: *Theoretical Advanced Study Institute in Elementary Particle Physics: Physics of the Large and the Small*. Jan. 2011. arXiv: 1101.5205 [astro-ph.CO].

[Sch22]   Schutz, K. *An axion dark matter-induced echo of supernova remnants.* https://indico. cern.ch/event/922783/contributions/4892630/. 2022.

[Sei58]   Seitz, F. "On the Theory of the Bubble Chamber." In: *The Physics of Fluids* vol. 1, no. 1 (Jan. 1958), pp. 2–13.

[Sik08]   Sikivie, P. "Axion Cosmology." In: *Lect. Notes Phys.* vol. 741 (2008). Ed. by Kuster, M., Raffelt, G., and Beltran, B., pp. 19–50. arXiv: astro-ph/0610440.

[Sik83]   Sikivie, P. "Experimental Tests of the "Invisible" Axion." In: *Phys. Rev. Lett.* vol. 51 (16 Oct. 1983), pp. 1415–1417.

[Sik85]   Sikivie, P. "Detection Rates for 'Invisible' Axion Searches." In: *Phys. Rev. D* vol. 32 (1985). [Erratum: Phys.Rev.D 36, 974 (1987)], p. 2988.

[Sik87]   Sikivie, P. "Erratum: Detection rates for "invisible"-axion searches." In: *Phys. Rev. D* vol. 36 (3 Aug. 1987), pp. 974–974.

[Sil+17]   Silva-Feaver, M. et al. "Design Overview of DM Radio Pathfinder Experiment." In: *IEEE Trans. Appl. Supercond.* vol. 27, no. 4 (2017), p. 1400204. arXiv: 1610.09344 [astro-ph.IM].

[Sir+17]   Sirunyan, A. M. et al. "Particle-flow reconstruction and global event description with the CMS detector." In: *JINST* vol. 12, no. 10 (2017), P10003. arXiv: 1706.04965 [physics.ins-det].

[SM09]   Sciolla, G. and Martoff, C. J. "Gaseous Dark Matter Detectors." In: *New J. Phys.* vol. 11 (2009), p. 105018. arXiv: 0905.3675 [astro-ph.IM].

[Smi+23]   Smith-Orlik, A. et al. "The impact of the Large Magellanic Cloud on dark matter direct detection signals." In: *JCAP* vol. 10 (2023), p. 070. arXiv: 2302.04281 [astro-ph.GA].

[SN13]   Saha, K. and Naab, T. "Spinning dark matter haloes promote bar formation." In: *Monthly Notices of the Royal Astronomical Society* vol. 434, no. 2 (July 2013), pp. 1287–1299. eprint: https://academic.oup.com/mnras/article-pdf/434/2/1287/18496031/stt1088.pdf.

[sna12]   snagator. *Earth.* [Online; accessed 4-April-2024]. 2012.

[Spa14]   Spada, Francesca R. "AMS-02 on the International Space Station." In: *EPJ Web of Conferences* vol. 70 (2014), p. 00026.

[Spe88]   Spergel, D. N. "Motion of the Earth and the detection of weakly interacting massive particles." In: *Physical Review D* vol. 37, no. 6 (1988), p. 1353.

[Spy+12]   Spyrou, A. et al. "First Observation of Ground State Dineutron Decay: ^{16}Be." In: *Phys. Rev. Lett.* vol. 108 (10 Mar. 2012), p. 102501.

[SSD19]   Sahoo, S., Seth, S., and Das, M. "The threshold of gamma-ray induced bubble nucleation in superheated emulsion." In: *Nuclear Instruments and Methods in Physics Research Section A: Accelerators, Spectrometers, Detectors and Associated Equipment* vol. 931 (2019), pp. 44–51.

[Sti23]   Stifter, K. "Sub-GeV dark matter searches with SENSEI." https://indico.cern.ch/event/ 1188759/contributions/5222760/. UCLA Dark Matter Conference 2023. Mar. 2023.

[Sun+22]   Sun, Y. et al. "Axion dark matter-induced echo of supernova remnants." In: *Phys. Rev. D* vol. 105, no. 6 (2022), p. 063007. arXiv: 2110.13920 [hep-ph].

[Sun+24]   Sun, Y. et al. "Looking in the axion mirror: An all-sky analysis of stimulated decay." In: *Phys. Rev. D* vol. 109, no. 4 (2024), p. 043042. arXiv: 2310.03788 [hep-ph].

[SVZ80]   Shifman, M., Vainshtein, A., and Zakharov, V. "Can confinement ensure natural CP invariance of strong interactions?" In: *Nuclear Physics B* vol. 166, no. 3 (1980), pp. 493–506.

[SW01]   Smith, D. and Weiner, N. "Inelastic dark matter." In: *Phys. Rev. D* vol. 64 (4 July 2001), p. 043502.

[SW67]   Sachs, R. K. and Wolfe, A. M. "Perturbations of a Cosmological Model and Angular Variations of the Microwave Background." In: *Astrophysical Journal* vol. 147 (Jan. 1967), p. 73.

[SZ16]   Schutz, K. and Zurek, K. M. "Detectability of Light Dark Matter with Superfluid Helium." In: *Phys. Rev. Lett.* vol. 117, no. 12 (2016), p. 121302. arXiv: 1604.08206 [hep-ph].

[t H71]   't Hooft, G. "Predictions for neutrino - electron cross-sections in Weinberg's model of weak interactions." In: *Phys. Lett. B* vol. 37 (1971), pp. 195–196.

[Tar24]   Tarek Saab, Enectali Figueroa. *Dark Matter Limit Plotter*. Last accessed: April 4, 2024. 2024.

[Tav+18]   Tavani, M. et al. "Science with e-ASTROGAM: A space mission for MeVGeV gamma-ray astrophysics." In: *JHEAp* vol. 19 (2018). Ed. by De Angelis, A. et al., pp. 1–106. arXiv: 1711.01265 [astro-ph.HE].

[TES22]   TESSERACT Collaboration. *SNOWMASS 2021 Letter of Interest: The TESSERACT Dark Matter Project.* https://www.snowmass21.org/docs/files/summaries/CF/SNOWMASS21-CF1_CF2-IF8-120.pdf. 2022.

[TG79]   Tremaine, S. and Gunn, J. E. "Dynamical Role of Light Neutral Leptons in Cosmology." In: *Phys. Rev. Lett.* vol. 42 (6 Feb. 1979), pp. 407–410.

[Tif+17]   Tiffenberg, J. et al. "Single-electron and single-photon sensitivity with a silicon Skipper CCD." In: *Phys. Rev. Lett.* vol. 119, no. 13 (2017), p. 131802. arXiv: 1706.00028 [physics.ins-det].

[Tol39]   Tolman, R. C. "Static Solutions of Einstein's Field Equations for Spheres of Fluid." In: *Phys. Rev.* vol. 55 (4 Feb. 1939), pp. 364–373.

[Tom+23]   Tomsick, J. A. et al. "The Compton Spectrometer and Imager." In: *PoS* vol. ICRC2023 (2023), p. 745. arXiv: 2308.12362 [astro-ph.HE].

[Tov+00]   Tovey, D. et al. "A new model-independent method for extracting spin-dependent cross section limits from dark matter searches." In: *Physics Letters B* vol. 488, no. 1 (2000), pp. 17–26.

[Trz+19]   Trzaska, W. H. et al. "Cosmic-ray muon flux at Canfranc Underground Laboratory." In: *Eur. Phys. J. C* vol. 79, no. 8 (2019), p. 721. arXiv: 1902.00868 [physics.ins-det].

[Tum+23]   Tumasyan, A. et al. "A search for decays of the Higgs boson to invisible particles in events with a top-antitop quark pair or a vector boson in proton-proton collisions at $\sqrt{s} = 13$ TeV." In: *Eur. Phys. J. C* vol. 83, no. 10 (2023), p. 933. arXiv: 2303.01214 [hep-ex].

[use19]   user121799. *Starparalax Help*. TeX Stack Exchange. URL:tex.stackexchange.com/questions/479330 (version: 2019-03-14). 2019. eprint: tex.stackexchange.com/questions/479330.

[Vah+20]   Vahsen, S. E. et al. "CYGNUS: Feasibility of a nuclear recoil observatory with directional sensitivity to dark matter and neutrinos." In: (Aug. 2020). arXiv: 2008.12587 [physics.ins-det].

[Vin+15]   Vinyoles, N. et al. "New axion and hidden photon constraints from a solar data global fit." In: *Journal of Cosmology and Astroparticle Physics* vol. 2015, no. 10 (Oct. 2015), pp. 015–015.

[Vit+18]   Vittino, A. et al. "DRAGON2 : A novel code for Cosmic-Ray transport in the Galaxy." In: *Nuclear and Particle Physics Proceedings* vol. 297-299 (2018). Cosmic Ray Origin - Beyond the Standard Models, pp. 135–142.

[WCW80]   Walsh, D., Carswell, R. F., and Weymann, R. J. "0957 + 561 A, B: twin quasistellar objects or gravitational lens?" In: *Nature* vol. 279, no. 5712 (May 1980), pp. 381–384.

[Wd10]   Weber, M. and de Boer, W. "Determination of the local dark matter density in our Galaxy." In: *A&A* vol. 509 (2010), A25.

[Wei78]   Weinberg, S. "A New Light Boson?" In: *Phys. Rev. Lett.* vol. 40 (4 Jan. 1978), pp. 223–226.

[WFP24]   Woodley, W., Fedynitch, A., and Piro, M.-C. "Cosmic ray muons in laboratories deep underground." In: *Phys. Rev. D* vol. 110 (6 Sept. 2024), p. 063006.

[Wid+02]   Widenhorn, R. et al. "Temperature dependence of dark current in a CCD." In: *Sensors and Camera Systems for Scientific, Industrial, and Digital Photography Applications III*. Ed. by Sampat, N. et al. Vol. 4669. International Society for Optics and Photonics. SPIE, 2002, pp. 193–201.

[Wik20a]   Wikimedia Commons. *File:Bullet cluster lensing.jpg — Wikimedia Commons, the free media repository*. [Online; accessed 8-June-2023]. 2020.

[Wik20b]   Wikimedia Commons. *File:Gravitational-lensing-angles.png — Wikimedia Commons, the free media repository*. [Online; accessed 19-September-2024]. 2020.

[Wik20c]   Wikimedia Commons. *File:ILC SchemeTDR.jpg — Wikimedia Commons, the free media repository*. [Online; accessed 8-February-2024]. 2020.

[Wik20d]   Wikimedia Commons. *File:NEDM P&T violation.png — Wikimedia Commons, the free media repository*. [Online; accessed 29-April-2024]. 2020.

[Wik20e]   Wikimedia Commons. *File:Stanford-linear-accelerator-usgs-ortho-kaminski-5900.jpg — Wikimedia Commons, the free media repository*. [Online; accessed 28-January-2024]. 2020.

[Wik21]   Wikimedia Commons. *File:Rotation curve of spiral galaxy Messier 33 (Triangulum).png — Wikimedia Commons, the free media repository*. [Online; accessed 25-August-2022]. 2021.

[Wik22a]   Wikimedia Commons. *File:ETH-BIB-Zwicky, Fritz (1898-1974)-Portr 01030.tif — Wikimedia Commons, the free media repository*. [Online; accessed 27-July-2024]. 2022.

[Wik22b]   Wikimedia Commons. *File:NEDM-History-wiki.png — Wikimedia Commons, the free media repository*. [Online; accessed 21-August-2024]. 2022.

[Wik23a]   Wikimedia Commons. *File:Bullet cluster.jpg — Wikimedia Commons, the free media repository*. [Online; accessed 8-June-2023]. 2023.

[Wik23b]   Wikimedia Commons. *File:Cern-accelerator-complex.svg — Wikimedia Commons, the free media repository*. [Online; accessed 8-February-2024]. 2023.

[Wik23c]   Wikimedia Commons. *File:DNA Structure+Key+Labelled.pn NoBB.png — Wikimedia Commons, the free media repository*. [Online; accessed 21-February-2024]. 2023.

[Wik23d]   Wikimedia Commons. *File:Measuring a qubit leaves no room for error.jpg — Wikimedia Commons, the free media repository*. [Online; accessed 9-March-2024]. 2023.

[Wik23e]   Wikimedia Commons. *File:QSO B0957+0561.jpg — Wikimedia Commons, the free media repository*. [Online; accessed 21-September-2024]. 2023.

[Wik23f]   Wikimedia Commons. *File:Scenario for Mysterious Light Beams from Active Galaxy (2020-58-4781).tif — Wikimedia Commons, the free media repository*. [Online; accessed 4-April-2024]. 2023.

[Wik23g]   Wikimedia Commons. *File:Shapenoise.svg — Wikimedia Commons, the free media repository*. [Online; accessed 8-June-2023]. 2023.

[Wik23h]   Wikimedia Commons. *File:Spherical Harmonics.png — Wikimedia Commons, the free media repository*. [Online; accessed 4-April-2024]. 2023.

[Wik23i]   Wikimedia Commons. *File:Vera Rubin.jpg — Wikimedia Commons, the free media repository*. [Online; accessed 27-July-2024]. 2023.

[Wik23j]   Wikipedia contributors. *List of galaxy groups and clusters — Wikipedia, The Free Encyclopedia.* https://en.wikipedia.org/w/index.php?title=List_of_galaxy_groups_and_clusters&oldid=1189949309. [Online; accessed 28-April-2024]. 2023.

[Wik24a]   Wikimedia Commons. *File:1e0657 scale.jpg — Wikimedia Commons, the free media repository.* [Online; accessed 20-September-2024]. 2024.

[Wik24b]   Wikimedia Commons. *File:CAST-Experiment.jpg — Wikimedia Commons, the free media repository.* [Online; accessed 5-April-2024]. 2024.

[Wik24c]   Wikimedia Commons. *File:CMB Timeline300 no WMAP.jpg — Wikimedia Commons, the free media repository.* [Online; accessed 4-April-2024]. 2024.

[Wik24d]   Wikimedia Commons. *File:KEK Belle II Detector (1).jpg — Wikimedia Commons, the free media repository.* [Online; accessed 5-September-2024]. 2024.

[Wik24e]   Wikimedia Commons. *File:Standard Model of Elementary Particles.svg — Wikimedia Commons, the free media repository.* [Online; accessed 4-April-2024]. 2024.

[Wil78]   Wilczek, F. "Problem of Strong P and T Invariance in the Presence of Instantons." In: *Phys. Rev. Lett.* vol. 40 (5 Jan. 1978), pp. 279–282.

[WM17]   Westerdale, S. and Meyers, P. D. "Radiogenic Neutron Yield Calculations for Low-Background Experiments." In: *Nucl. Instrum. Meth. A* vol. 875 (2017), pp. 57–64. arXiv: 1702.02465 [physics.ins-det].

[woo06]   wooptoo. *World map.* [Online; accessed 12-October-2024]. 2006.

[Wor+22]   Workman, R. L. et al. "Review of Particle Physics." In: *PTEP* vol. 2022 (2022), p. 083C01.

[WS10]   Wantz, O. and Shellard, E. P. S. "Axion Cosmology Revisited." In: *Phys. Rev. D* vol. 82 (2010), p. 123508. arXiv: 0910.1066 [astro-ph.CO].

[Wu+14]   Wu, X. et al. "PANGU: A High Resolution Gamma-ray Space Telescope." In: *Proc. SPIE Int. Soc. Opt. Eng.* vol. 9144 (2014). Ed. by Takahashi, T., Herder, J.-W. A. den, and Bautz, M., 91440F. arXiv: 1407.0710 [astro-ph.IM].

[Wu+19]   Wu, T. et al. "Search for Axionlike Dark Matter with a Liquid-State Nuclear Spin Comagnetometer." In: *Phys. Rev. Lett.* vol. 122, no. 19 (2019), p. 191302. arXiv: 1901.10843 [hep-ex].

[Wu16]   Wu, X. "PANGU: A High Resolution Gamma-Ray Space Telescope." In: *PoS* vol. ICRC2015 (2016), p. 964.

[Yak+20]   Yakabe, R. et al. "First limits from a 3D-vector directional dark matter search with the NEWAGE-0.3b detector." In: *PTEP* vol. 2020, no. 11 (2020), 113F01. arXiv: 2005.05157 [hep-ex].

[Yel02]   Yellin, S. "Finding an upper limit in the presence of an unknown background." In: *Phys. Rev. D* vol. 66 (3 Aug. 2002), p. 032005.

[Zha+13]   Zhang, L. et al. "THE GRAVITATIONAL POTENTIAL NEAR THE SUN FROM SEGUE K-DWARF KINEMATICS." In: *The Astrophysical Journal* vol. 772, no. 2 (July 2013), p. 108.

[Zhu+05]   Zhu, J. et al. "Study on the muon background in the underground laboratory of KIMS." English. In: *Kao Neng Wu Li Yu Ho Wu Li/High Energy Physics and Nuclear Physics* vol. 29, no. 8 (Aug. 2005), pp. 721–726.

[ZM14]   Zhang, C. and Mei, D. .-. "Measuring Muon-Induced Neutrons with Liquid Scintillation Detector at Soudan Mine." In: *Phys. Rev. D* vol. 90, no. 12 (2014), p. 122003. arXiv: 1407.3246 [physics.ins-det].

[Zur24]   Zurek, K. M. "Dark Matter Candidates of a Very Low Mass." In: (Jan. 2024). arXiv: 2401.03025 [hep-ph].

[Zwi33]   Zwicky, F. "Die Rotverschiebung von extragalaktischen Nebeln." In: *Helvetica Physica Acta* vol. 6 (1933), pp. 110–127.

[Zwi37]   Zwicky, F. "On the Masses of Nebulae and of Clusters of Nebulae." In: *Astrophysical Journal* vol. 86 (Oct. 1937), p. 217.

Index

www.ingramcontent.com/pod-product-compliance
Lightning Source LLC
Chambersburg PA
CBHW081456190326
41458CB00015B/5266